NOBEL LECTURES IN CHEMISTRY

1971–1980

NOBEL LECTURES

Including Presentation Speeches
And Laureates' Biographies

PHYSICS

CHEMISTRY

PHYSIOLOGY OR MEDICINE

LITERATURE

PEACE

ECONOMIC SCIENCES

NOBEL LECTURES

INCLUDING PRESENTATION SPEECHES AND LAUREATES' BIOGRAPHIES

CHEMISTRY

1971–1980

EDITOR-IN-CHARGE

TORE FRÄNGSMYR

Uppsala University, Sweden

EDITOR

STURE FORSÉN

University of Lund, Sweden

World Scientific
Singapore • New Jersey • London • Hong Kong

Edge, NJ 07661

UK office: 73 Lynton Mead, Totteridge, London N20 8DH

Reprinted 1994

NOBEL LECTURES IN CHEMISTRY (1971–1980)

ISBN 981-02-0786-7
ISBN 981-02-0787-5 (pbk)

Printed in Singapore.

FOREWORD

Since 1901 the Nobel Foundation has published annually "Les Prix Nobel" with reports from the Nobel Award Ceremonies in Stockholm and Oslo as well as the biographies and Nobel lectures of the laureates. In order to make the lectures available to people with special interests in the different prize fields the Foundation gave Elsevier Publishing Company the right to publish in English the lectures for 1901–1970, which were published in 1964–1972 through the following volumes:

Physics 1901–1970	4 vols.
Chemistry 1901–1970	4 vols.
Physiology or Medicine 1901–1970	4 vols.
Literature 1901–1967	1 vol.
Peace 1901–1970	3 vols.

Elsevier decided later not to continue the Nobel project. It is therefore with great satisfaction that the Nobel Foundation has given World Scientific Publishing Company the right to bring the series up to date beginning with the Prize lectures in Economics in 2 volumes 1969–1990. Thereafter the lectures in all the other prize fields will follow.

The Nobel Foundation is very pleased that the intellectual and spiritual message to the world laid down in the laureates' lectures, thanks to the efforts of World Scientific, will reach new readers all over the world.

Lars Gyllensten
Chairman of the Board

Stig Ramel
Executive Director

Stockholm, June 1991

PREFACE

It has been said that the very concept of "interdisciplinary research" points to the inadequacy of the traditional classification of natural science into different specific disciplines. I take some pride in the fact that the members of the Nobel Committee for Chemistry and the Royal Swedish Academy of Sciences have over the years held a rather broadminded view of what should be regarded as chemistry. This attitude is well documented by a listing of the awards presented during the eight decades that have elapsed since the inauguration of the Nobel Prize in 1901.

The awards in chemistry with which this volume is concerned, spanning the years 1971 to 1980, also reflect that long tradition. They cover exciting developments and fundamental discoveries in the fields of small molecules with only a handful of atoms, organic and metallo-organic compounds, synthetic and naturally occurring macromolecules, as well as complex chemical systems.

Reading through and editing the lectures in this book has been a great treat. Although I have been fairly familiar with the work leading to the awards, following the accounts presented here by the laureates themselves has been exceptionally fascinating and educational. I expect that other readers will share my experience.

The material in this volume has been taken from the original publication, "Les Prix Nobel". Some errors have been corrected and the lists of references have been updated, where necessary, in consultation with the laureates.

Sture Forsén
Editor

CONTENTS

Chemistry 1971

GERHARD HERZBERG

for his contributions to the knowledge of electronic structure and geometry of molecules, particularly free radicals

THE NOBEL PRIZE FOR CHEMISTRY

Speech by Professor STIG CLAESSON of the Royal Academy of Sciences
Translation from the Swedish text

Your Majesty, Your Royal Highnesses, Ladies and Gentlemen,

This year's Nobel Prize winner in Chemistry, Dr. Gerhard Herzberg, is generally considered to be the world's foremost molecular spectroscopist and his large institute in Ottawa is the indisputed center for such research. It is quite exceptional, in the field of science, that a single individual, however distinguished, in this way can be the leader of a whole area of research of general importance. A noted English chemist has also said that the only institutions that have previously played such a role were the Cavendish laboratory in Cambridge and Bohr's institute in Copenhagen.

Herzberg began as a physicist and his first contributions to molecular spectroscopy were published at the end of the 1920's. In such investigations one measures how molecules absorb light-energy—also outside the visible region —i.e. in the ultraviolet and infrared. Since light-energy is packaged as quanta, these measurements can provide accurate information about energy contents in molecules. From this information their size, shape and other properties can be derived. Such calculations must be based on the description of matter given by quantum mechanics. The development of this subject during the 1920's and 30's is regarded as one of the most exciting periods in the history of physical science. Herzberg's elegant experimental investigations combined with his theoretical insight into their interpretation contributed to the progress of quantum mechanics while being decisive for the rapid development of molecular spectroscopy.

One may now ask why Herzberg—originally a physicist and even famous as an astrophysicist—finally was awarded the Nobel prize in chemistry.

The explanation is that around 1950 molecular spectroscopy had progressed so far that one could begin to study even complicated systems of great chemical interest. This is brilliantly demonstrated by Herzberg's pioneering investigations of free radicals. Knowledge of their properties is of fundamental importance to our understanding of how chemical reactions proceed.

For a chemical reaction to occur the original molecules must in some way break up into fragments which rearrange to form the new molecules. These fragments, or intermediates, are called free radicals.

Free radicals are very difficult to study due to their short life-times— measured in millionth's of a second. Herzberg therefore had ample opportunity to repeatedly demonstrate his exceptional experimental skill when the necessary spectroscopic technique was worked out.

Herzberg has so far performed extensive precision determinations of the properties of over thirty free radicals among which are to be found the radicals methyl and methylene—well known from organic chemistry. Among his exciting discoveries may be mentioned that radicals drastically change their shape with increasing energy. For example, methylene is linear in its ground state but bent in states of higher energy. Many of the most important results were only achieved after several years' work and some of the most exciting as late as at the end of the 1960's. One can therefore note that this year's prize is truly an award for contributions of great current interest.

Dr. Herzberg,

I have tried to explain your great contributions to molecular spectroscopy and particularly your pioneering work on free radicals. The ideas and results presented by you—not least regarding the quantum mechanical aspects of the interpretation of molecular properties—have influenced scientific progress in almost all branches of chemistry.

On behalf of the Royal Academy of Sciences I beg you to accept our congratulations and ask you to receive your Nobel Prize from the hands of His Majesty the King.

Gerhard Herzberg

GERHARD HERZBERG

Gerhard Herzberg was born in Hamburg, Germany, on 25 December, 1904. He was married in 1929 to Luise Herzberg née Oettinger and has two children. He was widowed in 1971.

Herzberg received his early training in Hamburg and subsequently studied physics at the Darmstadt Institute of Technology where in 1928 he obtained his Dr. Ing. degree under H. Rau (a pupil of W. Wien). From 1928 to 1930 he carried out post-doctorate work at the University of Göttingen under James Franck and Max Born and the University of Bristol. In 1930 he was appointed Privatdozent (lecturer) and senior assistant in the Physics Department of the Darmstadt Institute of Technology.

In August 1935 Herzberg was forced to leave Germany as a refugee and took up a guest professorship at the University of Saskatchewan (Saskatoon, Canada), for which funds had been made available by the Carnegie Foundation. A few months later he was appointed research professor of physics, a position he held until 1945. From 1945 to 1948 Herzberg was professor of spectroscopy at the Yerkes Observatory of the University of Chicago. He returned to Canada in 1948 and was made Principal Research Officer and shortly afterwards Director of the Division of Physics at the National Research Council. In 1955, after the Division had been divided into one in pure and one in applied physics, Herzberg remained Director of the Division of Pure Physics, a position which he held until 1969 when he was appointed Distinguished Research Scientist in the recombined Division of Physics.

Herzberg's main contributions are to the field of atomic and molecular spectroscopy. He and his associates have determined the structures of a large number of diatomic and polyatomic molecules, including the structures of many free radicals difficult to determine in any other way (among others, those of free methyl and methylene). Herzberg has also applied these spectroscopic studies to the identification of certain molecules in planetary atmospheres, in comets, and in interstellar space.

Herzberg has been active as President or Vice President of several international commissions dealing with spectroscopy. He was also Vice

President of the International Union of Pure and Applied Physics from
1957 to 1963. He held the offices of President of the Canadian Association of Physicists for the year 1956—57 and President of the Royal Society
of Canada for the year 1966—67.

Herzberg was elected a Fellow of the Royal Society of Canada in 1939
and of the Royal Society of London in 1951. He was Bakerian Lecturer
of the Royal Society of London in 1960 and received a Royal Medal from
the Society in 1971. He was George Fischer Baker Non-Resident Lecturer in Chemistry at Cornell University in 1968, and Faraday Med-
allist and Lecturer of the Chemical Society of London in 1970. He is
Honorary Member or Fellow of a number of scientific societies, including
the American Academy of Arts and Sciences, the Optical Society of
America and the Chemical Society. He is also a Foreign Associate of the
National Academy of Sciences in Washington and a member of the
Pontifical Academy of Sciences. He is a Companion of the Order of
Canada. He has received many other medals and awards and holds
Honorary Degrees from a number of universities in Canada and abroad,
including one from the University of Stockholm.

In addition to many papers describing original results Herzberg has
published the following books:

Atomic Spectra and Atomic Structure
 1st ed., Prentice Hall, New York, 1937
 2nd ed., Dover Publications, New York, 1944
 Russian translation, Moscow, 1948
 Italian translation, Torino, 1961
 Japanese translation, 1964
Molecular Spectra and Molecular Structure. 1. Spectra of Diatomic Molecules
 1st ed., Prentice Hall, New York, 1939
 2nd ed., D. Van Nostrand, Company, Inc., New York, 1950
 Russian translation, Moscow, 1949
 Hungarian translation, Budapest, 1956
*Molecular Spectra and Molecular Structure. 2. Infrared and Raman Spectra
 of Polyatomic Molecules*
 Published by D. Van Nostrand Company, Inc., New York, 1945
 Russian translation, Moscow, 1949
 Hungarian translation, Budapest, 1960
*Molecular Spectra and Molecular Structure. 3. Electronic Spectra and Electronic
 Structure of Polyatomic Molecules*
 Published by D. Van Nostrand Company, Inc., New York, 1966.
 Russian translation, Moscow, 1969
*The Spectra and Structures of Simple Free Radicals: An Introduction to Molecular
 Spectroscopy*
 Published by Cornell University Press, 1971

SPECTROSCOPIC STUDIES OF MOLECULAR STRUCTURE

Nobel Lecture, December 11, 1971

by

G. HERZBERG

Division of Physics, National Research Council of Canada, Ottawa, Canada
K1A OR6

A. INTRODUCTION

The citation for the 1971 Nobel Prize in Chemistry reads "for contribution to the knowledge of electronic structures and geometry of molecules, especially free radicals" and therefore implies that the Prize has been awarded for a long series of studies extending practically over my whole scientific life. I shall try to present in this lecture a few of what I consider to be the more significant results of this work.

It was recognized very early in the development of spectroscopy that the study of molecular spectra is one of the most important tools for the determination of molecular structures. When I began my scientific work it had already been firmly established that band spectra are molecular spectra (in contrast to line spectra which are atomic spectra) and that a band system such as the one shown in Fig. 1 represents all transitions between the vibrational and rotational levels of two electronic states, i.e. corresponds to a single line or a single multiplet in an atomic spectrum.

It was also well-known how, by determining the spacing between appropriate bands in a band system, we obtain the vibrational intervals in the upper and lower electronic states, which are simply and directly related to the vibrational frequencies of the molecule in these states and how, by determining the separations of appropriate lines in a given band, we obtain the rotational intervals in these states which are directly related to the moments of inertia and therefore to the internuclear distances (or in other words the geometrical structures) in the two states.

The fine structure of the rotational lines tells us something about the nature of the electronic states involved, whether they are singlet, doublet, triplet, ... states while the nature and number of branches tells us, in the case of linear polyatomic molecules and diatomic molecules, whether the electronic states are of the type Σ^+ or Σ^-, or Π, or Δ.

In the years 1925—28 the work of James Franck (2) and Birge and Sponer (3) had established how in suitable spectra where a sufficient number of vibrational levels or the limit of an absorption continuum has been observed the dissociation energy can be accurately determined.

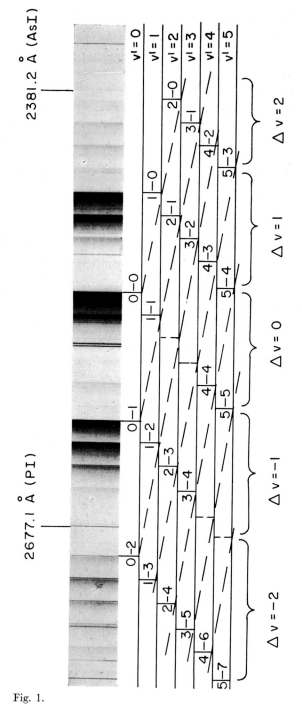

Fig. 1.
Band system of the PN molecule after Curry, Herzberg and Herzberg (1).

B. Diatomic Molecules, Radicals and Ions

The spectra of a large number of diatomic molecules, radicals and ions (~ 600) have been studied by various investigators in the past 50 years. For many of these molecules several band systems have been found and therefore several (sometimes many) electronic states have been established allowing a detailed interpretation in terms of molecular orbital theory [see Mulliken's Nobel lecture (4)]. If Rydberg series of electronic states are observed (so far only in relatively few cases) it is possible to determine the ionization potential in much the same way as for atoms.

In the following I shall describe the results only with regard to a few of the many diatomic systems studied in our laboratory.

(1) H_2, D_2, HD, H_2^+

The simplest molecular systems, H_2 and H_2^+ are even now the only ones for which *ab initio* calculations of very high precision can be and have been carried out [Kolos and Roothaan (5), Kolos and Wolniewicz (6), Hunter and Pritchard (7)]. It therefore appeared of considerable interest to improve as much as possible the experimental accuracy of some of the molecular constants. We have attempted to do this for the dissociation energy, the vibrational and rotational intervals in the ground state and for the ionization potential.

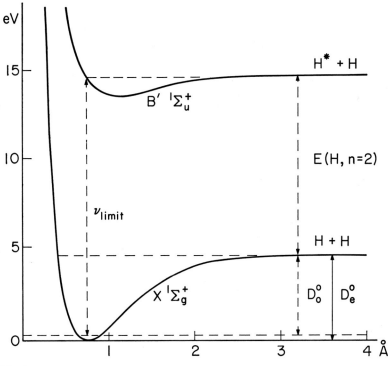

Fig. 2.
Potential functions of the ground state ($X^1\Sigma_g^+$) and the second excited state ($B'\ ^1\Sigma_u^+$) of H_2 showing the relation between the absorption limit and the dissociation energy in the ground state: $D_0^0 = v_{\text{limit}} - E(H, n = 2)$.

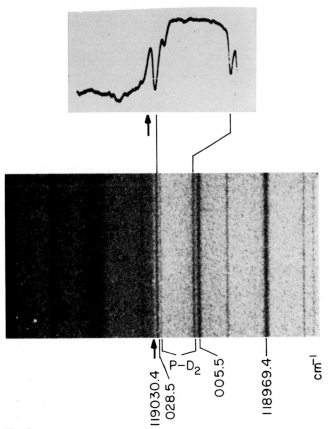

Fig. 3.
Section of far ultraviolet absorption spectrum of D_2 at low temperature showing the absorption limit corresponding to $J'' = 0$ after Herzberg (8).

The dissociation energy has been determined from the long wavelength limit of the continuum that joins onto the discrete absorption bands which correspond to transitions from the ground state to an excited electronic state designated B' in Fig. 2. This state dissociates into one normal and one excited $(n = 2)$ H atom; therefore the limit of the continuous absorption spectrum occurs at an energy equal to the sum of the dissociation energy of the ground state and the excitation energy of the H atom. As an example Fig. 3 shows a small section of the far ultraviolet absorption spectrum of D_2 near 840 Å under very high resolution taken at liquid nitrogen temperature where the absorption limit is clearly visible (for H_2 the corresponding limit is overlapped by an absorption line which makes it difficult to obtain as precise a limit as for D_2). Subtracting the excitation energy of D (or H) from the wave number of the limit and including a very small correction for the rotational barrier at $J = 1$ in the upper state we have obtained the dissociation energies given in the last column of Table 1 which should be compared with the theoretical values in the second column. The agreement between theory and experiment is clearly very good.

Table 1. Calculated and observed dissociation energies of hydrogen

	Theor. (a)	Obs. (8)
$D_0{}^0(H_2)$	36 117.9	$<$36 118.3
		$>$36 116.3
$D_0{}^0(HD)$	36 405.5	$\begin{cases} 36\ 406.6 \\ 36\ 405.8 \end{cases}$
$D_0{}^0(D_2)$	36 748.2	36 748.9$\pm$0.4 cm^{-1}
	cm^{-1}	cm^{-1}

(a) From Kolos and Wolniewicz (6) but including small non-adiabatic corrections according to Bunker (9).

Precise values of the vibrational intervals ΔG and the rotational constants B_v have been obtained for low v values from the Raman spectrum [Stoicheff (10)] and the infrared quadrupole spectrum [Herzberg (11), Rank and collaborators (12)], and for higher v values from the vacuum ultraviolet emission spectrum [Herzberg and Howe (13), Bredohl and Herzberg (14)]. In Table 2 the observed ΔG values of H$_2$ are compared with the theoretical values obtained from the Kolos-Roothaan-Wolniewicz potential (6). The slight systematic differences are in all probability due to the neglect of non-adiabatic corrections [see Poll and Karl (15) and Bunker (9)]. In Fig. 4 the deviations are plotted for H$_2$, HD and D$_2$. Similar very small systematic differences are also found for the rotational constants B_v as shown in Fig. 5. There is little doubt that these small discrepancies, of the order of 100 p.p.m., will be fully accounted for once the non-adiabatic corrections have been accurately evaluated.

If one disregards adiabatic and non-adiabatic corrections one obtains from the observed B_v values an internuclear distance for the equilibrium position of

Table 2. Observed and calculated vibrational quanta in the ground state of H$_2$

v	$\Delta G(v+^1/_2)$	
	obs.	theor.
0	4161.14	4162.06
1	3925.98	3926.64
2	3695.24	3696.14
3	3468.01	3468.68
4	3241.56	3242.24
5	3013.73	3014.49
6	2782.18	2782.82
7	2543.14	2543.89
8	2292.96	2293.65
9	2026.26	2026.81
10	1736.66	1737.13
11	1414.98	1415.54
12	1049.18	1048.98
13	621.96	620.16

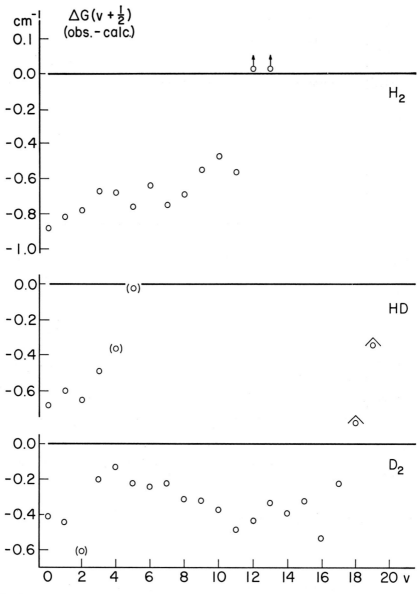

Fig. 4.
Deviations of the observed vibrational quanta of H₂, HD, and D₂ from those obtained from theory.

0.74139 and 0.74156 Å for H_2 and D_2 while the theoretical value (the same for both) is 0.74140 Å.

Finally the detailed study of the Rydberg series of H_2 near 800 Å, which is made somewhat complicated by interesting perturbations, by preionization effects and small pressure shifts, has led to the following experimental value for the ionization potential of H_2 [Herzberg and Jungen (16)]

$$\text{I.P. } (H_2) = 124\ 417.2 \pm 0.4 \text{ cm}^{-1}$$

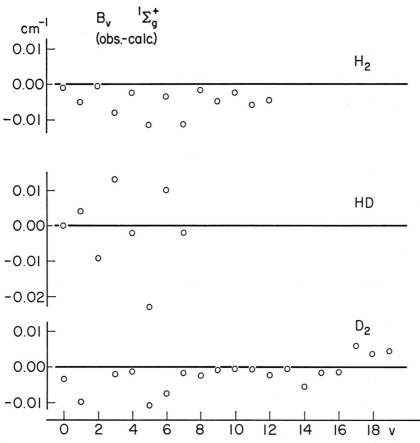

Fig. 5.
Deviations of the observed rotational constants B_v of H_2, HD, and D_2 from those obtained from theory.

and a less detailed study by Takezawa (17) of D_2, after a pressure shift correction, to the value

$$\text{I.P. } (D_2) = 124\ 745.6 \pm 0.6 \text{ cm}^{-1}$$

while the theoretical values, including relativistic and Lamb shift corrections, are [Hunter and Pritchard (7), Jeziorski and Kołos (18), Bunker (9)]

$$\text{I.P.}_{\text{theor}}(H_2) = 124\ 417.3 \text{ cm}^{-1}$$

$$\text{I.P.}_{\text{theor}}(D_2) = 124\ 745.2 \text{ cm}^{-1}$$

The agreement is well within the error of measurement, i.e. within about 3 p.p.m.

From the ionization potentials and the dissociation energies of H_2 and D_2 we obtain, according to the general relation

$$D_0{}^0(X_2{}^+) = \text{I.P.}(X) + D_0{}^0(X_2) - \text{I.P.}(X_2)$$

the following values for the dissociation energies of the ions

$$D_0{}^0{}_{obs.}(H_2{}^+) \leqslant 21\ 379.9 \pm 0.4\ \text{cm}^{-1}$$

$$D_0{}^0{}_{obs.}(D_2{}^+) = 21\ 711.9 \pm 0.6\ \text{cm}^{-1}$$

These values may be compared with the theoretical values

$$D_0{}^0{}_{theor.}(H_2{}^+) = 21\ 379.3\ \text{cm}^{-1}$$

$$D_0{}^0{}_{theor.}(D_2{}^+) = 21\ 711.6\ \text{cm}^{-1}$$

which are considered to be accurate to ± 0.2 cm^{-1}. That the agreement is again very satisfactory is not surprising since $D_0{}^0(H_2{}^+, D_2{}^+)$ are determined by $D_0{}^0(H_2, D_2)$ and I.P.(H_2, D_2) and for both of the latter quantities very good agreement between theory and experiment was found.

(2) O_2

The study of forbidden electronic transitions, that is, transitions forbidden by the normal selection rules, has greatly aided in the understanding of the electronic structure of diatomic (and polyatomic) molecules. The O_2 molecule is particularly rich in such forbidden transitions. The lowest electron configuration ... $\pi_u{}^4 \pi_g{}^2$ gives rise to three states $^3\Sigma_g{}^-$, $^1\Delta_g$ and $^1\Sigma_g{}^+$ of which the first forms the ground state of the molecule. As was first recognized by Van Vleck (19) the forbidden transitions from the ground state to the $^1\Delta_g$ and $^1\Sigma^+$ states can only occur as magnetic dipole (or much more weakly as quadrupole) radiation. Because of the long absorption path in the atmosphere, these forbidden transitions are very prominent in the solar spectrum observed from the ground. In emission they occur strongly in the red and infrared spectrum of the night sky. Even the transition $^1\Sigma_g{}^+$—$^1\Delta_g$ which can occur only as electric quadrupole radiation has been observed [Noxon (20)].

Another group of forbidden transitions in the near ultraviolet corresponds to transitions to states of the configuration ... $\pi_u{}^3 \pi_g{}^3$ which gives rise to the six states $^1\Sigma_u{}^+$, $^1\Sigma_u{}^-$, $^1\Delta_u$, $^3\Sigma_u{}^+$, $^3\Sigma_u{}^-$, $^3\Delta_u$. Only one of these states can combine with the ground state in an allowed transition, namely, $^3\Sigma_u{}^-$—$^3\Sigma_g{}^-$; this transition, the well-known Schuman-Runge bands, limits the transparency of air in the ultraviolet. Forbidden transitions to three of the other states, $^1\Sigma_u{}^-$, $^3\Sigma_u{}^+$ and $^3\Delta_u$ have been observed [Herzberg (21)], of which $^3\Sigma_u{}^+$—$^3\Sigma_g{}^-$ is the best known and is quite prominent in the light of the night sky. Fig. 6 shows a potential diagram of the lower electronic states of O_2. The fact that all transitions to the ground state from non-repulsive states arising from normal atoms ($^3P + {}^3P$) are forbidden accounts for the observation that in the upper atmosphere the radiative recombination of O atoms is very slow indeed. On the other hand the weak continuous absorption joining onto the $^3\Sigma_u{}^+$—$^3\Sigma_g{}^-$ absorption system ($\lambda < 2440$ Å) gives rise to the production of free O atoms even at fairly low altitudes and thus accounts for the formation of the ozone layer. Thus the chemistry of the upper atmosphere is greatly affected by the forbidden nature of these transitions i.e. by the electronic structure of the O_2 molecule. In Table 3 some of the molecular constants of O_2 are summarized and compared with those of S_2. In S_2 the analogues of the forbidden transitions of O_2 have not yet been observed.

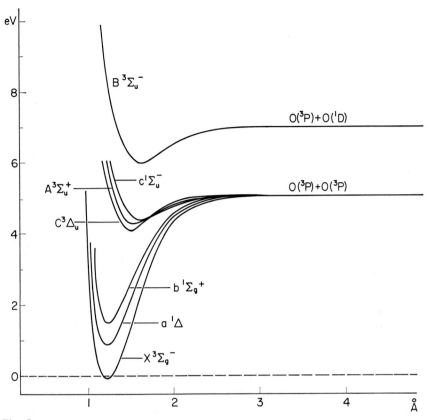

Fig. 6.
Potential functions of the ground state and the lower excited states of the O₂ molecule.
The state C $^3\Delta_u$ should have been designated A' $^3\Delta_u$ to be in accord with Table 3.

Table 3. Molecular constants of the lower states of the O₂ and S₂ molecules

			$T_0(\text{cm}^{-1})$	$\omega_e(\text{cm}^{-1})$	$D_0(\text{eV})$	$r_e(\text{Å})$
O₂	B	$^3\Sigma_u^-$	49 363.1	700.36	0.9627	1.6049
	A	$^3\Sigma_u^+$	35 007.2	799.07	0.7752	1.5214
	A'	$^3\Delta_u$	34 320	(611.2)	0.8604	(1.49)
	c	$^1\Sigma_u^-$	32 664.1	794.29	1.0657	1.5174
	b	$^1\Sigma_g^+$	13 120.9	1432.69	3.4887	1.2269
	a	$^1\Delta_g$	7 882.4	1509.3	4.1382	1.2156
	X	$^3\Sigma_g^-$	0	1580.36	5.1155	1.2075
S₂	B	$^3\Sigma_u^-$	31 689	434.0	1.5828	2.1702
	A	$^3\Sigma_u^+$	(22 682)	481.4	(1.55)	
	A'	$^3\Delta_u$	(21 855)	488.4	(1.66)	2.1506
	c	$^1\Sigma_u^-$				2.15
	b	$^1\Sigma_g^+$		(699.7)		
	a	$^1\Delta_g$	(5,100)	701.94	(3.73)	1.8978
	X	$^3\Sigma_g^-$	0	725.65	4.3662	1.8892

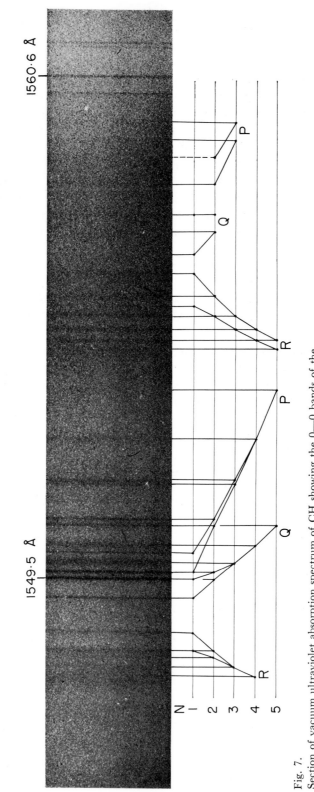

Fig. 7.
Section of vacuum ultraviolet absorption spectrum of CH showing the 0—0 bands of the
$E\ ^2\Pi—X\ ^2\Pi$ and $F\ ^2\Sigma—X\ ^2\Pi$ transitions after Herzberg and Johns (28).

Many studies of forbidden transitions in other molecules have been made but will not be discussed here [see Herzberg (22), (23)].

(3) CH and CH⁺

The visible and near ultraviolet emission spectrum of the CH radical has been known ever since spectra of an ordinary Bunsen burner have been taken. That the well-known bands at 4314 and 3889 Å really belong to CH was established of course only after the theory of diatomic spectra was sufficiently developed [Heurlinger (24), (25), Hulthén (26)]. The band at 3145 Å later discovered by Hori (27) in electric discharges has the same lower state as the other two, the $^2\Pi$ ground state of the molecule. While the ground state has the electron configuration $\sigma^2\,\sigma^2\,\sigma^2\,\pi$ the three excited states have the configuration . . . . $\sigma\pi^2$. This configuration, in addition to $^2\Sigma^-$, $^2\Delta$ and $^2\Sigma^+$ also gives rise to a $^4\Sigma^-$ state which however has not yet been observed even though it is expected to be a stable and fairly low-lying state.

The CH radical is a very reactive radical which under most conditions has a very short life time. It is for this reason that its absorption spectrum has been observed and studied in detail only fairly recently [Herzberg and Johns (28)]. This study has revealed a number of new transitions in the vacuum ultra-violet of which Fig. 7 gives an example. Indeed a whole Rydberg series was found which allowed an accurate determination of the ionization potential (= 10.64 eV). In addition the value for the dissociation energy was slightly refined in this work and much evidence of predissociation in all excited states except the $^2\Delta$ state was obtained.

Fig. 8. shows a diagram of the observed electronic states of CH and Table 4 gives some of the principal molecular constants.

The CH radical was the first molecule recognized in the interstellar medium [Swings and Rosenfeld (29), McKellar (30)]. The life time of CH under the action of the interstellar radiation field is relatively short (about 30 years) because of the predissociation that has been established to occur in all absorption systems of CH except the longest wavelength one ($^2\Delta$—$^2\Pi$). It is therefore somewhat surprising that CH is present in sufficient concentration to show its absorption spectrum.

The ion CH⁺ has also been observed, first in interstellar absorption [Adams (31)] and then in emission in the laboratory [Douglas and Herzberg (32)]. The observed transition is the expected $\sigma\pi$ $^1\Pi$—σ^2 $^1\Sigma$ transition. The molecular constants are included in Table 4. Fig. 9 shows a laboratory spectrum. The interstellar lines [R (0) lines coming from the lowest rotational level] are marked. In spite of much effort we have not yet been able to observe this spectrum *in absorption* in the laboratory.

(4) C₂⁻

A few years ago, in an attempt to observe absorption spectra of CH_4^+, CH_3^+, CH_2^+ and CH⁺, using flash discharges through CH_4, we observed a new very simple spectrum shown in Fig. 10 [Herzberg and Lagerqvist (33)]. The analysis was very easy and showed immediately and conclusively that the carrier of this

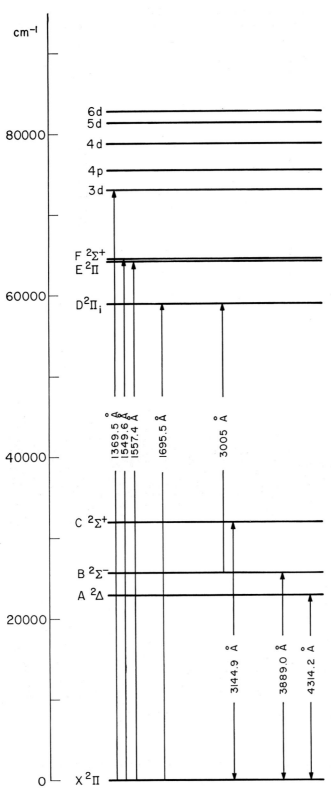

Fig. 8.
Energy level diagram
of the electronic states
of the CH radical showing
the observed transitions.
The Rydberg transitions
are not marked.

Table 4. Molecular constants of the known electronic states of CH and CH$^+$

State		T_o	$\Delta G(1/2)$	B_o	r_o(Å)
CH$^+$	$^1\Pi$	109 446	1642.16	11.428	1.2596
	$^1\Sigma^+$	85 850	2739.70	13.932	1.1409
CH	$^2\Delta$				
3d	$^2\Pi$	72 960			
	$^2\Sigma^+$				
3p	$^2\Pi_r$	64 532		12.6	1.20
	$^2\Sigma^+$			12.17	1.221
D	$^2\Pi_i$	58 981	(2800)	13.7	1.12
C	$^2\Sigma^+$	31 778	2613	14.25	1.128
B	$^2\Sigma^-$	25 698	1795	12.64	1.198
A	$^2\Delta$	23 217	2737	14.58	1.115
X	$^2\Pi$	0	2732.5	14.19	1.130
		cm^{-1}	cm^{-1}	cm^{-1}	

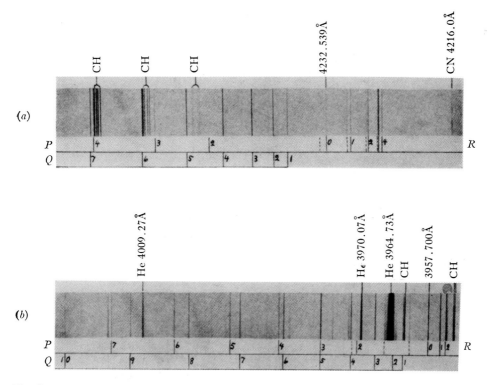

Fig. 9.
Two emission bands of CH$^+$ after Douglas and Herzberg (32). The interstellar lines are the R(0) lines.

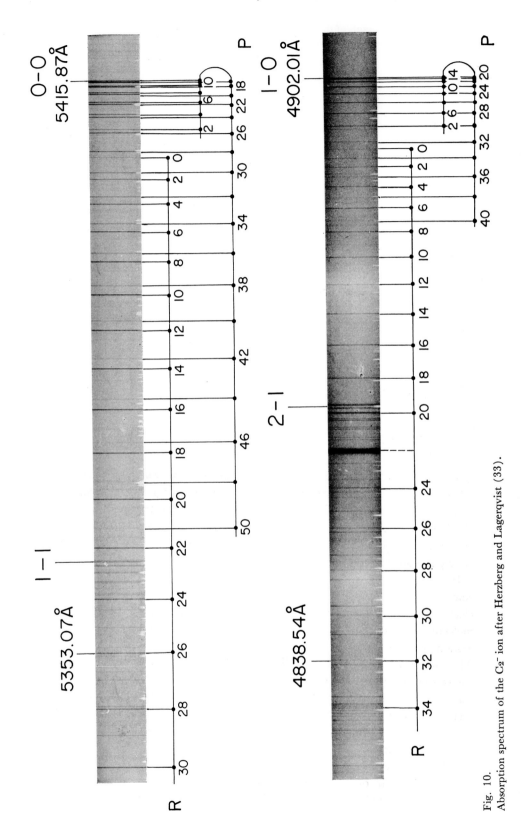

Fig. 10.

Absorption spectrum of the C₂⁻ ion after Herzberg and Lagerqvist (33).

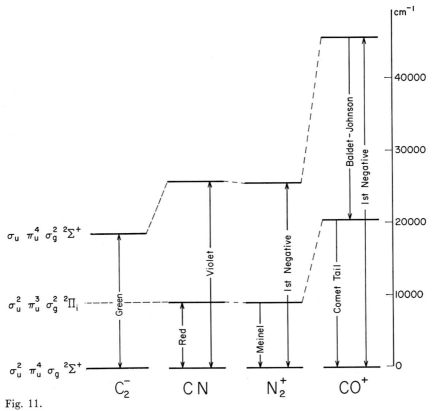

Fig. 11.
Observed electronic states of 13-electron systems.

spectrum must be the C_2 radical or one of its ions. It seemed impossible to fit this transition into the known system of energy levels of C_2. Therefore and because of the similarity with N_2^+ we suggested that the spectrum belongs to the C_2^- ion (i.e. represents the first discrete spectrum of a negative molecular ion) even though the required doublet structure of the spectrum was not very clearly recognizable.

In Fig. 11 the observed electronic energy levels of N_2^+, CN and CO^+ are compared with those of the new transition. All these molecules or ions have 13 electrons. Comparisons such as this historically formed the starting point of molecular orbital theory. Here the comparison serves to strengthen the suggestion that the new transition belongs to a 13 electron system, that is, C_2^-. This suggestion was strikingly confirmed first by the work of Milligan and Jacox (34) on the analogous spectrum observed in a matrix since it is considerably strengthened by the addition of an electron donor like Cs; and recently even more conclusively by the work of Lineberger and Patterson (35) who photo-ionized a C_2^- beam by a two-photon absorption of a tunable laser beam and found, as shown in Fig. 12, a photoionization exactly at the wave lengths of the new absorption bands.

Meinel (36) in our laboratory has recently observed the spectrum of C_2^+. Other diatomic molecules, radicals and ions studied in our laboratory are listed

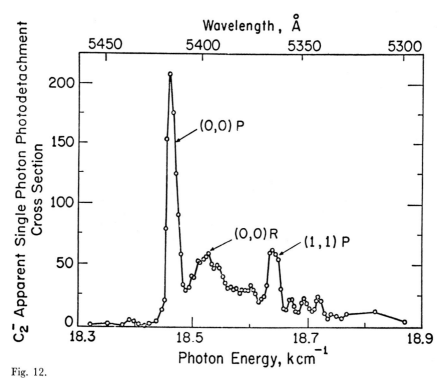

Fig. 12.

Photoionization spectrum of a C_2^- beam produced by two-photon absorption of a tunable laser beam after Lineberger and Patterson (35).

in Table 5. For all these systems, information about their geometrical structure (internuclear distances), their vibrational frequencies in various electronic states, and about their electronic structure has been obtained. Much work on these molecules and many other molecules which we have not studied has proceeded in many laboratories throughout the world notably here in Stockholm in Professor Lagerqvist's laboratory.

Table 5. Diatomic molecules, radicals and ions studied at the National Research Council of Canada

---

H_2, He_2, B_2, C_2, N_2, O_2, F_2, Mg_2, Si_2, P_2, S_2, Cl_2, I_2
LiH, BH, CH, NH, OH, SiH, PH, SH, HCl, HBr, CrH, CuH
BN, CN, PN
CO, NO, PO, SO, IO
BS, CS, SnS
BF, NF, NaF, SiF, PF, KF, RbF
BCl, CCl, SiI, NCl, AlCl
NBr, AlBr, CP, AlC
CH^+, NH^+, SiH^+, PH^+
C_2^-, C_2^+, N_2^+, N_2^{++}, P_2^+, CO^+, CN^+, NO^+

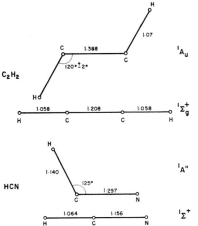

Fig. 13.

Geometrical structures of C_2H_2 and HCN in their ground and first excited states. [Ingold and King (36 a), Innes (36 b), Herzberg and Innes (36 c).

C. Polyatomic molecules, radicals and ions

(1) General remarks

In polyatomic molecules and radicals there are in general several geometrical parameters which are required to describe their structures (rather than one, r_e, for diatomic molecules). For a large number of stable molecules these parameters have been determined in their ground states by the techniques of infrared, Raman and microwave spectroscopy. In excited electronic states these parameters can only be obtained by a detailed study of electronic band systems in absorption or emission. In many cases it has been found that the shape (point group) of the molecule in an excited state is different from its shape in the ground state. For example the molecules C_2H_2 and HCN, well-known to be linear in their ground states are found to be strongly bent in their first (singlet) excited states as shown in Fig. 13. It is clear that this behaviour throws an interesting light on the way in which the electronic structure determines molecular shape.

For polyatomic free radicals and ions one is dependent both for the ground states and the excited states on the study of electronic spectra to obtain the shapes and the geometrical parameters. Only for a few radicals have microwave and infrared spectra been obtained in the gaseous state. Not infrequently, just as for stable molecules, electronic spectra of radicals are diffuse or even continuous in which case no information about geometrical structure can be obtained. The BH_3 radical may be such a case. An additional difficulty in the interpretation of free radical spectra is the problem of identification, that is the problem to which particular radical an observed spectrum belongs. As an example of these difficulties I should like to discuss the history of the discovery of the spectra of CH_2 and C_3.

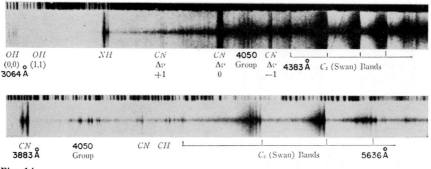

OH   OH   NH   CN   CN   **4050** CN
(0,0) $_o$ (1,1)   Δv   Δv   Group Δv   **4383 Å** C_2 (Swan) Bands
3064 Å   +1   0   —1

CN $_o$ **4050**   CN CH
3883 Å   Group C_2 (Swan) Bands   **5636 Å**

Fig. 14.
Spectrum of comet Cunningham (1940 c) after Swings (37).

(2) CH_2 and C_3

In 1941 the Belgian astronomer Swings wrote to me about a problem that had arisen in the interpretation of the spectra of comets. Fig. 14 shows the spectrum of a comet. In this spectrum the emission bands of CN, C_2, CH, NH, OH can be clearly seen, but in addition there is a group of bands near 4050 Å whose origin nobody had been able to identify at that time. On the basis of the structure of this spectrum, I thought I could eliminate the possibility that it was due to a *diatomic* free radical. Rather, the 4050 group appeared to me like a perpendicular band of a nearly symmetric top molecule, and because of the wide spacing of the subbands I concluded (38) that it must be due to a nonlinear molecule XH_2 with a bond angle of the order of 140°. The most likely identification appeared to be CH_2, particularly since at that time Mulliken (39) had just predicted a spectrum of CH_2 to occur in the region 4000—4500 Å. Since CH was known to be present in comets, the identification of the 4050 group as due to CH_2 seemed eminently reasonable.

On the basis of these considerations I proceeded to do some laboratory experiments. I tried the obvious, passing a discharge through methane (CH_4) in the hope of obtaining in this discharge a spectrum of CH_2. While the continuous discharge through methane showed only well-known features such as CH and H_2, I noticed that the colour of the discharge in the first instant after it was turned on was slightly different from the later colour. I therefore took a spectrum with the discharge turned on and off repeatedly. On this spectrum, in addition to the bands of CH, a new feature appeared precisely at 4050 Å which agreed in almost every detail with the 4050 group as observed in comets. This agreement is shown in Fig. 15. Thus, for the first time [Herzberg (40)], the 4050 group of comets had been observed in the laboratory, and this had been done by choosing conditions suggested by the assumption that the spectrum was due to CH_2. Therefore, it was perhaps excusable that I felt confirmed in my belief that this spectrum was due to CH_2.

However, in 1949 Monfils and Rosen (41) at Liège repeated our experiment but replaced the hydrogen by deuterium. The spectrum that they observed was identical in every detail with the spectrum that I had observed,

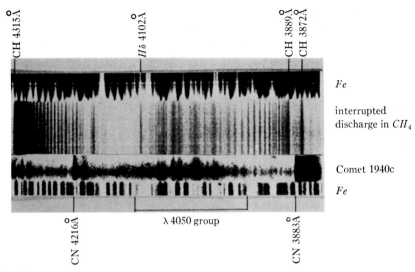

Fig. 15.
The $\lambda 4050$ group in the laboratory and in a comet after Herzberg (40).

whereas of course small isotope shifts would have been expected had this spectrum been due to CH_2. Dr. Douglas and I at Ottawa immediately repeated this experiment using much higher resolution and confirmed the result of the Belgian physicists, thus establishing without doubt that neither the cometary spectrum nor the laboratory spectrum was due to CH_2. Douglas (42) then proceeded to find the true carrier of this spectrum by using methane with C^{13} in it [supplied by the late K. Clusius]. He observed that the main emission band at 4050 Å in a 50—50 mixture of $C^{12}H_4$ and $C^{13}H_4$ was replaced by six bands, showing immediately that the molecule responsible for this spectrum must have three carbon atoms in it. Further consideration of the fine structure of this band left no doubt that the spectrum must be due to the free C_3 radical. At the time when Douglas established this conclusion, the C_3 radical had not even been postulated in any chemical reaction, but since that time it has been found to be one of the dominant constituents of carbon vapour as obtained by the evaporation of graphite.

The question now arose, if the 4050 group is not due to CH_2, where is the true spectrum of CH_2—or does such a spectrum not exist? It was almost ten years after the identification of the carrier of the 4050 group before a spectrum of CH_2 was found. From photochemical evidence it was well known that there are two molecules which on photolysis give CH_2—namely, ketene (CH_2CO) and diazomethane (CH_2N_2). Since the latter compound is rather explosive, we began by studying the continuous photolysis of ketene and, when that failed, by turning to flash photolysis, which had in the meantime been developed [Norrish and Porter (43)]. Even though we extended our search into the vacuum ultraviolet, we did not find a spectrum of CH_2. As a last resort we decided that we should try diazomethane in spite of its hazardous properties. Almost the first absorption spectrum of flash-photolyzed diazomethane showed

1414.5 Å

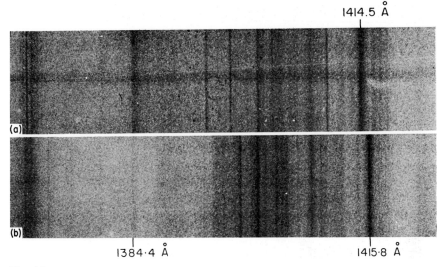

(a)

(b)

1384·4 Å   1415·8 Å

Fig. 16.
Absorption spectra of flash-photolyzed (a) diazomethane and (b) deuterated diazomethane after Herzberg (44).

a new transient feature (of a lifetime of about 10 μsec), which turned out to
be the spectrum of CH_2. We did not observe this particular feature in the
flash photolysis of ketene because ketene itself absorbs strongly at this
particular wavelength.

We were fortunate in being able to obtain immediately, with the help of
Dr. Leitch at Ottawa, a quantity of deuterated diazomethane, and in this way
were able to verify, as shown by Fig. 16, that the new feature, at 1415 Å, actually does shift when hydrogen is substituted by deuterium. Thus, at least we
were sure that the molecule or radical responsible for this feature contained
hydrogen, but of course this observation did not yet prove that the radical
was CH_2.

Spectra taken at high resolution in partially or fully deuterated diazomethane
showed three different positions for the principal band depending on the
deuterium concentration. These spectra, reproduced in Fig. 17, show for two
of the isotopes a clear and simple fine structure. For CD_2 there is in addition
a characteristic intensity alternation indicating a symmetric position of the two
D atoms. The simple structure of the bands suggests that the molecule is
linear. According to molecular orbital theory the ground state of linear CH_2
must be

$$. . . \sigma_g{}^2 \; \pi_u{}^2 \; {}^3\Sigma_g{}^-$$

and in agreement with that the odd lines are strong in the CD_2 band. We
concluded therefore that the ground state of CH_2 is a triplet state and that
the molecule is linear even though the triplet splitting was not resolved.

During the last year several theoretical calculations (45), (46) as well as
electron-spin resonance studies (47), (48) of CH_2 and CD_2 in inert matrices

CH$_2$

1414.5 Å

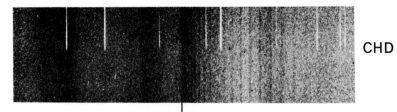

CHD

1415.5 Å

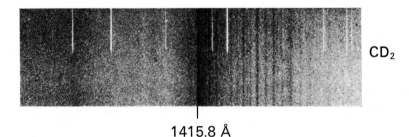

CD$_2$

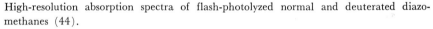

1415.8 Å

Fig. 17.
High-resolution absorption spectra of flash-photolyzed normal and deuterated diazo-methanes (44).

have strongly suggested that CH$_2$ in its lowest triplet state is not linear. If that is so, the molecule is a symmetric top and in addition to the K = 0 subband there should be several other subbands with $K \neq 0$. In 1960 the non-observation of these subbands was considered as strong evidence that the molecule is linear. However in view of the new ESR evidence one must consider the possibility that the $K \neq 0$ subbands are so strongly predissociated that they escape observation [Herzberg and Johns (49)]. Indeed a calculation of the bond angle from the observed B_o values of CD$_2$ and CHD leads to 136° and a bond distance of 1.078 Å. We believe now that this is the correct structure for the ground state of CH$_2$.

In addition to the VUV spectrum a series of bands with complicated fine structure were found [(44), (50)] in the red and photographic infrared of which Fig. 18 gives an example. This spectrum was analyzed as that of an asymmetric top and corresponds to a second modification of CH$_2$ with zero spin (singlet CH$_2$). Its life time is shorter than that of the first (triplet)

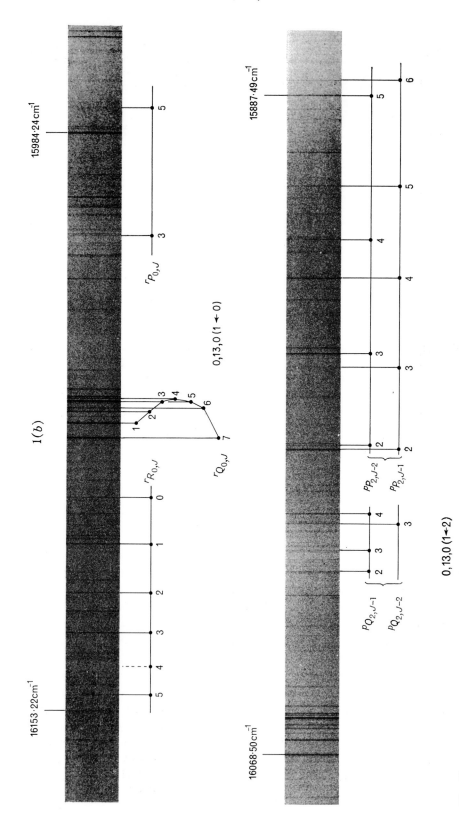

Fig. 18.
Two subbands (1—0 and 1—2) of the 0 13 0—0 0 0 band of the red absorption system of CH_2.

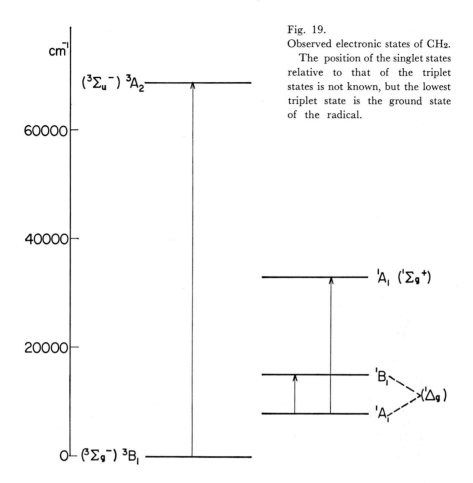

Fig. 19.

Observed electronic states of CH₂.

The position of the singlet states relative to that of the triplet states is not known, but the lowest triplet state is the ground state of the radical.

modification suggesting that the latter represents the ground state. In Fig. 19 an energy level diagram of the observed states is shown. The energy difference of the lowest singlet and triplet state is uncertain. It should be noted that the observed low-lying states are precisely those predicted from the electron configuration $(^3\Sigma_g^-, \, ^1\Delta_g, \, ^1\Sigma_g^+$ for the linear conformation.

In Fig. 20 the geometrical structure of CH₂ in the lowest states is illustrated graphically.

(3) Other dihydrides

Several years before the spectrum of CH₂ was observed Dressler and Ramsay (51) at Ottawa observed and analyzed the spectrum of the NH₂ radical obtained by the flash photolysis of NH₃. The spectrum is quite complicated since the molecule is an asymmetric top and its analysis by Dressler and Ramsay represented a very considerable accomplishment. The experience gained by them was of great help in the analysis of the singlet spectrum of CH₂ referred

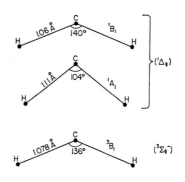

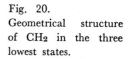

Fig. 20.
Geometrical structure of CH₂ in the three lowest states.

to earlier. Fig. 21 shows the structure of NH_2 derived from the spectrum.

More recently Dr. Johns and I (52) obtained the spectrum of BH_2 by the flash photolysis of BH_3CO. This spectrum lies in the same region as that of NH_2. Fig. 22 shows a section of the spectrum showing the B^{10}—B^{11} isotope shift and Fig. 23 shows the resulting structure. Table 6 summarizes the information of observed bond angles in these and other dihydrides.

Several years before these structures were spectroscopically established Walsh (53) gave some rules for the prediction of such structures on the basis of certain semi-empirical assumptions about the molecular orbitals in these systems. In the Walsh diagram for XH_2 molecules shown in Fig. 24 the predicted energies of the lowest orbitals are plotted as a function of the bond angle (going from 90° to 180°). The orbitals designated $2a_1$ and $1b_2$ (or $2\sigma_g$ and $1\sigma_u$) favour slightly the linear conformation while the orbital $3a_1$ which arises as one component of $1\pi_u$ strongly favours the bent conformation. The other orbital $1b_1$ arising from $1\pi_u$ favours neither conformation. In the ground state of BH_2 the electron configuration is

$$(2a_1)^2 \ (1b_2)^2 \ 3a_1$$

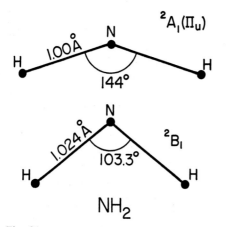

Fig. 21.
Geometrical structure of NH₂ in the two known electronic states.

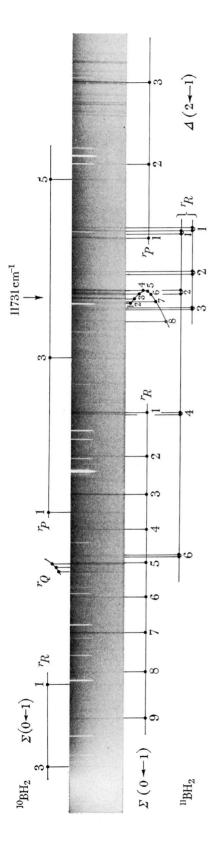

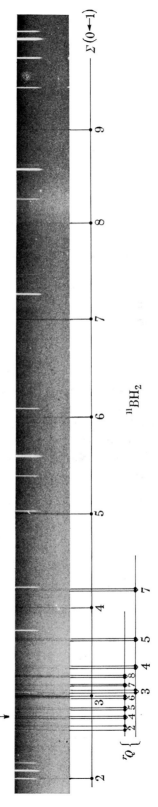

Fig. 22.

Subbands of the BH$_2$ band near 8520 Å after Herzberg and Johns (52).
Lines of ^{10}BH$_2$ are marked above, those of ^{11}BH$_2$ below the spectra.

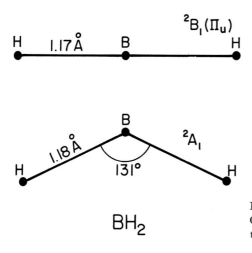

Fig. 23.
Geometrical structure of BH₂ in the two known electronic states.

and therefore in this state the molecule is bent (see Table 6); but in the first excited state

$$(2a_1)^2 \, (1b_2)^2 \, 1b_1$$

there is no tendency for bending and in agreement with observation the molecule is predicted to be linear. In a similar way the other molecules of Table 6 can be treated: Two electrons in the $3a_1$ orbital always lead to strong bending with an angle of about $105°$ as in the ground state of H_2O.

(4) Triatomic monohydrides
A number of triatomic radicals with one H atom have been studied. The first was HCO of which Fig. 25 shows one of the absorption bands [Herzberg and Ramsay (54)]. Here again there is a striking change of shape in the electronic transition as shown in Fig. 26. Moreover this is another case (the first to be recognized) in which only one K value, here $K'' = 1$, appears since in the upper state (where the molecule is linear) all levels with $l \neq 0$ are strongly predissociated.

The spectra of HNO [Dalby (55)] and HCF [Merer and Travis (56)] shown in Figs. 27 and 28 are examples for cases in which the radicals are bent in both upper and lower state and where no predissociation occurs, i.e. several

Table 6. Bond angles in the ground states of triatomic dihydrides

BH₂	CH₂	NH₂	H₂O
131°	136° (triplet)	103.3°	105.2°
	104° (singlet)		

AlH₂	SiH₂	PH₂	H₂S
119°	92.1° (singlet)	91.5°	92.2°

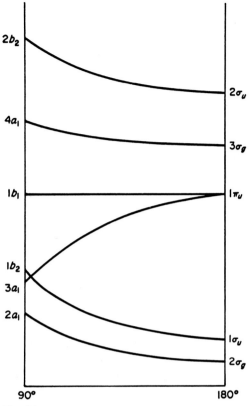

$2b_2$

$4a_1$

$1b_1$

$1b_2$

$3a_1$

$2a_1$

$2\sigma_u$

$3\sigma_g$

$1\pi_u$

$1\sigma_u$

$2\sigma_g$

90° 180°

Fig. 24.
Walsh diagram for XH₂ molecules.
   The variation of the orbital energies in going from a bond angle of 90° to 180° is shown.

subbands of different K are observed. The geometrical structures resulting from the analysis of these spectra are shown in Figs. 29 and 30.

   The most recent radicals studied in this group are HNF [Woodman (57)] and HSiI [Billingsley (58)] for which Figs. 31 and 32 show the resulting structures. HNF is one of the first examples of an asymmetric top in which the spin doubling is fully resolved and analyzed.

(5) Triatomic non-hydrides

   Among triatomic non-hydrides the C_3 radical has already been discussed. It is linear in both the upper and the lower state of the only known electronic transition which is of type $^1\Pi_u\text{—}^1\Sigma_g^+$. In the excited state ($^1\Pi_u$) there is a strong interaction between electronic and vibrational angular momentum (Renner-Teller effect) leading to considerable anomalies in the spacing of the vibrational levels. In the ground state ($^1\Sigma_g^+$) the bending frequency is surprisingly small, only 63 cm⁻¹. The corresponding force constant is 1/100 of that in CO_2 i.e. the molecule is very floppy. The electron configurations of the two states are $(1\pi_u)^4 (3\sigma_u) 1\pi_g)$ and $\ldots (1\pi_u)^4 (3\sigma_u)^2$.

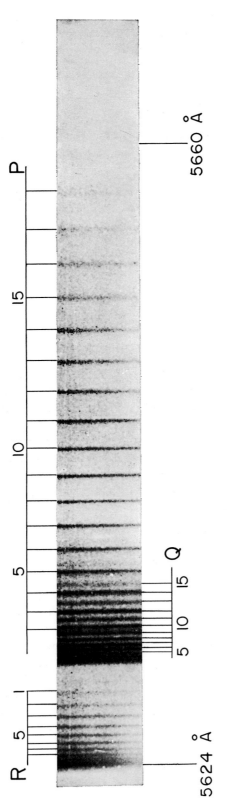

Fig. 25.

The 0 11 0—0 0 0 band of the A—X system of HCO at 5624 Å after Herzberg and Ramsay (54).

Only the 0—1 subband is observed. Each line is an unresolved doublet.

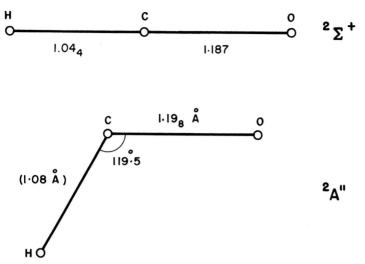

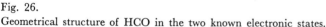

Fig. 26.
Geometrical structure of HCO in the two known electronic states.

Radicals with one more electron which has to go into a π_g orbital are CCN and CNC observed by Merer and Travis (59) at Ottawa. The NCN molecule with two electrons in the π_g orbital and therefore with a $^3\Sigma_g^-$ ground state, was studied by Travis and myself (60), the similar molecule CCO by Devillers and Ramsay (61). We have looked for the corresponding spectrum of the ion N_3^+ which has the same number of electrons. Ledbetter (62) found a complicated absorption band in flash discharges through nitrogen which is not otherwise identifiable, but no proof for its belonging to N_3^+ has yet been obtained. Several radicals and ions with three electrons in the π_g orbital have been studied by various investigators.

Table 7 summarizes the molecular constants for the ground states of this group of molecules. Note the strong increase of the bending frequency ν_2 as a function of the number of π_g electrons.

(6) CH₃ and CH₄⁺

The spectrum of the methyl radical was easier to obtain than that of methylene (44). First by the flash photolysis of $Hg(CH_3)_2$ and later that of many other methyl containing compounds a spectrum consisting of two diffuse peaks near 2160 Å was obtained as shown in Fig. 33 a. For CD_3 a simple (even though still diffuse) fine structure is observed (Fig. 33 b) which yields the conclusion that CD_3 is planar (or very nearly planar) in its ground state and that the CD distance is 1.069 Å. In addition several Rydberg series have been observed in the region 1500—1280 Å which yield accurate values for the ionization potential [I.P.(CH_3) = 9.843eV].

The electron configurations of the two lowest states of CH_3 assuming D_{3h} symmetry are

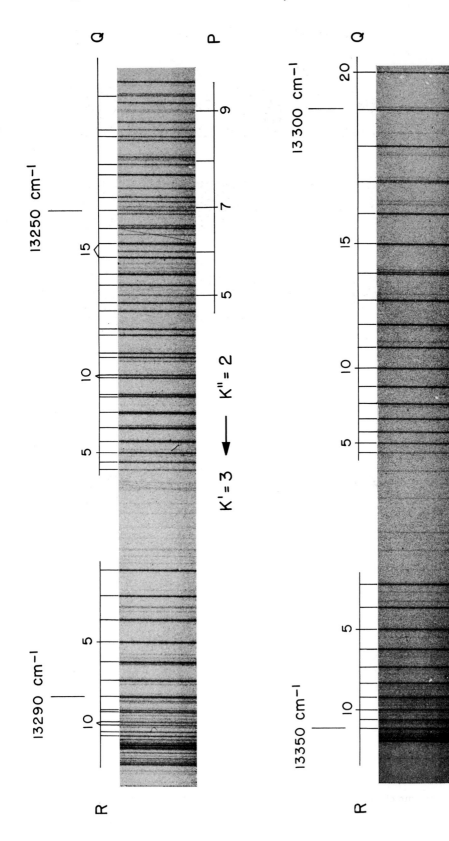

Fig. 27.
Two subbands of the 000—000 band of the Ã—Ã system of the HNO radical after
Dalby (55).

SUB-BANDS OF HCF

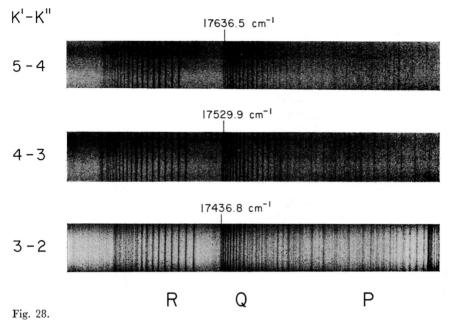

$K'-K''$

17636.5 cm⁻¹

5 - 4

17529.9 cm⁻¹

4 - 3

17436.8 cm⁻¹

3 - 2

R   Q   P

Fig. 28.
Subbands of the 000—000 band of the $\tilde{A}$—$\tilde{X}$ system of the HCF radical after Merer and Travis (56).

$$\ldots (1e')^4 (1a_2'') \; ^2A_2''$$

$$\ldots (1e')^3 (1a_2'')^2 \; ^2E'$$

The transition between these two states should lie in the visible region but for planar CH_3 it is a forbidden transition and so far has not been observed. The upper states of all observed transitions are Rydberg states of the type

$$\ldots (1e')^4 \; nsa_1'; \; nda_1'; \; nde''$$

The spectrum of SiH_3 is of course expected to be very similar to that of CH_3, but in spite of considerable effort we have not yet been able to observe

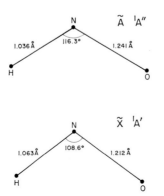

Fig. 29.
Geometrical structure of the HNO radical in the two known electronic states.

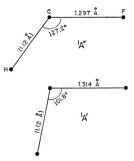

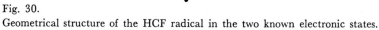

Fig. 30.
Geometrical structure of the HCF radical in the two known electronic states.

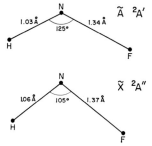

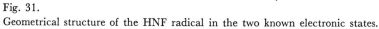

Fig. 31.
Geometrical structure of the HNF radical in the two known electronic states.

Table 7. Molecular constants in the ground states of linear triatomic non-hydrides

Molecule	State	ΔG cm^{-1}	B_0 cm^{-1}	r_0 Å	N
C_3	$^1\Sigma_g^+$	63.1	0.4305	1.277	12
CCN	$^2\Pi_r$	(325)	0.3981		13
CNC	$^2\Pi_g$	321	0.4535	1.245	13
NCN	$^3\Sigma_g^-$	(423)	0.3968	1.232	
CCO	$^3\Sigma^-$	379.4	0.3851		14
NCO	$^2\Pi_i$	(539)	0.3894		
N_3	$^2\Pi_g$		0.4312	1.182	15
BO_2	$^2\Pi_g$	464	0.3292	1.265	
CO_2	$^1\Sigma_g^+$	667.4	0.3902	1.162	16

N = number of valence electrons

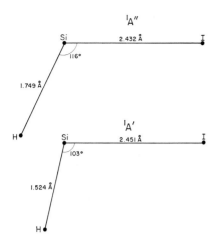

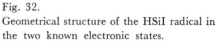

Fig. 32.
Geometrical structure of the HSiI radical in the two known electronic states.

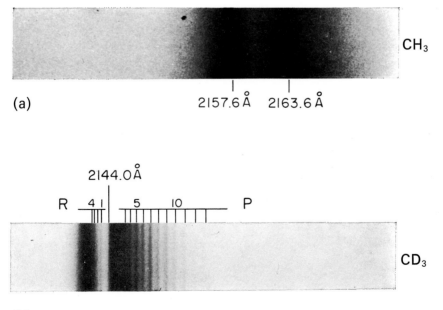

(a)

2157.6 Å   2163.6 Å

2144.0 Å

R   4 | 5   10   P

(b)

Fig. 33.
Absorption bands of (a) CH₃ and (b) CD₃ near 2150 Å after Herzberg (44).

it. Thus it is still an open question whether SiH_3 is planar or non-planar in its ground state although the latter appears more likely.

A molecular system with the same number of electrons as CH_3 is the CH_4^+ ion. If CH_4^+ were tetrahedral like CH_4 its ground state would have the electron configuration

$$(1a_1)^2 \ (2a_1)^2 \ (1f_2)^5 \ {}^2F_2$$

that is, would be a triply degenerate state. However, according to the Jahn-Teller theorem, in such a triply degenerate state the equilibrium conformation cannot be the symmetrical tetrahedral one. Recent theoretical calculations [see for example Dixon (63)] suggest, that CH_4^+ has a D_{2d} structure and that the electron configurations of the two lowest electronic states derived from ${}^2F_2$ are

$$(1a_1)^2 \ (2a_1)^2 \ (1e)^4 \ (1b_2) \ \ {}^2B_2$$

$$\cdots\cdots\cdots (1e)^3 \ (1b_2)^2 \ \ {}^2E$$

If, instead, CH_4^+ had C_{3v} symmetry (i.e. one C-H bond longer than the other three) the states ${}^2A_1$ and ${}^2E$ with electron configurations even more similar to those of the two low-lying states of CH_3 would result. But now the transition between these two low-lying states is not forbidden. That these two low-lying states of CH_4^+ exist has been conclusively shown by photo-electron spectroscopy. Fig. 34 shows the photo-electron spectrum as observed by Price (64). The two maxima at the right correspond to photoelectrons with CH_4^+ left in one or the other of the two states. The allowed transition between these two states would lie in the visible region but has not yet been observed.

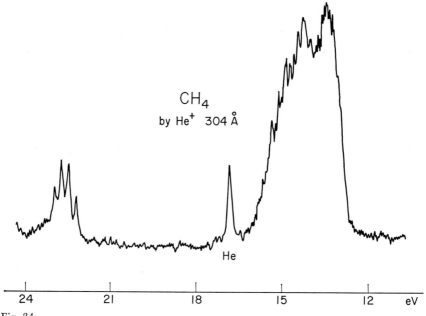

CH$_4$

by He$^+$ 304 Å

He

| 24 | 21 | 18 | 15 | 12 | eV |

Fig. 34.

Photo-electron spectrum of CH$_4$ obtained with the He$^+$ line at 304 Å by Price (64).

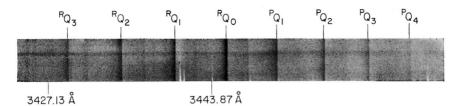

Fig. 35.
The 0—0 band of the Ä—Ä system of the HNCN radical at 3440 Å in absorption after Herzberg and Warsop (65).

One reason (in addition to its intrinsic interest) why we have spent consider-able effort to observe this spectrum is it possible astrophysical importance. It is possible that the diffuse interstellar lines (which represent the last major identification problem in astronomical spectroscopy) may be due to this transition in CH_4^+, the diffuseness being due to predissociation of the upper state (the dissociation energy of CH_4^+ is only about 1 eV). Since CH_4 is almost certainly present in the interstellar medium it would not be surprising if a sufficient stationary concentration of CH_4^+ arises for the appearance of the ^2E—$^2B_2$ transition.

(7) HNCN

As a final example I should like to mention a radical which, like many others, was observed in the flash photolysis of diazomethane, namely HNCN [Herz-berg and Warsop (65)]. As shown in Fig. 35 a particularly simple spectrum arises, a textbook example of a perpendicular band of a nearly symmetric top, consisting of a series of almost equidistant subbands. The large separa-tion of the subbands indicates that the heavy atoms lie very nearly on a straight line while the H atom lies off that line. Isotope effects with D, C^{13} show that only one hydrogen and one carbon atom is present. Fig. 36 shows the derived structure; the position of the C atom cannot be accurately deter-mined from the present data.

D. Conclusion

In the preceding discussion I have not described any of the techniques used in our work since the principal method, the flash-photolysis technique, was described by Norrish and Porter in their Nobel lectures (43). We have more recently developed two methods for the study of spectra of molecular ions: the flash discharge technique which is closely related to the flash photolysis technique and the flash radiolysis (or pulse radiolysis) technique in which a powerful pulse of electrons is sent through the absorption cell and produces many ions. Both methods have been described in more detail in my Faraday lecture (66). The results obtained with the second method have as yet been very sparse but we hope to overcome some of the difficulties and obtain absorption spectra of ions such as $C_2H_2^+$, HCN^+, CH_4^+, CH_3^+, CH_2^+, H_3O^+, H_2O^+, ... $C_6H_6^+$ and many others about which nothing is known at present. The

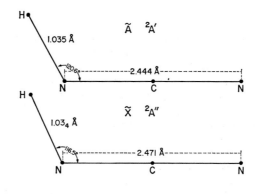

Fig. 36.
Geometrical structure of HNCN in the two known electronic states.

information so obtained would be of immense help for a deeper understanding of the electronic structure of these ions as well as the corresponding neutral molecules. The new methods of tunable lasers as shown by Patel and his collaborators (67) promise to give infrared spectra of radicals and ions about which up to now very little is known. For example the vibrational frequencies in the ground state of CH_2 are unknown at present. These few remarks may suffice to indicate that a great deal remains to be done in the study of the geometric and electronic structure of radicals and molecular ions.

ACKNOWLEDGEMENT

The work described in this lecture could not have been accomplished without the effective help and independent work of many collaborators. The list of references gives the names of most of them. I should like to single out particularly the constant support by most valuable criticisms and suggestions which I have had from Dr. A. E. Douglas, now my successor as director of the Physics Division of the National Research Council. I should also like to acknowledge the devoted help I have had for twenty years from Mr. J. Shoosmith who carried out all the experimental work with great efficiency, energy and skill.

In receiving the Nobel Prize I am deeply aware of the debt I owe to all these collaborators and technical assistants for their continued help and support.

REFERENCES

1. Curry, J., Herzberg, L. and Herzberg, G., Z. Physik **86**, 348 (1933).
2. Franck, J., Trans. Faraday Soc. **21**, 536 (1925).
3. Birge, R. T. and Sponer, H., Phys. Rev. **28**, 259 (1926).
4. Mulliken, R. S., Les Prix Nobel en 1966, Stockholm 1967.
5. Kolos W. and Roothaan, C. C. J., Rev. Mod. Phys. **32**, 219 (1960).
6. Kolos W. and Wolniewicz, L., J. Chem. Phys. **48**, 3672 (1968); **49**, 404 (1968).
7. Hunter, G. and Pritchard, H. O., J. Chem. Phys. **46**, 2153 (1967).
8. Herzberg, G., J. Mol. Spec. **33**, 147 (1970).
9. Bunker, P. R., J. Mol. Spec. (in press).
10. Stoicheff, B. P., Can. J. Phys. **35**, 730 (1957).
11. Herzberg, G., Can. J. Research A, **28**, 144 (1950).
12. Fink, O., Wiggins, T. A. and Rank, D. H., J. Mol. Spec. **18**, 384 (1965); Foltz, T. V., Rank, D. H. and Wiggins, T. A., J. Mol. Spec. **21**, 203 (1966).
13. Herzberg, G. and Howe, L. L., Can. J. Phys. **37**, 636 (1959)
14. Bredohl, H. and Herzberg, G., to be published.
15. Poll, J. D. and Karl, G., Can. J. Phys. **44**, 1467 (1966).
16. Herzberg, G. and Jungen, C., J. Mol. Spec. (in press).
17. Takezawa, S., Abstracts, Symposium on Molecular Structure and Spectroscopy, Columbus, Ohio (1971).
18. Jeziorski, B. and Kołos, W., Chem. Phys. Letters **3**, 677 (1969).
19. Van Vleck, J. H., Astrophys. J. **80**, 161 (1934).
20. Noxon, J. F., Can. J. Phys. **39**, 1110 (1961).
21. Herzberg, G., Naturwiss. **20**, 577 (1932); Can. J. Phys. **30**, 185 (1952); **31**, 657 (1953).
22. Herzberg, G., Trans. Roy. Soc. Canada **46**, 1 (1952).
23. Herzberg, G., Mémoires de la Soc. Roy. des Sciences de Liège, ser. V, **17**, p. 121. (1969).
24. Heurlinger, Diss. Lund (1918).
25. Heurlinger and Hulthén, E., Z. wiss. Photographie, Photophys. und Photochem. **18**, 241 (1919)
26. Hulthén, E., Z. Phys. **11**, 284 (1922); Diss. Lund (1923).
27. Hori, T., Z. Physic, **59**, 91 (1929).
28. Herzberg, G. and Johns, J. W. C., Astrophys. J. **158**, 399 (1969).
29. Swings, P. and Rosenfeld, L., Astrophys. J. **86**, 483 (1937).
30. McKellar, A., Publ. Astron. Soc. Pac. **52**, 307 (1940); Publ. Dominion Astrophys. Observ. **7**, 251 (1941).
31. Adams, W. S., Astrophys. J. **93**, 11 (1941).
32. Douglas, A. E. and Herzberg, G., Astrophys. J. **94**, 381 (1941); Can. J. Res. A **20**, 71 (1942).
33. Herzberg, G. and Lagerqvist, A., Can. J. Phys. **46**, 2363 (1968).
34. Milligan, D. E. and Jacox, M. E., J. Chem. Phys. **51**, 1952 (1969).
35. Lineberger, W. C. and Patterson, T. A., in press.
36. Meinel, H., Can. J. Phys. **50**, 158 (1972).
36a. Ingold, C. K. and King, G. W., J. Chem. Soc., p. 2702 (1953).
36b. Innes, K. K., J. Chem. Phys. **22**, 863 (1954).
36c. Herzberg, G. and Innes, K. K., Can. J. Phys. **35**, 842 (1957).
37. Swings, P., Publ. Astron. Soc. Pac. **54**, 123 (1942).
38. Herzberg, G., Rev. Mod. Phys. **14**, 195 (1942).
39. Mulliken, R. S., quoted in (38).
40. Herzberg, G., Astrophys. J. **96**, 314, (1942).
41. Monfils, A. and Rosen, B., Nature **164**, 713 (1949).
42. Clusius, K. and Douglas, A. E., Can. J. Phys. **32**, 319 (1954).
43. Norrish, R. G. W. and Porter, G., Nature **164**, 658 (1949); also Les Prix Nobel en 1967, Stockholm 1968.

44. Herzberg, G., Proc. Roy. Soc. **262 A,** 291 (1961).
45. Harrison, J. F. and Allen, J. C., J. Amer. Chem. Soc. **91,** 807 (1969); **93,** 4112 (1971).
46. Bender, C. F., Schaefer, H. F. and O'Neil, V., J. Amer. Chem. Soc. **92,** 4984 (1970); J. Chem. Phys. 55, 162 (1971).
47. Bernheim, R. A., Bernard, H. W., Wang, P. S., Wood, L. S. and Shell, P. S., J. Chem. Phys. **53,** 1280 (1970).
48. Wasserman, E., Jager, W. A., Kuch, V. J. and Hutton, R. S., Chem. Phys. Let. **7,** 409 (1970); J. Amer. Chem. Soc. **92,** 7491 (1970).
49. Herzberg, G. and Johns, J. W. C., J. Chem. Phys. **54,** 2276 (1971).
50. Herzberg, G. and Johns, J. W. C., Proc. Roy. Soc. **295 A,** 107 (1966).
51. Dressler, K. and Ramsay, D. A., Phil. Trans. **251 A,** 553 (1959).
52. Herzberg, G. and Johns, J. W. C., Proc. Roy. Soc. **298 A,** 142 (1967).
53. Walsh, A. D., J. Chem. Soc. p. 2260, 2266, 2288, 2296 (1953).
54. Herzberg, G. and Ramsay, D. A., Proc. Roy. Soc. **233 A,** 34 (1955).
55. Dalby, F. W., Can. J. Phys. **36,** 1336 (1958).
56. Merer, A. J. and Travis, D. N., Can. J. Phys. **44,** 1541 (1966).
57. Woodman, C. M., J. Mol. Spec. **33,** 311 (1970).
58. Billingsley, J., to be published.
59. Merer, A. J. and Travis, D. N., Can. J. Phys. **43,** 1795 (1965); **44,** 353 (1966).
60. Herzberg, G. and Travis, D. N., Can. J. Phys. **42,** 1658 (1964).
61. Devillers, C. and Ramsay, D. A., Can. J. Phys. **49,** 2839 (1971).
62. Ledbetter, J., unpublished.
63. Dixon, R. N., Mol. Phys. **20,** 113 (1971).
64. Price, W. C., in "Molecular Spectroscopy" ed. Hepple, P. W., Inst. Petroleum, London 1968, p. 221.
65. Herzberg, G. and Warsop, P. A., Can. J. Phys. **41,** 286 (1963).
66. Herzberg, G., Quart. Revs. Chem. Soc. **25,** 201 (1971).
67. Patel, C. K. N., paper presented at Esfahan Symposium on Fundamental and Applied Laser Physics (1971).

Chemistry 1972

CHRISTIAN B. ANFINSEN

for his work on ribonuclease, especially concerning the connection between the amino acid sequence and the biologically active confirmation

STANFORD MOORE and WILLIAM H. STEIN

for their contribution to the understanding of the connection between chemical structure and catalytic activity of the active centre of the ribonuclease molecule

THE NOBEL PRIZE FOR CHEMISTRY

Speech by Professor Bo MALMSTRÖM of the Royal Academy of Sciences
Translation from the Swedish text

Your Royal Highnesses, Ladies and Gentlemen,
The key substances of life are called enzymes. Everything we humans undertake—if we sit here enjoying the splendour of a Nobel ceremony, if we perform work, or even if we simply feel joy or sorrow—occurs by means of enzyme reactions. The phenomenon described as life is a network of coupled enzymatic processes. In chemical terminology the enzymes are catalysts, i.e. substances which accelerate chemical reactions without being consumed themselves. The concept of catalysis was introduced about 150 years ago by the great Swedish chemist Jöns Jacob Berzelius, who also, with astonishing intuition, suggested that the tissues of a living organism have catalytic activity. Around the turn of the century scientists started to associate this catalytic effect with specific substances, enzymes. This year's three Nobel Prize winners in Chemistry, Christian B. Anfinsen, Stanford Moore and William H. Stein, have performed fundamental studies with the enzyme ribonuclease making it possible for us now to approach the problem of enzymatic activity on a molecular level.

From a chemical point of view enzymes are proteins. These are built up of 20 different amino acids which are linked together into long chains. Despite the fact that proteins have only 20 building blocks, there are thousands of enzymes, each with its specific properties. This large degree of variation becomes possible because the number and sequence of the amino acids in the chain can be varied. Ribonuclease was the first enzyme for which the complete amino acid sequence was determined thanks to contributions from Anfinsen and from Moore and Stein.

Every living organism has its own characteristic pattern of enzymes. It can also produce a copy of itself, and this progeny has the same enzymes. An important question concerns the source of the information which has to be passed on from generation to generation for the enzyme pattern to be preserved. We know, thanks to contributions which have led to earlier Nobel Prize awards, that a specific molecule, called DNA, serves as the carrier of the traits of inheritance. These traits are expressed by DNA controlling the synthesis of enzymes. DNA accomplishes this by determining the sequence of the amino acids making up a particular protein molecule. An active enzyme does not, however, consist just of a long chain of amino acids linked together, but the chain is folded in space in a way which gives the molecule a globular form. What is the source of the information responsible for this specific folding of the peptide chain? It is this question in particular which has been the concern of Anfinsen's investigations. In a series of elegant experiments he showed that the necessary information is inherent in the linear sequence of amino acids in the peptide chain, so that no further genetic information than that found in DNA is necessary.

The contributions of Moore and Stein concern another fundamental question regarding ribonuclease, namely the basis for its catalytic activity. The reacting substances, the substrates, are bound to an enzyme in what is generally called its active site. In the complex so formed there is an interaction between enzyme and substrate leading to a changed reactivity of the substrate. Knowledge about the structure of an enzyme is of little help in understanding this interaction if it is not possible to find the active site and to determine the chemical groups in it. Moore and Stein discovered as an important principle that the active site contains amino acids with an anomalously high reactivity compared to the same amino acids in free form. This high reactivity is of direct importance for the catalytic activity of the enzyme, but Moore and Stein also found it possible to utilize it to label two amino acids in the active site by chemical modification. In this way the position of these amino acids in the long peptide chain could be unambigously determined. Through these investigations Moore and Stein were able to provide a detailed picture of the active site of ribonuclease long before the three-dimensional structure of the enzyme had been determined.

Dr. Anfinsen,

I have tried to explain your pioneering investigations showing that the linear sequence of amino acids in the enzyme ribonuclease determines the biologically active conformation of this enzyme. This finding has profound implications for our understanding of the way in which active enzyme molecules are formed in living cells.

Drs. Moore and Stein,

I have attempted to summarize your fundamental contributions to our understanding of the relationship between chemical structure and catalytic activity in the enzyme ribonuclease. In particular, I have stressed your studies leading to the localization of two specific histidine residues in the active site of the enzyme. It is for these pioneering experiments that the Royal Academy of Sciences has decided to award this year's Nobel Prize in Chemistry to you together with Dr. Anfinsen.

Drs. Anfinsen, Moore and Stein,

On behalf of the Royal Academy of Sciences I wish to convey to you our warmest congratulations, and I now ask you to receive your Prizes from the hands of His Royal Highness the Crown Prince.

Christian B. Anfinsen

CHRISTIAN B. ANFINSEN

Born in Monessen, Pennsylvania, March 26, 1916 Dr. Anfinsen obtained a B.A. degree from Swarthmore College in 1937 and an M.S. in organic chemistry in 1939 from the University of Pennsylvania. He spent the year 1939—40 as a Visiting Investigator at the Carlsberg Laboratory in Copenhagen. In 1943, he received a Ph.D. from Harvard Medical School in biochemistry and spent the next seven years at Harvard Medical School; first as Instructor and then as Assistant Professor of Biological Chemistry. During this time, he spent a year (1947—48) as a Senior Fellow of the American Cancer Society working with Dr. Hugo Theorell at the Medical Nobel Institute. Dr. Anfinsen left Harvard in 1950 to become Chief of the Laboratory of Cellular Physiology and Metabolism in the National Heart Institute of the National Institutes of Health. He was again at Harvard Medical School as Professor of Biological Chemistry in 1962—63 and then returned to the National Institutes of Health to assume his present position.

In Anfinsen's early work, he and Steinberg studied the non-uniform labelling in newly synthesized proteins—a technique with later permitted Dintzis, Canfield and others to determine that proteins are synthesized sequentially from the amino-terminal and *in vivo*, and to calculate the rate at which amino acids are polymerized.

In the mid 1950's Anfinsen began to concentrate on the problem of the relationship between structure and function in enzymes. On the basis of studies on ribonuclease with Sela and White, he proposed that the information determining the tertiary structure of a protein resides in the chemistry of its amino acid sequence. Investigations on reversible denaturation of several proteins served to verify this proposal experimentally. It was demonstrated that, after cleavage of disulfide bonds and disruption of tertiary structure, many proteins could spontaneously refold to their native forms. This work resulted in general acceptance of the "thermodynamic hypothesis". Studies on the rate and extent of renaturation *in vitro* led to the discovery of a microsomal enzyme which catalyzes sulfhydryl-disulfide interchange and thereby accelerates, *in vitro*, the refolding of denatured proteins containing disulfide bonds. In the presence of this enzyme the rate of renaturation approaches that sufficient to account for folding of newly completed polypeptide chains during protein biosynthesis. These findings have given important impetus to studies on the organic synthesis of proteins, since they demonstrate that, under physiological conditions of environment, attainment of the native structure rests solely upon the correct sequential polymerization of the amino acids.

In addition to his research activities, Dr. Anfinsen is an editor of Advances in Protein Chemistry, served on the Editorial Board of the Journal of Biological Chemistry and wrote "The Molecular Basis of Evolution" which was published in 1959. He is active as a member of the Board of Governors of the Weizmann Institute of Science in Rehovot, Israel, and was elected President of the American Society of Biological Chemists for the Academic Year 1971—72. His honors include a Rockefeller Foundation Public Service Award in 1954, a Guggenheim Fellowship in 1958, election to the National Academy of Sciences in 1963 and he Royal Danish Academy in 1964, and Honorary Doctor of Science degrees from Swarthmore College (1965), Georgetown University (1967), and New York Medical College (1969).

In recent years, Anfinsen has devoted himself primarily to comprehensive investigations of an extracellular nuclease of *Staphylococcus aureus*. He and his colleagues have determined the sequence of its 149 amino acids and have described its fundamental enzymological, physical, and immunological properties. They have used an extensive range of spectroscopic and chemical techniques, including new methods of affinity labeling and cross-linking, to delineate the identity and relationship of amino acids in its active site. Dr. Anfinsen has collaborated closely with a crystallographic group at M.I.T., under Professor F. A. Cotton, which has determined the three-dimensional structure of nuclease at high resolution.

Membership and activities:
Board of Governors, Weizmann Institute of Science, Rehovot, Israel
American Society of Biological Chemists (President, 1971—72)
National Academy of Sciences
Royal Danish Academy
Honors:
1954 Rockefeller Foundation Public Service Award
1957 International Union of Pure and Applied Chemistry
 Travel Grant to attend Symposium on Proteins, Paris France
1958 Guggenheim Fellowship for Travel to do research at the Weizmann
 Institute of Science
1959 National Science Foundation Travel Award to attend Conference on
 Genetic Specificity of Proteins, Copenhagen
1964 NIH Lecture
 Kelly Lecture (Purdue University)
1965 D.Sc. (Hon.)—Swarthmore College
1966 Harvey Lecture
1967 D.Sc. (Hon.)—Georgetown University
1969 D.Sc. (Hon.)—New York Medical College
 Honorary Fellow, Weizmann Institute of Science
 Leon Lecture (University of Pennsylvania)
1970 EMBO Lecturer for Sweden
 Visiting Fellow, All Souls College, Oxford
1972 Jubilee Lecture

STUDIES ON THE PRINCIPLES THAT GOVERN THE FOLDING OF PROTEIN CHAINS

Nobel Lecture, December 11, 1972

by

CHRISTIAN B. ANFINSEN

National Institutes of Health
Bethesda, Maryland

The telegram that I received from the Swedish Royal Academy of Sciences specifically cites ". . . studies on ribonuclease, in particular the relationship between the amino acid sequence and the biologically active conformation . . ." The work that my colleagues and I have carried out on the nature of the process that controls the folding of polypeptide chains into the unique three-dimensional structures of proteins was, indeed, strongly influenced by observations on the ribonuclease molecule. Many others, including Anson and Mirsky (1) in the '30s and Lumry and Eyring (2) in the '50s, had observed and discussed the reversibility of denaturation of proteins. However, the true elegance of this consequence of natural selection was dramatized by the ribonuclease work, since the refolding of this molecule, after full denaturation by reductive cleavage of its four disulfide bonds (Figure 1), required that only one of the 105

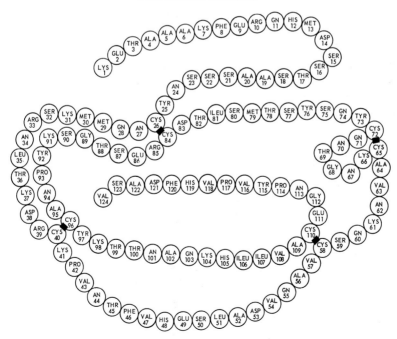

Fig. 1.
The amino acid sequence of bovine pancreatic ribonuclease (3, 4, 5).

possible pairings of eight sulfhydryl groups to form four disulfide linkages take place. The original observations that led to this conclusion were made together with my colleagues, Michael Sela and Fred White, in 1956—1957 (6). These were in actuality, the beginnings of a long series of studies that rather vaguely aimed at the eventual total synthesis of the protein. As we all know, Gutte and Merrifield (7) at the Rockefeller Institute, and Ralph Hirschman and his colleagues at the Merck Research Institute (8), have now accomplished this monumental task.

The studies on the renaturation of fully denatured ribonuclease required many supporting investigations (9, 10, 11, 12) to establish, finally, the generality which we have occasionally called (13) the "thermodynamic hypothesis". This hypothesis states that the three-dimensional structure of a native protein in its normal physiological milieu (solvent, pH, ionic strength, presence of other components such as metal ions or prosthetic groups, temperature, etc.) is the one in which the Gibbs free energy of the *whole system* is lowest; that is, that the native conformation is determined by the totality of interatomic interactions and hence by the amino acid sequence, in a *given environment*. In terms of natural selection through the "design" of macromolecules during evolution, this idea emphasized the fact that a protein molecule only makes stable, structural sense when it exists under conditions similar to those for which it was selected—the so-called physiological state.

After several years of study on the ribonuclease molecule it became clear to us, and to many others in the field of protein conformation, that proteins devoid of restrictive disulfide bonds or other covalent cross linkages would make more convenient models for the study of the thermodynamic and kinetic aspects of the nucleation, and subsequent pathways, of polypeptide chain folding. Much of what I will review will deal with studies on the flexible and convenient staphylococcal nuclease molecule, but I will first summarize some of the older, background experiments on bovine pancreatic ribonuclease itself.

Support for the "thermodynamic hypothesis."

An experiment that gave us a particular satisfaction in connection with the translation of information in the linear amino acid sequence into native conformation involved the rearrangement of so-called "scrambled" ribonuclease (12). When the fully reduced protein, with 8 SH groups, is allowed to reoxidize under denaturing conditions such as exist in a solution of 8 molar urea, a mixture of products is obtained containing many or all of the possible 105 isomeric disulfide bonded forms (schematically shown at the bottom right of Figure 2). This mixture is essentially inactive — having on the order of 1% the activity of the native enzyme. If the urea is removed and the "scrambled" protein is exposed to a small amount of a sulfhydryl group-containing reagent such as mercaptoethanol, disulfide interchange takes place and the mixture eventually is converted into a homogeneous product, indistinguishable from native ribonuclease. This process is driven entirely by the free energy of conformation that is gained in going to the stable, native structure. These experi-

IN VITRO

UNFOLDING
(Urea + Mercaptoethanol)

REFOLDING

Fig. 2.
Schematic representation of the reductive denaturation, in 8 molar urea solution containing 2-mercaptoethanol, of a disulfide-cross linked protein. The conversion of the extended, denatured form to a randomly cross linked, "scrambled" set of isomers is depicted in the lower right portion of the figure.

ments, incidentally, also make unlikely a process of obligatory, progressive folding during the elongation of the polypeptide chain, during biosynthesis, from the NH$_2$- to the COOH-terminus. The "scrambled" protein appears to be essentially devoid of the various aspects of structural regularity that characterize the native molecule.

A disturbing factor in the kinetics of the process of renaturation of reduced ribonuclease, or of the "unscrambling" experiments described above, was the slowness of these processes, frequently hours in duration (11). It had been established that the time required to synthesize the chain of a protein like ribonuclease, containing 124 amino acid residues, in the tissues of a higher organism would be approximately 2 minutes (14, 15). The discrepancy between the *in vitro* and *in vivo* rates led to the discovery of an enzyme system in the endoplasmic reticulum of cells (particularly in those concerned with the secretion of extracellular, SS-bonded proteins) which catalyzes the disulfide interchange reaction and which, when added to solutions of reduced ribonuclease or to protein containing randomized SS bonds, catalyzed the rapid formation of the correct, native disulfide pairing in a period less than the requisite two minutes (16, 17). The above discrepancy in rates would not have been observed in the case of the folding of non-crosslinked structures and, as discussed below, such motile proteins as staphylococcal nuclease or myoglobin can undergo virtually complete renaturation in a few a seconds or less.

The disulfide interchange enzyme subsequently served as a useful tool for the examination of the thermodynamic stability of disulfide-bonded protein structures. This enzyme, having a molecular weight of 42,000 and containing three half-cystine residues, one of which must be in the SH form for activity (18, 19), appears to carry out its rearranging activities on a purely random basis. Thus, a protein whose SS bonds have been deliberately broken and re-formed in an incorrect way, need only be exposed to the enzyme (with its essential half-cystine residue in the pre-reduced, SH form) and interchange of disulfide bonds occurs until the native form of the protein substrate is reached. Presumably, SS bonds occupying solvent-exposed, or other ther-modynamically unfavorable positions, are constantly probed and progressively replaced by more favorable half-cystine pairings, until the enzyme can no longer contact bonds because of steric factors, or because no further net decrease in conformational free energy can be achieved. Model studies on ribonuclease derivatives had shown that, when the intactness of the genetic message represented by the linear sequence of the protein was tampered with by certain cleavages of the chain, or by deletions of amino acids at various points, the added disulfide interchange enzyme, in the course of its "probing", discovered this situation of thermodynamic instability and caused the random reshuffling of SS bonds with the formation of an inactive cross-linked network of chains and chain fragments (e.g., (20)). With two naturally occurring proteins, insulin and chymotrypsin, the interchange enzyme did, indeed, induce such a randomizing phenomenon (21). Chymotrypsin, containing three SS-bonded chains, is known to be derived from a single-chained precursor, chy-motrypsinogen, by excision of two internal bits of sequence. The elegant studies of Steiner and his colleagues subsequently showed that insulin was also derived from a single-chained precursor, proinsulin (Figure 3), which is converted to

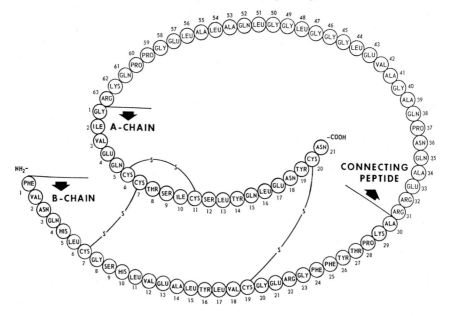

Fig. 3.
The structure of porcine proinsulin (R. E. Chance, R. M. Ellis and W. W. Bromer, Science, *161*, 165 (1968).

the two-chained form, in which we normally find the active hormone, by removal of a segment from the middle of the precursor strand after formation of the 3 SS bonds (22). In contrast, the multichained immune globulins are *not* scrambled and inactivated by the enzyme, reflecting the fact that they are normal products of the disulfide bonding of 4 preformed polypeptide chains.

FACTORS CONTRIBUTING TO THE CORRECT FOLDING OF POLYPEPTIDE CHAINS.
The results with the disulfide interchange enzyme discussed above suggested that the correct and unique translation of the genetic message for a particular protein backbone is no longer possible when the linear information has been tampered with by deletion of amino acid residues. As with most rules, however, this one is susceptible to many excpetions. First, a number of proteins have been shown to undergo reversible denaturation, including disulfide bond rupture and reformation, after being shortened at either the NH$_2$- or COOH-terminus (23). Others may be cleaved into two (24, 25, 26), or even three, fragments which, although devoid of detectable structure alone in solution, recombine through noncovalent forces to yield biologically active structures with physical properties very similar to those of the parent protein molecules. Richards and his colleagues (24) discovered the first of these recombining systems, ribonuclease-S (RNase-S), which consists of a 20 residue fragment from the NH$_2$-terminal end held by a large number of noncovalent interactions to the rest of the molecule, which consists of 104 residues and all four of the disulfide bridges. The work by Wyckoff, Richards and their associates on the three-dimensional structure of this two-fragment complex (27) and on the identification of many of the amino acid side chains that are essential for complementation is classical, as are studies by Hofmann (28) and Scoffone (29) and their colleagues on semi-synthetic analogues of this enzyme derivative. Studies in our own laboratory (30) showed that the 20 residue "RNase-S-peptide"

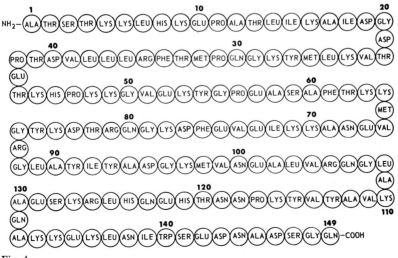

Fig. 4.
Covalent structure of the major extracellular nuclease of *Staphylococcus aureus* (32, 33).

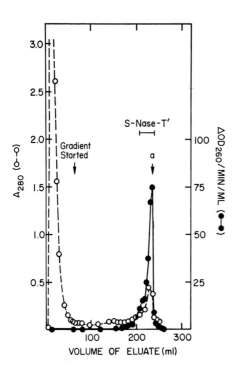

Fig. 5.

Isolation of semisynthetic nuclease-T on a phosphocellulose column following "functional" purification by trypsin digestion in the presence of calcium ions and thymidine-3'5'-diphosphate (41).

fragment could be reduced by 5 residues at its COOH-terminus without loss of enzymic activity in the complex, or of its intrinsic stability in solution.

Other examples of retention of native structural "memory" have been found with complexing fragments of the staphylococcal nuclease molecule (25, 31). This calcium-dependent, RNA and DNA cleaving enzyme (Figure 4) consists of 149 amino acids and is devoid of disulfide bridges and SH groups (32, 33). Although it exhibits considerable flexibility in solution as evidenced by the ready exchange of labile hydrogen atoms in the interior of the molecule with solvent hydrogen atoms (34), only a very small fraction of the total population deviates from the intact, native format at any moment. Spectral and hydrodynamic measurements indicate marked stability up to temperatures of approximately 55°. The protein is greatly stabilized, both against hydrogen exchange (34) and against digestion by proteolytic enzymes (35) when calcium ions and the inhibitory ligand, 3'5'-thymidine diphosphate (pdTp), are added. Trypsin, for example, then only cleaves at very restricted positions — the loose amino terminal portion of the chain and a loop of residues that protrudes out from the molecule as visualized by X-ray crystallography. Cleavage occurs between lysine residues 5 and 6 and, in the sequence -Pro-Lys-Lys-Gly- (residues 47 through 50), between residues 48 and 49 or 49 and 50 (25). The resulting fragments (6—48) and (49—149) or (50—149), are devoid of detectable structure in solution (36). However, as in the case of RNase-S, when they are mixed in stoichiometric amounts regeneration of activity (about 10 %) and of native structural characteristics occurs (the complex is called nuclease-T). Nuclease-T has now been shown (37a) to be closely isomorphous with native nuclease (37b). Thus the cleavages

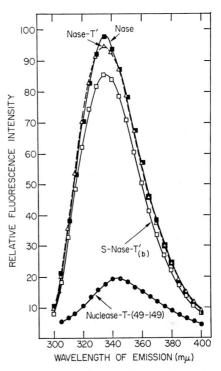

Fig. 6.
Use of fluorescence measurements to determine the relative hydrophobicity (presumably reflecting "nativeness" in the case of nuclease) of the molecular environment of the single tryptophan residue in this protein (39, 41).

and deletions do not destroy the geometric "sense" of the chain. Recently it was shown that residue 149 may be removed by carboxypeptidase treatment of nuclease, and that residues 45 through 49 are dispensable, the latter conclusion the result of solid phase-synthetic studies (38) on analogues of the fragment, (6—47).

Earlier studies by David Ontjes (39) had established that the rapid and convenient solid-phase method developed by Merrifield (40) for peptide synthesis could be applied to the synthesis of analogues of the (6—47) fragment of nuclease-T. The products, although contaminated by sizeable amounts of "mistake sequences" which lack an amino acid residue due to slight incompleteness of reaction during coupling, could be purified by ordinary chromatographic methods to a stage that permitted one to make definite conclusions about the relative importance of various components in the chain. Taking advantage of the limited proteolysis that occurs when nuclease is treated with trypsin in the presence of the stabilizing ligands, calcium and pdTp, Chaiken (41) was able to digest away those aberrant synthetic molecules of (6—47) that did not form a stable complex with the large, native fragment, (49—149). After digestion of the complex, chromatography on columns of phosphocellulose (Figure 5) yielded samples of semisynthetic nuclease-T that were essentially indistinguishable from native nuclease-T. For example, the large enhancement of fluorescence of the single tryptophan residue in nuclease (located at position 140 in the (50—149) fragment) upon addition of the native (6—49) fragment was also shown when, instead, synthetic (6—47) peptide isolated from semisynthetic nuclease-T that had been purified as described above was added (Figure 6).

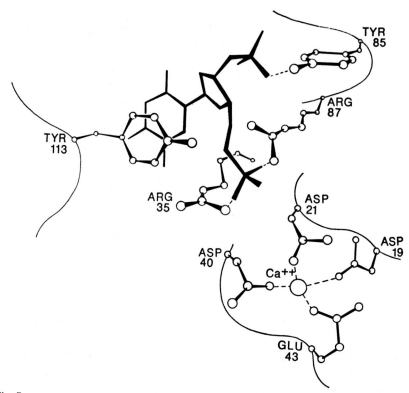

Fig. 7.

Amino acid residues in the sequence of nuclease that are of particular importance in the catalytic activity and binding of substrate and calcium ions (42).

The dispensability, or replaceability, of a number of residues to the stability of the nuclease-T complex was established by examining the fluorescence, activity, and stability to enzymatic digestion, of a large number of semisynthetic analogues (42). As illustrated in Figure 7, interaction with the calcium atom required for nuclease activity normally requires the participation of four dicarboxylic amino acids. Although the *activities* of complexes containing synthetic (6—47) fragments in which one of these had been replaced with an asparagine or glutamine residue were abolished (with one partial exception— asparagine at position 40), three dimensional structure and complex stability was retained for the most part. Similarly, replacement of arginine residue 35 with lysine yielded an inactive complex, but nevertheless one with strong three dimensional similarity to native nuclease-T.

A second kind of complementing system of nuclease fragments (31) consists of tryptic fragment (1—126) and a partially overlapping section of the sequence, (99—149), prepared by cyanogen bromide treatment of the native molecule (shown schematically in Figure 8). These two peptides form a complex with about 15 % the activity of nuclease itself which is sufficiently stable in the presence of pdTp and calcium ions to exhibit remarkable resistance to digestion by trypsin. Thus, many of the overlapping residues in the complex,

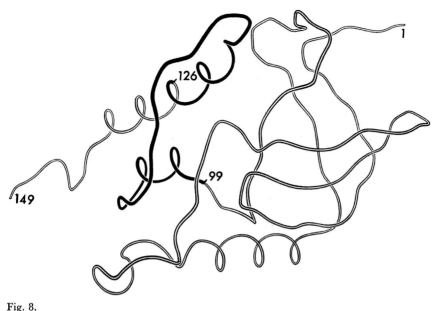

Fig. 8.
A schematic view of the three dimensional structure of staphylococcal nuclease (37b, 53).

(1—126): (99—149), may be "trimmed" away with the production of a derivative, (1—126): (111—149). Further degradation of each of the two components, the former with carboxypeptidases A and B and the latter with leucine
aminopeptidase, permits the preparation of (1—124): (114—149) which is as
active, and as structurally similar to native nuclease, (as evidenced by estimates of hydrodynamic, spectral, and helical properties) as the parent, un-
degraded complex. A number of synthetic analogues of the (114—149)
sequence have been prepared (43), which also exhibit activity and "native"
physical properties when added to (1—126). I will discuss below the manner
in which these complexing fragments have been useful in devising experiments
to study the processes of nucleation and folding of polypeptide chains.

MUTABILITY OF INFORMATION FOR CHAIN FOLDING.
Biological function appears to be more a correlate of macromolecular geometry
than of chemical detail. The classic chemical and crystallographic work on the
large number of abnormal human hemoglobins, the species variants of cytochrome *c* and other proteins from a very large variety of sources, and the isolation of numerous bacterial proteins after mutation of the corresponding genes
have made it quite clear that considerable modification of protein sequence
may be made without loss of function. In those cases where crystallographic
studies of three-dimensional structure have been made, the results indicate
that the geometric problem of "designing," through natural selection, molecules that can subserve a particular functional need can be solved in many
ways. Only the *geometry* of the protein and its active site need be conserved,
except, of course, for such residues as actually participate in a unique way in
a catalytic or regulatory mechanism (44). Studies of model systems have led

to similar conclusions. In our own work on ribonuclease, for example, it was shown that fairly long chains of poly-D, L-alanine could be attached to eight of the eleven amino groups of the enzyme without loss of enzyme activity (45). Furthermore, the polyalanylated enzyme could be converted to an extended chain by reduction of the four SS bridges in 8 M urea and this fully denatured material could then be reoxidized to yield the active, correctly folded starting substance. Thus, the chemistry of the protein could be greatly modified, and its capacity to refold after denaturation seemed to be dependent only on *internal* residues and not those on the outside, exposed to solvent. This is, of course, precisely the conclusion reached by Perutz and his colleagues (46), and by others (47) who have reviewed and correlated the data on various protein systems. Mutation and natural selection are permitted a high degree of freedom during the evolution of species, or during accidental mutation, but a limited number of residues, destined to become involved in the internal, hydrophobic core of proteins, must be carefully conserved (or at most replaced with other residues with a close similarity in bulk and hydrophobicity).

THE COOPERATIVITY REQUIRED FOR FOLDING AND STABILITY OF PROTEINS.

The examples of non-covalent interaction of complementing fragments of proteins quoted above give strong support to the idea of the essentiality of cooperative interactions in the stability of protein structure. As in the basic rules of languages, an incomplete sentence frequently conveys only gibberish. There appears to exist a very fine balance between stable, native protein structure and random, biologically meaningless polypeptide chains.

A very good example of the inadequacy of an incomplete sequence comes from our observations on the nuclease fragment, (1—126). This fragment contains all of the residues that make up the active center of nuclease. Nevertheless, this fragment, representing about 85 % of the total sequence of nuclease, exhibits only about 0.12 % the activity of the native enzyme (48). The further

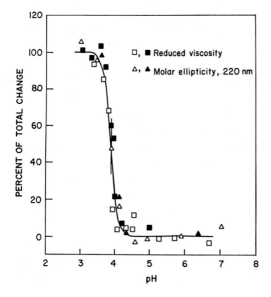

Fig. 9.
Changes in reduced viscocity and molar ellipticity at 220 nm during the acid-induced transition from native to denatured nuclease.
☐ and ■, Reduced viscosity; △ and ▲, molar ellipticity at 220 nm. ☐ and △, Measurements made during the addition of acid; ■ and ▲, measurements made during the addition of base.
A. N. Schechter, H. F. Epstein and C. B. Anfinsen, unpublished results.

addition of 23 residues during biosynthesis, or the addition, *in vitro*, of residues 99—149 as a complementing fragment (31), restores the stability required for activity to this unfinished gene translation.

The transition from incomplete, inactive enzyme, with random structure, to competent enzyme, with unique and stable structure, is clearly a delicately balanced one. The sharpness of this transition may be emphasized by experiments of the sort illustrated in Figure 9. Nuclease undergoes a dramatic change from native globular structure to random disoriented polypeptide over a very narrow range of pH, centered at pH 3.9. The transition has the appearance of a "two-stage" process—either all native or all denatured—and, indeed two-state mathematical treatment has classically been employed to describe such data. In actuality, it has been possible to show, by NMR and spectrophotometric-experiments (49), that one of the 4 histidines and one tyrosine residue of the 7 in nuclease become disoriented before the general and sudden disintegration of organized structure. However, such evidences of a stepwise denaturation and renaturation process are certainly not typical of the bulk of the cooperatively stabilized molecule.

The experiments in Figure 9, involving measurements of intrinsic vicosity and helix-dependent circular dichroism, are typical of those obtained with most proteins. In the case of nuclease, not only is the transition from native to denatured molecule during transfer from solution at pH 3.2 to 6.7 very abrupt, but the process of renaturation occurs over a very short time period. I will not discuss these stop-flow kinetic experiments (50) in detail in this lecture. In brief, the process can be shown to take place in at least two phases; an initial rapid nucleation and folding with a half-time of about 50 milliseconds and a second, somewhat slower transformation with a half time of about 200 milliseconds. The first phase is essentially temperature independent (and therefore possibly entropically driven) and the second temperature dependent.

NUCLEATION OF FOLDING.

A chain of 149 amino acid residues with two rotatable bonds per residue, each bond probably having 2 or 3 permissible or favored orientations, would be able to assume on the order of 4^{149} to 9^{149} different conformations in solution. The extreme rapidity of the refolding makes it essential that the process take place along a limited number of "pathways", even when the statistics are severely restricted by the kinds of stereochemical ground rules that are implicit in a so-called Ramachandran plot. It becomes necessary to postulate the existence of a limited number of allowable initiating events in the folding process. Such events, generally referred to as nucleations, are most likely to occur in parts of the polypeptide chain that can participate in conformational equilibria between random and cooperatively stabilized arrangements. The likelihood of a requirement for cooperative stabilization is high because, in aqueous solution, ionic or hydrogen bonded interactions would not be expected to compete effectively with interactions with solvent molecules and anything less than a sizeable nucleus of interacting amino acid side chains would probably have a very short lifetime. Furthermore, it is important to stress that the

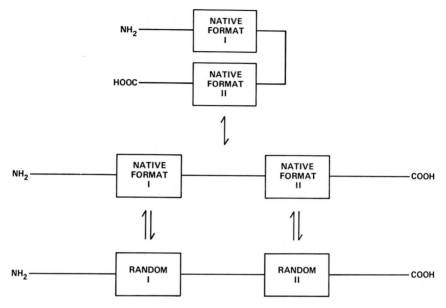

Fig. 10.
How protein chains might fold (see the text for a discussion of this fairly reasonable, but subjective proposal).

amino acid sequences of polypeptide chains designed to be the fabric of protein molecules only make functional sense when they are in the three dimensional arrangement that characterizes them *in the native protein structure*. It seems reasonable to suggest that portions of a protein chain that can serve as nucleation sites for folding will be those that can "flicker" in and out of the conformation that they occupy in the final protein, and that they will form a relatively rigid structure, stabilized by a set of cooperative interactions. These nucleation centers, in what we have termed their "native format", (Figure 10) might be expected to involve such potentially self-dependent substructures as helices, pleated sheets or beta-bends.

Unfortunately, the methods that depend upon hydrodynamic or spectral measurements are not able to detect the presence of these infrequent and transient nucleations. To detect the postulated "flickering equilibria" and to determine their probable lifetimes in solution requires indirect methods that will record the brief appearance of individual "native format" molecules in the population under study. One such method, recently used in our laboratory in a study of the folding of staphylococcal nuclease and its fragments, employs specific antibodies against restricted portions of the amino acid sequence (51).

Figure 8 depicts the three dimensional pattern assumed by staphylococcal nuclease in solution. Major features involving organized structure are the three-stranded antiparallel pleated sheet approximately located between residues 12 and 35, and the three alpha-helical regions between residues 54—67, 99—106 and 121—134. Antibodies against specific regions of the nuclease molecule were prepared by immunization of goats with either polypeptide

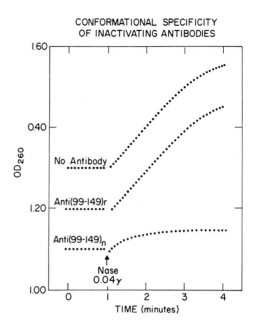

CONFORMATIONAL SPECIFICITY
OF INACTIVATING ANTIBODIES

Fig. 11.
Inhibition of nuclease activity by anti-[$(99-149)_n$], and lack of inhibition by anti-[$(99-149)_r$] made against the peptide $(99-149)$, presumably in a random conformation (51).

fragments of the enzyme or by injection of the intact, native protein with subsequent fractionation of the resulting antibody population on affinity chromatography columns consisting of agarose bearing the covalently attached peptide fragment of interest (51, 52). In the former manner there was prepared, for example, an antibody directed against the polypeptide, residues 99—149, known to exist in solution as a random chain without the extensive helicity that characterizes this portion of the nuclease chain when present as part of the intact enzyme. Such an antibody preparation is referred to as anti-$(99-149)_r$, the subscript indicating the disordered state of the antigen.

When, on the other hand, a fraction of anti-native nuclease serum, isolated on an agarose-nuclease column, was further fractionated on agarose-$(99-149)$, a fraction was obtained which was specific for the sequence $(99-149)$ but presumably only when this bit of sequence occupied the "native format". This latter conclusion is based on the observation that the latter fraction, termed anti-$(99-149)_n$ (the subscript "n" referring to the native format) exhibited a strong inhibitory effect on the enzymic activity of nuclease whereas anti-$(99-126)_r$ or anti-$(99-149)_r$ were devoid of such an effect (see Figure 11). This conclusion was further supported by the observation that the conformation--stabilizing ligands, pdTp and calcium ions, showed a market inhibitory effect on the precipitability of nuclease by anti-$(1-126)_r$ and anti-$(99-149)_r$ but had little effect, if any on such precipitability by anti-$(1-149)_n$ (51). This finding reinforced the idea that many of the antigenic determinants recognized by the anti-*fragment* antibodies are present only in the "unfolded" or "non-native" conformation of nuclease. Analysis of the reaction between anti-$(99-149)$ and nuclease could be shown by measurements of changes in the

INHIBITION OF ANTIBODY-INDUCED INACTIVATION

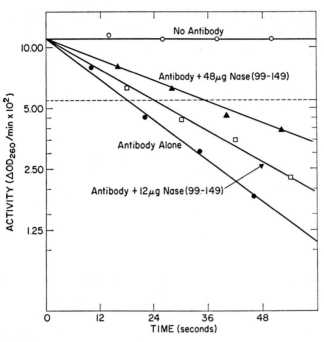

Fig. 12.
Semilogarithmic plot of activity *vs.* time for assays of 0.05 μg of nuclease in the presence of:
○—○ no antibody, ●—● 6 μg of anti-(99—126)$_n$, □—□ 6 μg of anti-(99—126)$_n$ plus
12 μg of (99—149) and ▲—▲ 6 μg of anti-(99—126)$_n$ plus 48 μg of (99—149). The dotted
line represents one-half of the initial activity.

kinetics of inhibition of enzyme activity (Fig. 12), to be extremely rapid:
k_{off}, on the other hand, is negligibly small.

The system may be described by two simultaneous equilibria, the first
concerned with the "flickering" of fragment (99—149), which we shall term
"P", from random to native "format", and the second with the association of
anti-(99—149)$_n$, which we shall term, simply, "Ab" with fragment P in its
native format: i.e. P_n.

$$P_r \rightleftharpoons P_n \quad K_{conf.} = \frac{[P_n]}{[P_r]}$$

$$Ab + P_n \rightleftharpoons AbP_n \quad K_{assoc.} = \frac{[AbP_n]}{[Ab] \cdot [P_n]}$$

$$K_{conf.} = \frac{[AbP_n]}{K_{assoc.} \cdot [Ab] \cdot [P_r]}$$

Two equilibria involving fragment (99—149) of nuclease with the corresponding equilibrium-constant expressions.

The amount of unbound antibody in the second equilibrium may be estimated
from measurements of the kinetics of inactivation of the digestion of denatured

Table 1

Studies of the equilibrium between the peptide fragment (99 — 149) in its random form [99 — 149)$_r$] and in the form this fragment assumes in the native structure of nuclease (99 — 149)$_n$]

Abbreviations: P, fragment (99 — 149); Ab, antibody.

$$K_{conf.} = \frac{[AbP_n]}{K_{assoc.} \cdot [Ab] \cdot [P_T]}$$

[Ab]$_{total\ sites}$ (μM)	[P$_T$] (μM)	t$_{1/2}$(s)	[Ab]$_{free\ sites}$ (μM)	[Ab]$_{bound\ sites}$ (μM)	K$_{conf.}$	% of P$_T$ as P$_n$
0.076	0	18	0.076	0	—	—
0.076	0.65	20	0.068	0.0080	2.20×10^{-4}	0.022
0.076	2.0	24	0.057	0.019	2.02×10^{-4}	0.020
0.076	2.6	27	0.051	0.025	2.29×10^{-4}	0.023
0.076	7.8	35	0.039	0.037	1.47×10^{-4}	0.015
0.076	6.5	33	0.042	0.034	1.51×10^{-4}	0.015

$$K_{conf.} = (2.0 \pm 0.4) \times 10^{-4}$$

DNA substrate by a standard amount of nuclease added to the preincubated mixture of fragment (99 — 149) and anti-(99 — 149)$_n$. Making the assumption that the affinity of anti-(99 — 149)$_n$ for (99 — 149)$_n$ in its folded (P) form is the same as that determined for this antigenic determinant in native nuclease, the value for the term K$_{conf.}$ may be calculated from measureable parameters. A series of typical values shown in Table 1, suggests that approximately 0.02% of fragment (99 — 149) exists in the native format at any moment. Such a value, although low, is very large relative to the likelihood of a peptide fragment of a protein being found in its native format on the basis of chance alone.

Empirical considerations of the large amount of data now available on correlations between sequence and three dimensional structure (54), together with an increasing sophistication in the theoretical treatment of the energetics of polypeptide chain folding (55) are beginning to make more realistic the idea of the *a priori* prediction of protein conformation. It is certain that major advances in the understanding of cellular organization, and of the causes and control of abnormalities in such organization, will occur when we can predict, in advance, the three dimensional, phenotypic consequences of a genetic message.

BIBLIOGRAPHY

1. Anson, M. L., Advan. Protein Chem. 2, 361 (1945).
2. Lumry, R. and Eyring, H., J. Phys. Chem. 58, 110 (1954).
3. Hirs, C. H. W., Moore, S. and Stein, W. H., J. Biol. Chem. 235, 633 (1960).
4. Potts, J. T., Berger, A., Cooke, J. and Anfinsen, C. B., J. Biol. Chem. 237, 1851 (1962).
5. Smyth, D. G., Stein, W. H. and Moore, S., J. Biol. Chem. 238, 227 (1963).
6. Sela, M., White, F. H. and Anfinsen, C. B., Science 125, 691 (1957).
7. Gutte, B. and Merrifield, R. B., J. Biol. Chem. 246, 1922 (1971).
8. Hirschmann, R., Nutt, R. F., Veber, D. F., Vitali, R. A., Varga, S. L., Jacob, T. A., Holly, F. W. and Denkewalter, R. G., J. Am. Chem. Soc. 91, 507 (1969).
9. White, F. H., Jr. and Anfinsen, C. B., Ann. N. Y. Acad. Sci. 81, 515 (1959).

10. White, F. H., Jr., J. Biol. Chem. *236*, 1353 (1961).
11. Anfinsen, C. B., Haber, E., Sela, M. and White, F. H., Jr., Proc. Nat. Acad. Sci. U. S. *47*, 1309 (1961).
12. Haber, E. and Anfinsen, C. B., J. Biol. Chem. *237*, 1839 (1962).
13. Epstein, C. J., Goldberger, R. F. and Anfinsen, C. B., Cold Spring Harbor Symp. Quant. Biol. *28*, 439 (1963).
14. Dintzis, H. M., Proc. Nat. Acad. Sci. U. S. *47*, 247 (1961).
15. Canfield, R. E. and Anfinsen, C. B., Biochemistry *2*, 1073 (1963).
16. Goldberger, R. F., Epstein, C. J. and Anfinsen, C. B., J. Biol. Chem. *238*, 628 (1963).
17. Venetianer, P. and Straub, F. B., Biochim. Biophys. Acta *67*, 166 (1963).
18. Fuchs, S., DeLorenzo, F. and Anfinsen, C. B., J. Biol. Chem. *242*, 398 (1967).
19. DeLorenzo, F., Goldberger, R. F., Steers, E., Givol, D. and Anfinsen, C. B., J. Biol. Chem. *241*, 1562 (1966).
20. Kato, I. and Anfinsen, C. B., J. Biol. Chem. *244*, 5849 (1969).
21. Givol, D., DeLorenzo, F., Goldberger, R. F. and Anfinsen, C. B., Proc. Nat. Acad. Sci. U. S. *53*, 766 (1965).
22. Steiner, D. F., Trans. N. Y. Acad. Sci. Ser. II *30*, 60 (1967).
23. Anfinsen, C. B., Developmental Biology Supplement *2*, 1 (1968), Academic Press Inc. U.S.A., 1968.
24. Richards, F. M., Proc. Nat. Acad. Sci. U. S. *44*, 162 (1958).
25. Taniuchi, H., Anfinsen, C. B. and Sodja, A., Proc. Nat. Acad. Sci. U. S. *58*, 1235 (1967).
26. Kato, I. and Tominaga, N., FEBS Letters *10*, 313 (1970).
27. Wyckoff, H. W., Tsernoglou, D., Hanson, A. W., Knox, J. R., Lee, B. and Richards, F. M., J. Biol. Chem. *245*, 305 (1970).
28. Hofmann, K., Finn, F. M., Linetti, M., Montibeller, J. and Zanetti, G., J. Amer. Chem. Soc. *88*, 3633 (1966).
29. Scoffone, E., Rocchi, R., Marchiori, F., Moroder, L., Marzotto, A. and Tamburro, A. M., J. Amer. Chem. Soc. *89*, 5450 (1967).
30. Potts, J. T., Jr., Young, D. M. and Anfinsen, C. B., J. Biol. Chem. *238*, 2593 (1963).
31. Taniuchi, H. and Anfinsen, C. B., J. Biol. Chem. *246*, 2291 (1971).
32. Cone, J. L., Cusumano, C. L., Taniuchi, H. and Anfinsen, C. B., J. Biol. Chem. *246*, 3103 (1971).
33. Bohnert, J. L. and Taniuchi, H., J. Biol. Chem. *247*, 4557 (1972).
34. Schechter, A. N., Moravek. L. and Anfinsen, C. B., J. Biol. Chem. *244*, 4981 (1969).
35. Taniuchi, H., Moravek, L. and Anfinsen, C. B., J. Biol. Chem. *244*, 4600 (1969).
36. Taniuchi, H. and Anfinsen, C. B., J. Biol. Chem. *244*, 3864 (1969).
37a. Taniuchi, H., Davies, D. and Anfinsen, C. B., J. Biol. Chem. *247*, 3362 (1972).
   b. Arnone, A., Bier, C. J., Cotton, F. A., Hazen, E. E., Jr., Richardson, D. C., Richardson, J. S. and Yonath, A., J. Biol. Chem. *246*, 2302 (1971).
38. Sanchez, G. R., Chaiken, I. M. and Anfinsen, C. B., J. Biol. Chem., *248*, 3653 (1973).
39. Ontjes, D. and Anfinsen, C. B., J. Biol. Chem. *244*, 6316 (1969).
40. Merrifield, R. B., Science *150*, 178 (1965).
41. Chaiken, I. M., J. Biol. Chem. *246*, 2948 (1971).
42. Chaiken, I. M. and Anfinsen, C. B., J. Biol. Chem. *246*, 2285 (1971).
43. Parikh, I., Corley, L. and Anfinsen, C. B., J. Biol. Chem. *246*, 7392 (1971).
44. Fitch, W. M. and Margoliash, E., Evolutionary Biology, *4*, 67 (1970), Edited by Th. Dobzhansky, M. K. Hecht and W. C. Steere, Appleton-Century-Crofts, New York, 1970.
45. Cooke, J. P., Anfinsen, C. B. and Sela, M., J. Biol. Chem. *238*, 2034 (1963).
46. Perutz, M. F., Kendrew, J. C. and Watson, H. C., J. Mol. Biol. *13*, 669 (1965).
47. Epstein, C. J., Nature *210*, 25 (1966).
48. D. Sachs, H. Taniuchi, A. N. Schechter and A. Eastlake, unpublished work.
49. Epstein, H. F., Schechter, A. N., and Cohen, J. S. *Proc. Nat. Acad. Sci. U. S., 68*, 2042 (1971).

50. Epstein, H. F., Schechter, A. N., Chen, R. F. and Anfinsen, C. B., J. Mol. Biol. *60*, 499 (1971).
51. Sachs, D. H., Schechter, A. N., Eastlake, A. and Anfinsen, C. B., Proc. Nat. Acad. Sci. U. S., *69*, 3790, (1972).
52. Sachs, D. H., Schechter, A. N., Eastlake, A. and Anfinsen, C. B., J. Immunol. *109*, 1300 (1972).
53. Sachs, D. H., Schechter, A. N., Eastlake, A. and Anfinsen, C. B., Biochemistry, *11*, 4268 (1972).
54. Anfinsen, C. B. and Scheraga, H., Adv. in Prot. Chem. *29*, 205 (1975).
55. H. A. Scheraga, Chemical Reviews *71*, 195 (1971).

STANFORD MOORE

Stanford Moore was born in 1913 in Chicago, Illinois, and grew up in Nashville, Tennessee, where his father was a member of the faculty of the School of Law of Vanderbilt University. His developmental years were in a home environment which made the pursuit of knowledge an eagerly adopted undertaking. He had the opportunity to attend a high school administered by the George Peabody College for Teachers in Nashville. A skilled teacher of science, R. O. Beauchamp, kindled an interest in chemistry. Moore entered Vanderbilt University undecided between a career in chemistry or aeronautical engineering. The courses which he took in the engineering school presaged a concern for instrumentation. But a gifted professor of organic chemistry, Arthur Ingersoll, succeeded in presenting the study of molecular architecture as an even more appealing discipline. Moore graduated from Vanderbilt (B.A. 1935, *summa cum laude*) with a major in chemistry. The faculty recommended him for a Wisconsin Alumni Research Foundation Fellowship which took him to the University of Wisconsin where he received his Ph.D. in organic chemistry in 1938.

His thesis research was in biochemistry in the laboratory of Karl Paul Link. The first lessons that the young graduate student received from the skilled hands of his professor were in the microanalytical methods of Pregl for the determination of C, H, and N; Link had recently returned from Europe where he had studied in the laboratory of Fritz Pregl in Graz. This training from Link in microchemistry was especially valuable for a student who was later to be concerned with the quantitative analysis of proteins. Moore's thesis was on the characterization of carbohydrates as benzimidazole derivatives. The experience of bringing that work from the bench to the printed page under Link's guidance marked Moore's transition from a student to a productive scholar.

Karl Paul Link was a friend of Max Bergmann, who had recently arrived from Germany to lead a laboratory at the Rockefeller Institute for Medical Research in New York. Through that friendship, Moore was encouraged to join the Bergmann Laboratory in 1939, which was an internationally renowned center of research on the chemistry of proteins and enzymes. During Emil Fischer's last years Max Bergmann had been his senior research associate, and Bergmann had attracted to Rockefeller a group of versatile chemists who maintained a tradition of innovative research and high productivity. After nearly three valuable years in such company, which included William H. Stein, the advent of World War II drew Moore out of the laboratory to serve as a junior administrative officer in Washington for academic and industrial chemical projects administered by the Office of Scientific Research and Development. At the close of the war, he was on duty with the Operational Research Section attached to the Headquarters of the United States Armed Forces in the Pacific Ocean Area, Hawaii.

During the war years, the situation at the Rockefeller Institute had changed. The untimely death of Max Bergmann in 1944 had brought to a close the major chapter in biochemistry which the contributions of his laboratory comprised. Moore's decision to return to Rockefeller was influenced by Herbert Gasser, then the Director of The Rockefeller Institute, who offered to give modest space to Moore and Stein to pursue the theme of research which they had begun with Bergmann or any new lines of investigation that appealed to them. Thus began the collaboration that led to the development of quantitative chromatographic methods for amino acid analysis, their automation, and the utilization of such techniques, in cooperation with younger associates, in the researches in protein chemistry summarized in the Nobel Lecture by Moore and Stein in this volume.

The investigations were conducted in an atmosphere at Rockefeller that encouraged interdepartmental cooperation and international consultation that would expedite research. Interludes included Moore's tenure of the Francqui Chair at the University of Brussels in 1950, where, at the generous invitation of E. J. Bigwood, a laboratory of amino acid analysis was organized in the School of Medicine. Moore had the opportunity to round out the year in Europe with six months in England at the University of Cambridge where he shared part of a laboratory with Frederick Sanger during the time of the pioneering studies on insulin. In 1968, Moore was a Visiting Professor of Health Sciences at the Vanderbilt University School of Medicine.

Memberships and Activities: American Society of Biological Chemists (Treasurer, 1956—59; Editorial Board, 1950—60; President, 1966), American Chemical Society, hon. member Belgian Biochemical Society, foreign correspondent Belgian Royal Academy of Medicine, Biochemical Society (Great Britain), U.S. National Academy of Sciences (Chairman, Section of Biochemistry, 1970), American Academy of Arts and Sciences, Harvey Society, Chairman of Panel on Proteins of the Committee on Growth of the National Research Council (1947—49), Secretary of the Commission on Proteins of the International Union of Pure and Applied Chemistry (1953—57), Chairman of the Organizing Committee for the Sixth International Congress of Biochemistry (1964), President of the Federation of American Societies for Experimental Biology (1970).

Honors:
Docteur *honoris causa* from the Faculty of Medicine of the University of Brussels (1954) and from the University of Paris (1964). Award shared with William H. Stein: American Chemical Society Award in Chromatography and Electrophoresis, 1964; Richards Medal of the American Chemical Society, 1972; Linderstrøm-Lang Medal, Copenhagen, 1972.

Stanford Moore died in 1982.

William H Stein
(PMS)

WILLIAM H. STEIN

I was born June 25, 1911 in New York City, the second of three children, to Freed M. and Beatrice Borg Stein. My father was a business man who was greatly interested in communal affairs, particularly those dealing with health, and he retired quite early in life in order to devote his full time to such matters as the New York Tuberculosis and Health Association, Montefiore Hospital and others. My mother, too, was greatly interested in communal affairs and devoted most of her life to bettering the lot of the children of New York City. During my childhood, I received much encouragement from both of my parents to enter into medicine or a fundamental science.

My early education was at the Lincoln School of Teachers College of Columbia University in New York City, a school which was considered progressive for that time and which fostered in me an active interest in creative arts, music, and writing. There I had my first course in chemistry which proved to be an extremely valuable and interesting one. I left this school when I was about sixteen and went to an excellent preparatory school in New England, namely Phillips Exeter Academy, which was at the time, although it has changed since, a much more rigid and much more demanding educational experience than I had had at Lincoln. It was at Exeter that I was introduced to standards of precision of writing, and of work generally which I think has stood me in very good stead, and I believe that the combination of a progressive school and a more demanding school such as I enjoyed was an ideal preparation. From Exeter I went to Harvard where I had a very enjoyable, although not a very academically distinguished career, and graduated from the college in 1933 at the depths of the economic depression. I had majored in chemistry at college and decided to continue on at Harvard as a graduate student in that subject. This proved to be a rather unfortunate experience because my first graduate year was undistinguished, to say the very least. I was almost ready to abandon a career in science when it was suggested to me that I might enjoy biochemistry much more than straight organic chemistry.

The next year, I transferred to the Department of Biochemistry, then headed by the late Hans Clarke at the College of Physicians and Surgeons, Columbia University in New York. The department at Columbia was an eye-opener for me. Professor Clarke had succeeded in surrounding himself with a fascinating and active faculty and an almost equally stimulating group of graduate students. From both of these I learned a tremendous amount in a short time. My thesis involved the amino acid analysis of the protein elastin, which was then thought to play a role in coronary artery disease and I completed the requirements for my degree at Columbia late in 1937 and went directly to the laboratory of Max Bergmann at the Rockefeller Institute.

While still a graduate student, I had the extreme good fortune to marry, in 1936, Phoebe Hockstader who has been of enormous support to me ever since. We have three sons, William H. Jr., 35; David F., 33; Robert J., 28.

Bergmann was, I still feel, one of the very great protein chemists of this century and he, too, had the ability to surround himself with a most talented group of postdoctoral colleagues. In the laboratory at the time that I was there were, of course, Dr. Moore, and, in addition, Dr Joseph S. Fruton, Dr. Emil L. Smith, Dr. Klaus Hofmann, Dr. Paul Zamecnik, and many others. It was impossible not to learn a great deal about the business of research in protein chemistry from Bergmann, himself, and from the outstanding group he had around him.

The task of Moore and myself was to devise accurate analytical methods for the determination of the amino acid composition of proteins, because Bergmann firmly believed, as did we, that the amino acid analysis of proteins bore the same relationship to these macromolecules that elementary analysis bore to the chemistry of simpler organic substances. It was during this period in the mid-thirties that Bergmann and Fruton and their colleagues were working out the specificity of proteolytic enzymes, work which has had a profound effect upon our knowledge of how enzymes function and has made it possible to use these proteolytic enzymes as tools for the degradation and subsequent derivation of structure of protein molecules ever since.

Work on proteins was suspended during the war for other more pressing matters and Dr. Moore left the laboratory in order to be of assistance in Washington and elsewhere. Our entire group was engaged in working for the Office of Scientific Research and Development. Bergmann's death in 1944 robbed the world of a distinguished chemist and, of course, left the laboratory without a chief. The group continued to function until the end of the war at which time Moore and I had the very great good fortune to be asked by Dr. Herbert Gasser, Director of the Institute, to stay on at Rockefeller with the freedom to do anything we pleased in the biochemical field.

In the meantime, had come the remarkable developments in England on the separation of amino acids by paper chromatography by Martin and Synge and Sanger had started his classical work on the derivation of the structure of insulin. It was then, perhaps, not surprising that Moore and I resumed our collaboration, and following a suggestion of Synge began to try to separate amino acids on columns of potato starch. We were very fortunate in hitting upon a type of potato starch which was well-suited to our needs almost immediately, and from that day on began to work first on the amino acid analysis, and then on the structural analysis of proteins. From columns of potato starch, we progressed to columns of ion exchange resins, developed the automatic amino acid analyzer, and together with a group of very devoted and extremely skillful collaborators, began work on the structure of ribonuclease. These columns were also used for other purposes. In the course of the early work, we developed a drop-counting automatic fraction collector which is now a common instrument in most biochemical laboratories throughout the world.

I should like to emphasize that the development of methods grew out of a need rather than a particular desire to develop methods as ends in themselves. We needed to know the amino acid composition of proteins, we needed to be able to separate and analyze peptides in good yield, and we needed to be able to purify proteins chromatographically. Since there were no methods for doing any of these things at the time that we started, we had to devise them ourselves. We not only wanted to know what the amino acid sequence of an enzyme such as ribonuclease was, but we tried to find out as much as we could about what made it an enzyme and after we had taken that particular enzyme about as far as we thought we could profitably go, we turned to a number of others which have been listed in the Nobel Lecture.

During all of this time, we had the undeviating support of an enlightened administration at Rockefeller who believed in allowing us to do those things which we thought to be important, and, during the last years of this work, we also have had great financial assistance from the NIH. For this and particularly for the very large number of devoted and talented colleagues which we have had in the laboratory we shall be forever grateful.

During all of this time, each of us, naturally, developed interests outside of the laboratory. I, for example, became greatly concerned about the promulgation of scientific information and have been attached, in one way or another, to the Journal of Biological Chemistry for a matter of over fifteen years. During this time it has been my privilege to work with a knowledgeable and dedicated group of biochemists who have devoted themselves unselfishly to serving the interests of their fellow biochemists throughout the world.

Scientific Societies—National Academy of Sciences, American Academy of Arts and Sciences, American Society of Biological Chemists, Biochemical Society of London, American Chemical Society, American Association for the Advancement of Science, Harvey Society of New York.

I was a member of the Editorial Committee of the Journal of Biological Chemistry, which is an elective office, for six years and Chairman of this Committee for three, 1958—61. After the conclusion of my work on the Editorial Committee, I became a member of the Editorial Board of the Journal of Biological Chemistry in 1962, and then an Associate Editor from 1964 until 1968. I assumed the Editorship, succeeding John T. Edsall, in 1968, a post I was forced to relinquish by illness in 1971.

Other Activities—Member of the Council of the Institute of Neurological Diseases and Blindness of the NIH, 1961—66; Chairman of the U.S. National Committee for Biochemistry, 1968—69; Philip Schaffer Lecturer at Washington University at St. Louis, 1965; Harvey Lecturer, 1956; Phillips Lecturer at Haverford College, 1962; Visiting Professor at the University of Chicago, 1961; Visiting Professor at Harvard University, 1964; Member of Medical Advisory Board, Hebrew University-Hadassah Medical School, 1957—1970; Trustee, Montefiore Hospital.

Awards (shared with Stanford Moore): American Chemical Society Award in Chromatography and Electrophoresis, 1964; Richards Medal of the American Chemical Society, 1972; Kaj Linderstrøm-Lang Award, Copenhagen, 1972.

William H. Stein died in 1980.

THE CHEMICAL STRUCTURES OF PANCREATIC RIBONUCLEASE AND DEOXYRIBONUCLEASE

Combined text of the Nobel Lecture, December 11, 1972.

by

Stanford Moore and William H. Stein

The Rockefeller University, New York, New York

Introduction

In introducing this summary of experiments on two enzymes, we wish to indicate that the information is representative of what biochemists are obtaining about many proteins. An understanding of the host of reactions in which proteins participate in living cells requires information on the molecular architectures of a wide variety of proteins of different origins and different functions. Such information is coming from laboratories all over the world and draws upon a rich heritage of experience from many investigators. And such knowledge is fundamental to progress in medical research; the Nobel awards this year in Chemistry (concerning ribonuclease) and in Physiology or Medicine (concerning antibodies) both concern basic researches on the chemistry and the biology of proteins.

Occasionally (1) it has been educational to write the structural formula for ribonuclease in full, in terms of its 1,876 atoms of C, H, N, O, and S. Portrayal of the complete molecule with all of the atoms of the amino groups, carboxyl groups, hydroxyl groups, guanido groups, imidazole rings, phenolic groups, indole rings, aromatic, aliphatic, and thiother side chains, sulfhydryl groups, and disulfide bonds, helps in the visualization of the almost infinite number of ways in which such groups could be arranged. This characteristic of proteins makes it possible for nature to design catalysts for such a variety of specific reactions. There is no law that says that a nucleic acid or a polysaccharide could not be an enzyme. But it is understandable that the enzymes so far isolated have turned out to be proteins; a protein is equipped to participate, sometimes through cooperation with coenzymes, in the whole lexicon of organic reactions that require catalysis in the living cell.

Purification of Ribonuclease

The first step in the study of the structure of ribonuclease was, of course, its purification. Ribonuclease was first described in 1920 by Jones (2), who showed that there was present in beef pancreas a relatively heat-stable enzyme capable of digesting yeast nucleic acid. Dubos and Thompson (3) partially purified the enzyme some eighteen years later and in 1940 Kunitz (4) described the isolation of bovine ribonuclease in crystalline form after fractionation by ammonium sulfate precipitation. In order to be as certain as possible that we were beginning the structural study with a single molecular species, we under-

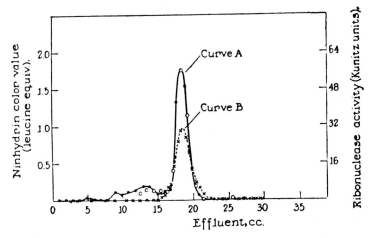

Fig. 1.

Chromatography of crystalline bovine pancreatic ribonuclease (Curve A) on the poly-methacrylic acid resin Amberlite IRC-50. Elution was with 0.2 M sodium phosphate buffer at pH 6.45. Curve B was obtained upon rechromatography of material from Peak A. (From (5).

took to apply the potential resolving power of ion exchange chromatography to ribonuclease (Fig. 1). While Werner Hirs, in our laboratory, was exploring the chromatographic purification of ribonuclease on the polymethacrylic resin Amberlite IRC-50 (5,6), Paléus and Neilands (7), in Stockholm, were studying cytochrome C on the same exchanger. These two proteins were the first molecules of their size to be thus purified. The best resolution for ribonuclease (Fig. 2) is now obtained (8) with an exchanger invented in Uppsala, a sulfoethyl cross-linked dextran, which was a development that grew from Porath and Flodin's (9) experiments on gel filtration and drew upon Sober and Peterson's (10) emphasis on the advantages of a carbohydrate matrix for the exchanger.

When pancreatic extracts were analyzed without prior fractionation, two peaks of enzymatic activity were observed by us (6) by ion exchange chromatography and by Martin and Porter (11) by partition chromatography. The major component, ribonuclease A, was selected for the first structural studies. (In later independent experiments, Plummer and Hirs (12, 13) isolated ribonuclease B in pure form from pancreatic juice and showed it to be the same as A but with the addition of a carbohydrate side chain attached to one asparagine residue.)

AMINO ACID ANALYSIS

The second step in the structural study of ribonuclease A was the determination of the empirical formula of the chromatographically homogeneous protein in terms of the constituent amino acids. Our appreciation of the importance of quantitative amino acid analysis began in the late 1930's when we had the special privilege of starting our postdoctoral studies in apprenticeship to Max

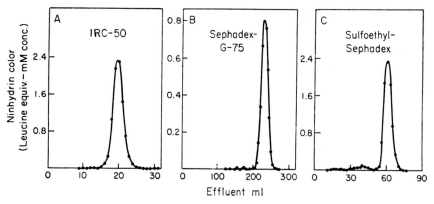

Fig. 2.

Behavior of IRC-50 purified ribonuclease on IRC-50 (A), Sephadex G-75 (B), and sulfoethyl-Sephadex (C) at pH 6.5. From (8).

Bergmann (14). In 1945 it was possible to take a new look at the subject in the light of the renaissance in chromatography stimulated by Martin and Synge in the early 1940's (15—17). In 1949, by combining a quantitative photometric ninhydrin method (18) with elution of amino acids from starch columns by alcohol: water eluents (19, 20) on an automatic fraction collector (21), we were able to analyze a protein hydrolysate in about two weeks by running three such chromatograms to resolve all overlaps. In the early 1950's, the process was speeded up to one week (Fig. 3) by turning to ion exchange chromatography on a sulfonated polystyrene resin (22, 23). In 1958, in co-operation with Darrel Spackman (25, 26) the process was automated (Fig. 4) to give recorded curves (Fig. 5) and the speed was increased to give an over-night run. Shorter columns and faster flow rates (27) permitted an analysis time of about 6 hours. Results from many academic and industrial laboratories have helped to make the procedures simpler and more rapid. Some recent users have adopted 2-hour systems (cf. (28)) and the ninhydrin reagent has been improved (29). In the 1970's, a number of industrially designed analyzers with increased automation have reduced the time for a complete analysis to about 1 hour and increased the sensitivity to the nanomole range. The sharing of knowledge among academic scientists and industrial designers of instruments and manufacturers of ion exchangers has played an important role in progress of biomedical research in this field.

In 1972 there are continuing developments that may make amino acid chromatography more ultramicro and more expeditious. These contributions include the introduction by Udenfriend and his colleagues of an analog of ninhydrin (30—32) that yields, at room temperature, a fluorescent product that can be detected at extremely low concentrations; there is also the continuing possibility that gas chromatography can give fully satisfactory results with amino acid derivatives.

The precision and the sensitivity of current procedures for amino acid analysis have been recently reviewed (33). The developmental research on

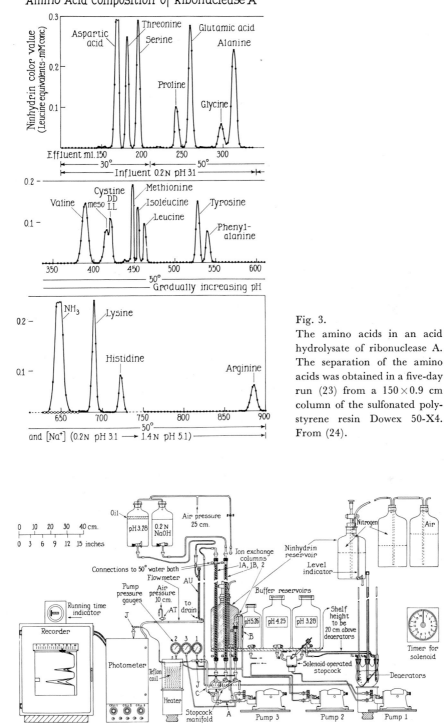

Fig. 3.
The amino acids in an acid hydrolysate of ribonuclease A. The separation of the amino acids was obtained in a five-day run (23) from a 150×0.9 cm column of the sulfonated polystyrene resin Dowex 50-X4. From (24).

Fig. 4.
Schematic diagram of automatic recording apparatus for the chromatographic analysis of mixtures of amino acids. From (26).

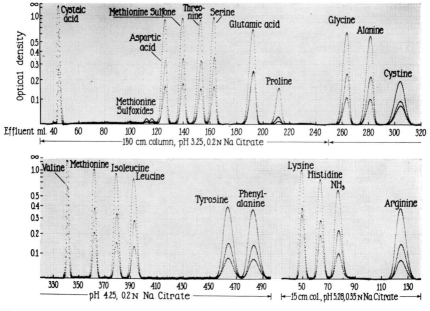

Fig. 5.
Chromatographic analysis of a mixture of amino acids automatically recorded in 22 hours
by the equipment shown in Fig. 4. From (26).

quantitative amino acid analysis has also yielded procedures for the isolation
of amino acids on a preparative scale (34, 35), for the determination of D-
and L-amino acids (36, 37), for the analysis of hydrolysates of foods (38, 39),
and for the determination of free amino acids in blood plasma (40), urine (41),
mammalian tissues (42), topics that extend beyond the scope of this lecture.
Specific discoveries from such studies include the findings by Harris Tallan
of 3-methylhistidine (43) and tyrosine-O-sulfate (44) in human urine, acetyl-
aspartic acid in brain (45), and cystathionine in human brain (46).

STRUCTURE OF RIBONUCLEASE
The empirical formula of bovine pancreatic ribonuclease (Table I), determined by the chromatographic methods applied during the structural study,
turned out to be that of a molecule containing 124 amino acid residues. From
the known mechanisms of protein biosynthesis, coupled with the susceptibility
of the peptide bonds to enzymatic hydrolysis, all of the residues are almost
certainly of the L-configuration. The calculated molecular weight is 13,683.
    The experience with amino acid chromatography led us to try to develop
column methods with sufficient resolving power for the separation of the peptides formed by the enzymatic hydrolysis of performic acid-oxidized ribonuclease (47), as in the chromatogram illustrated in Fig. 6 from the experiments
of Werner Hirs, who was the first postdoctoral associate to join our laboratory.
Fifteen young scholars began their postgraduate careers on the researches
summarized in this lecture. Each citation of their contributions connotes our

TABLE I
Amino Acid Composition of Ribonuclease A (47)

Amino Acid	Number of Residues per Molecule (mol. wt. 13,683)
Aspartic acid	15
Glutamic acid	12
Glycine	3
Alanine	12
Valine	9
Leucine	2
Isoleucine	3
Serine	15
Threonine	10
Half-cystine	8
Methionine	4
Proline	4
Phenylalanine	3
Tyrosine	6
Histidine	4
Lysine	10
Arginine	4
Total number of residues	124
Amide NH_3	17

recognition of the ideas, the hard work, and the enthusiasm that facilitate productive research; each of these biochemists shares the credit for the results reported on this occasion and we continue to be stimulated by their current independent accomplishments.

Werner Hirs, through gradient elution from Dowex 50-X2, obtained 100 % yields of the peptides that were completely liberated by tryptic hydrolysis. The elucidation of the sequences of amino acid residues in the peptides and the crossword puzzle-like ordering of the peptides followed many of the principles established by Sanger (48) in his pioneering determination of the structure of insulin, but with the larger molecule of 124 amino acid residues quantitative methods were particularly helpful in the interpretation of the results. A key chemical method in studies of molecules of this size has been the sequential degradation method developed by Pehr Edman (49) with phenyliso-thiocyanate as the reagent. Instead of determining the resulting phenylthio-hydantoins, we have generally used a subtractive procedure in which we utilize the amino acid analyzer to tell us which amino acid has been removed in each step.

The formula for ribonuclease (Fig. 7), largely developed by Werner Hirs (50, 51), Darrel Spackman (52) and Derek Smyth (53, 54), but drawing importantly upon the results of several key experiments by Christian B. Anfinsen and his associates (55—57), in Bethesda, is here written with the customary abbreviations. Ribonuclease was the first enzyme for which the

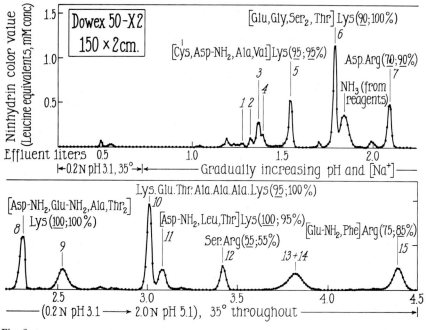

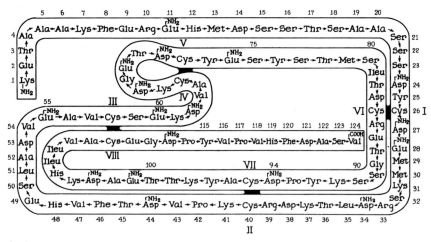

Fig. 6.

Chromatographic separation of the peptides in a tryptic hydrolysate of oxidized ribonuclease A. From (47).

sequence could be written and the determination of its structure was a logical sequel to Sanger's success with the hormone insulin.

The writing of such a two-dimensional formula is only the first step. Linderstrøm-Lang (58) referred to such a sequence as the primary structure of the protein. Catalysis is a three-dimensional operation which involves what Lang termed the secondary and tertiary structures of the chain.

Fig. 7.

The sequence of amino acid residues in bovine pancreatic ribonuclease A. From (54), based upon (50—57).

As chemists, we had made some predictions, through derivatization experiments, about residues that were folded together to form the active center of ribonuclease. Through Gerd Gundlach's studies on the inactiation of ribonuclease by alkylation with iodoacetate (59) and Arthur Crestfield's demonstration of the reciprocal alkylation of two essential histidine residues by iodoacetate at pH 5.5 (60), we concluded that the imidazole rings of histidine-119 and histidine-12 were at the active center and were about 5 Å apart. Robert Heinrikson, in our laboratory, from experiments on the alkylation of lysine at pH 8.5 (61), and drawing upon independent dinitrophenylation experiments by Hirs et al. (62), further concluded that the ε-amino group of lysine-41 was probably 7–10 Å from the imidazole ring of histidine-12 and somewhat further removed from histidine-119.

But the chemical approach does not begin to provide enough data to build an adequate model of an enzyme as a whole. The great advances in X-ray crystallography pioneered by Perutz (63) and Kendrew (64) have opened a whole new chapter in this regard, with knowledge of the sequence, at least in considerable part, being a pre-requisite for the solution of the X-ray problem in the present state of the art. We were waiting with great anticipation for the results of X-ray analysis of crystals of ribonuclease which came in 1967 through the researches of Kartha, Bello, and Harker (65) on RNase A and Wyckoff and Richards and their associates (66) on RNase S. In the S-form of the enzyme (67, 68), which is fully active, the chain has been cleaved primarily between the 20th and 21st residues by controlled proteolysis with subtilisin.

Examination of the model shows the approximate positions of the imidazole rings of histidines-12 and -119 and the ε-amino group of lysine-41 to be compatible with the chemical predictions. The substrate for RNase (Fig. 8) is ribonucleic acid, which, from the results of experiments in the laboratories of

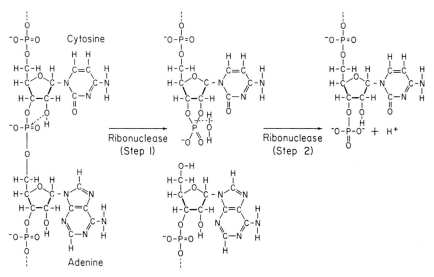

Fig. 8.
The action of ribonuclease on ribonucleic acid (reviewed in (69)).

Todd, of Cohn, and of Markham (reviewed in (69)) is cleaved at the 5'-phosphate ester bond following a pyrimidine-containing nucleotide to give, by transphosphorylation, the 2', 3'-cyclic phosphate which, in a second step, is hydrolyzed to the 3'-ester. The X-ray data show that the substrate fits in a trough on the surface of the protein with the phosphate moiety near the two imidazole rings of histidines-12 and -119 and with the pyrimidine ring tucked into a hydrophobic pocket close to the aromatic ring of phenylalanine-120

From this picture of the active site (reviewed in (70)), chemical and physical experimentation is progressing in a number of laboratories toward definition of the catalytic process in as explicit terms as possible with primary roles for one charged imidazole and one uncharged imidazole participating in the push-pull which results in transphosphorylation or hydrolysis of the phosphate ester bond.

These further experiments carry the subject into the third chapter in the history of ribonuclease, its chemical synthesis, which has grown from the many innovations in the methods for peptide synthesis in recent years. A preparation with 70 % of the activity of native ribonuclease has been synthesized through a major effort by Gutte and Merrifield (71, 72). An active RNase S-protein has been synthesized by Hirschmann, Denkewalter and associates (73). The yields in the syntheses are dependent upon a very important property of the disulfide bonds of RNase studied by White (74) and by Anfinsen and his associates (75) and reviewed particularly in terms of its special biological significance by Anfinsen (76). The reduced chain with 8 -SH groups folds to give the proper pairs of S-S bonds for the active conformation of the protein. The intramolecular forces that guide such a folding, and the similar forces that contribute to the specific aggregation of the chains of a protein with multiple subunits, such as hemoglobin, form a continuing subject of research.

The ability to synthesize RNase, or major parts thereof, opens new avenues for the identification of residues that may be essential for activity through the preparation of analogs of ribonuclease with substituitions at specific positions. Hofmann and Scoffone and their respective associates (reviewed in (70)) have done this at the amino end of the chain by synthetic variations in Richards' S-peptide. We have cooperated with Merrifield and his associates in the following recent series of experiments which illustrate the application of chemical surgery to the COOH end of the molecule and the use of synthetic replacements.

How can we examine the question of whether the proximity of the aromatic ring of phenylalanine-120 and the pyrimidine ring of the substrate leads to specific interaction between two six-membered rings in a way which is important in the binding of substrate to enzyme? Michael Lin, in our laboratory, was able to cut off, enzymatically (by pepsin according to Anfinsen (77) plus carboxypeptidase A) the last 6 residues from ribonuclease which include phenylalanine-120 and histidine-119. The resulting molecule (78) is completely inactive and does not bind substrate. Concurrently, Gutte and Merrifield had synthesized the 14-residue L-peptide from glutamic acid-111 through valine--124. When this synthetic fragment is mixed with the molecule missing residues

119 to 124, the added peptide is adsorbed and 90 % of the activity of the native enzyme is regained (79). The missing histidine is thus supplied for the active center; other residues in the peptide adsorb on the protein core in such a way as to re-form the binding site and the catalytic site. This result parallels in principle the earlier experiments of Richards and Vithayathil (68) on the removal and adsorption of the S-peptide at the NH_2 end.

When leucine or isoleucine is substituted for phenylalanine at position 120 in the synthetic peptide, the combination of peptide and protein has 10 % of the activity of ribonuclease and binds substrate as effectively as the native enzyme (80). In this way we conclude that the aromatic ring is not essential for binding of the pyrimidine ring or for activity. But the lower activity when leucine or isoleucine is substituted indicates that the aromatic ring of phenylalanine fits into the hydrophobic pocket more specifically than the aliphatic side chains and probably serves to orient histidine-119 more exactly in the delicate balance with histidine-12 that gives the active site its full catalytic power.

Another way to learn what residues can be varied without loss of activity is to study the changes in pancreatic ribonucleases from different species, as has been done for the enzymes from sheep (81), rat (82), and pig (83, 84).

DEOXYRIBONUCLEASE

A further way to gain insight into what makes ribonuclease so specific for its special substrate is to look at enzymes that hydrolyze similar substrates. In the past few years we have turned our attention to pancreatic deoxyribonuclease. This enzyme, which is about twice the size of ribonuclease, hydrolyzes DNA in the presence of bivalent cations, such as Mn^{++}, to give 5'-mononucleotides and larger fragments (85). Deoxyribonuclease first attracted special attention in the classic work of Avery, MacLeod, and McCarty (86, 87), who showed that the transforming principle of the pneumococcus could be destroyed by the action of the enzyme. McCarty's (88) experiments on the purification of the enzyme from pancreas were followed by those of Kunitz (89) and of

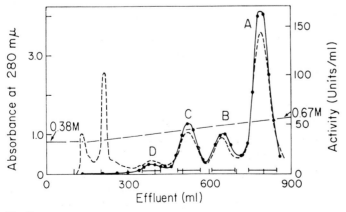

Fig. 9.
Chromatography of bovine pancreatic deoxyribonuclease, prepared by ammonium sulfate precipitation (89), on phosphocellulose at pH 4.7 with a sodium acetate buffer of increasing molarity. From (92).

                                                              Carb.
                                      10
Leu-Lys-Ile-Ala-Ala-Phe-Asn-Ile-Arg-Thr-Phe-Gly-Glu-Thr-Lys-Met-Ser-Asn-

20   30
Ala-Thr-Leu-Ala-Ser-Tyr-Ile-Val-Arg-Arg-Tyr-Asp-Ile-Val-Leu-Ile-Glu-Gln-Val-

           40   50
Arg-Asp-Ser-His-Leu-Val-Ala-Val-Gly-Lys-Leu-Leu-Asp-Tyr-Leu-Asn-Gln-Asp-Asp-

           60   70
Pro-Asn-Thr-Tyr-His-Tyr-Val-Val-Ser-Glu-Pro-Leu-Gly-Arg-Asn-Ser-Tyr-Lys-Glu-

              80   90
Arg-Tyr-Leu-Phe-Leu-Phe-Arg-Pro-Asn-Lys-Val-Ser-Val-Leu-Asp-Thr-Tyr-Gln-Tyr-

              100   110
Asp-Asp-Gly-Cys-Glu-Ser-Cys-Gly-Asn-Asp-Ser-Phe-Ser-Arg-Glu-Pro-Ala-Val-Val-

                                                     130
Lys-Phe-Ser-Ser-His-Ser-Thr-Lys-Val-Lys-Glu-Phe-Ala-Ile-Val-Ala-Leu-His-Ser-

                    140   150
Ala-Pro-Ser-Asp-Ala-Val-Ala-Glu-Ile-Asn-Ser-Leu-Tyr-Asp-Val-Tyr-Leu-Asp-Val-

                       160   170
Gln-Gln-Lys-Trp-His-Leu-Asn-Asp-Val-Met-Leu-Met-Gly-Asp-Phe-Asn-Ala-Asp-Cys-

                          180
Ser-Tyr-Val-Thr-Ser-Ser-Gln-Trp-Ser-Ser-Ile-Arg-Leu-Arg-Thr-Ser-Ser-Thr-Phe-

190   200
Gln-Trp-Leu-Ile-Pro-Asp-Ser-Ala-Asp-Thr-Thr-Ala-Thr-Ser-Thr-Asn-Cys-Ala-Tyr-

   210   220
Asp-Arg-Ile-Val-Val-Ala-Gly-Ser-Leu-Leu-Gln-Ser-Ser-Val-Val-Gly-Pro-Ser-Ala-

      230   240
Ala-Pro-Phe-Asp-Phe-Gln-Ala-Ala-Tyr-Gly-Leu-Ser-Asn-Glu-Met-Ala-Leu-Ala-Ile-

      250   257
Ser-Asp-His-Tyr-Pro-Val-Glu-Val-Thr-Leu-Thr

Fig. 10.
The sequence of amino acid residues in bovine pancreatic deoxyribonuclease A. From (93, 94).

Lindberg (90). Our studies began when Paul Price, as a graduate student, undertook the chromatographic purification of deoxyribonuclease. His initial studies showed that the enzyme, in the absence of bivalent metals, was extremely sensitive to proteolysis. Success in the purification depended upon keeping metals such as Ca^{++} present or adding diisopropyl phosphorofluoridate to inactivate the pancreatic proteases. He succeeded in resolving preparations of deoxyribonuclease into two active components on sulfoethyl-Sephadex (91) and Hans Salnikow subsequently obtained even higher resolving power (Fig. 9) with phosphocellulose (92). There are three main active components: DNase A is a glycoprotein, DNase B is a sialoglycoprotein, and DNase C is similar to A but with a proline residue substituted for one histidine. These three deoxyribonucleases were also present in the pancreatic juice from a single animal.

The determination of the chemical structure of DNase A was undertaken by

Hans Salnikow (93) and carried to completion this year by Ta-hsiu Liao (94). The working hypothesis for the structure of the molecule (Fig. 10) indicates a single chain of 257 residues with two disulfide bonds. The ordering of the tryptic and chymotryptic peptides in the reduced and carboxymethylated chain and the pairing of the half-cystine residues in the native enzyme were greatly facilitated by the cleavage of the molecule at the four methionine residues by the cyanogen bromide by the method of Gross and Witkop (95). Amino acid analyses at the nanomole level made possible sequence determinations on small amounts of peptides isolated by chromatography or paper electrophoresis.

Some of the special features of the structure can be discussed in reference to the diagram in Fig. 11. The carbohydrate side-chain, which contains 2 residues of N-acetylglucosamine and 2 to 6 residues of mannose (91, 92, 96) and which Brian Catley showed was attached via an aspartamidohexosamine linkage to a -Ser-Asn-Ala-Thr- sequence (96), is found at only one position in the chain, at residue 18. Tony Hugli studied the nitration of deoxyribonuclease (97) by tetranitromethane (98); the enzyme is inactivated by the modification of one tyrosine residue which turns out to be residue 62. Paul Price discovered that inactivation of DNase by iodoacetate in the presence of Cu^{++} and Tris buffer (99) is accompanied by carboxymethylation of one residue of histidine; from the sequences of a 3-carboxymethylhistidine-containing peptide and that of the protein, the essential imidazole ring is found to be in residue 131. DNase C (92) is the result of a mutation which causes one histidine to be replaced by a proline without any change in the activity. Hans Salnikow and Dagmar Murphy (100) have shown that this change occurs at position 118; the histidine at this position in DNase A is thus not essential for enzymatic activity.

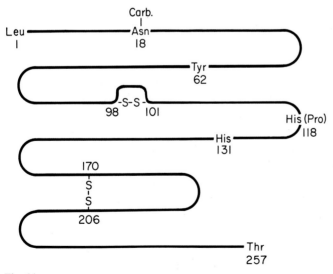

Fig. 11.
Diagram of special features of deoxyribonuclease A (94) and the substitution of Pro for His in deoxyribonuclease C (100).

The two disulfide bonds of deoxyribonuclease possess some unusual properties. Paul Price showed that even without the use of a denaturing agent both bonds are very easily reduced by mercaptans in the absence of calcium to give an inactive product. In the presence of calcium, one bond is stable and one bond is reduced (101) and the product is active. Ta-hsiu Liao has identified the non-essential disulfide bond as the one forming the small loop between residues 98 and 101 (94). When the larger loop, formed by half-cystines 170 and 206, is opened, the activity is lost.

The next step will be the correlation of the chemical evidence with the three-dimensional structure of the enzyme, if X-ray analysis of crystalline DNase A can be successfully accomplished.

CONCLUSION

In the course of studying enzymes of different functions, we have had the pleasure of cooperation with Kenji Takahashi in the identification of a carboxyl group of glutamic acid as part of the active site of ribonuclease T_1 (102). The essential -SH group and histidine residue of streptococcal proteinase have have been studied in collaboration with Stuart Elliott and Teh-yung Liu (103, 104). The esterification of carboxyl groups at the active center of pepsin was explored with T. G. Rajagopalan (105). There is a vast amount of basic information needed on various enzymes before biochemists can explan catalytic action in full detail. Enzyme chemistry today is in a stage of development that bears some similarity to that of organic chemistry at the beginning of this century. At that time there was great activity in documenting the properties of the myriad small organic compounds conceivable by man and nature. Today, in the polypeptide field, the list of determined structures is relatively small. The enzymes that have been studied first are those that can be prepared in gram quantities, such as ribonucleaase, trypsin, lysozyme, carboxypeptidase, and subtilisin. The experience in the determination of such structures is leading to ultramicromethods which will extend the range of structural studies to tissue enzymes that are present in very small amounts.

From the knowledge of the structures of a large series of enzymes, underlying principles of how nature designs catalysts for given purposes will evolve. And there will be practical dividends from such research on proteins. One example of research-in-progress can illustrate this possibility. The project developed in the following way: In the course of examining the importance of the three-dimensional configuration of RNase to its activity, George Stark had occasion to dissolve the enzyme in 8 M urea at 40° (106). In one of those experiments the RNase was not active after the urea was removed by dialysis, and it turned out that traces of cyanate in the urea solution had carbamylated the ε-NH_2 groups of the enzyme. The chemistry of the subject carries us back to Wöhler's (107) observations on the relationship between ammonium cyanate cyanate and urea in 1828. In 1970, Anthony Cerami and James Manning (108), two young investigators at the Rockefeller University, undertook to explore, fully on their own initiative, whether traces of cyanate in urea might have a role in the reported beneficial effect of urea on the sickling of erythrocytes of individuals carrying hemoglobin S. They have discovered that there is such an effect of cyanate on human erythrocytes, and that it is accompanied by carbamylation of the α-NH$_2$ groups of the valine residues of the α- and β-chains of hemoglobin S. The knowledge that a relatively simple chemical modification of hemoglobin S can restore nearly normal function to the deficient molecule opens the possibility that a genetic defect in man might be remedied, not by having to change the gene, but by redivatizing the protein.

Such results afford an example of the manner in which one finding leads to another in basic research and ultimately to possible benefits to man. When we consider biochemistry in 1972, it is important to realize how fragmentary is our knowledge of the molecular basis of life. Very few macromolecules can be discussed in the detail with which ribonuclease or hemoglobin can be defined. Such knowledge of structure-function relationships is basic to the rational approach to the intricate synergisms of living systems.

ACKNOWLEDGEMENTS

The researches from our laboratory on ribonuclease and deoxyribonuclease summarized in this lecture have been possible through financial backing by The Rockefeller University, The United States Public Health Service, and The National Science Foundation.

REFERENCES

1. Stein, W. H., and Moore, S. Scientific American 204, 81−92, 1961.
2. Jones, W., Am. J. Physiol. 52, 203−207, 1920.
3. Dubos, R. J., and Thompson, R. H. S., J. Biol. Chem. *124*, 501−510, 1938.
4. Kunitz, M., J. Gen. Physiol. *24*, 15−32, 1940.
5. Hirs, C. H. W., Stein, W. H., and Moore, S., J. Amer. Chem. Soc. *73*, 1893, 1951.
6. Hirs, C. H. W., Moore, S., and Stein, W. H., J. Biol. Chem. *200*, 493−506, 1953.
7. Paléus, S., and Neilands, J. B., Acta Chem. Scand. *4*, 1024−1030, 1950.
8. Crestfield, A. M., Stein, W. H., and Moore, S., J. Biol. Chem. *238*, 618−621, 1963.
9. Porath, J., and Flodin, P., Nature *183*, 1657−1659, 1959.
10. Sober, H. A., and Peterson, E. A., J. Amer. Chem. Soc. *76*, 1711−1712, 1954.
11. Martin, A. J. P., and Porter, R. R., Biochem. J. *49*, 215−218, 1951.
12. Plummer, T. H., Jr., and Hirs, C. H. W., J. Biol. Chem. *238*, 1396−1401, 1963.
13. Plummer, T. H., Jr., and Hirs, C. H. W., J. Biol. Chem. *239*, 2530−2538, 1964.
14. Moore, S., Stein, W. H., and Bergmann, M., Chem. Rev. *30*, 423−432, 1942.
15. Martin, A. J. P., and Synge, R. L. M., Biochem. J. *35*, 1358−1368, 1941.
16. Martin, A. J. P., Ann. Rev. Biochem. *19*, 517−535, 1950.
17. Synge, R. L. M., Analyst *71*, 256−258, 1946.
18. Moore, S., and Stein, W. H., J. Biol. Chem. *176*, 367−388, 1948.
19. Moore, S., and Stein, W. H., J. Biol. Chem. *178*, 53−77, 1949.
20. Stein, W. H., and Moore, S., J. Biol. Chem. *178*, 79−91, 1949.
21. Stein, W. H., and Moore, S., J. Biol. Chem. *176*, 337−365, 1948.
22. Moore, S., and Stein, W. H., J. Biol. Chem. *192*, 663−681, 1951.
23. Moore, S., and Stein, W. H., J. Biol. Chem. *211*, 893−906; *211*, 907−913, 1954.
24. Hirs, C. H. W., Stein, W. H., and Moore, S., J. Biol. Chem. *211*, 941−950, 1954.
25. Moore, S., Spackman, D. H., and Stein, W. H., Anal. Chem. *30*, 1185−1190, 1958.
26. Spackman, D. H., Stein, W. H., and Moore, S., Anal Chem. *30*, 1190−1206, 1958.
27. Spackman, D. H., Federation Proc. *22*, 244, 1963.

28. Spackman, D. H., In Methods in Enzymology (C. H. W. Hirs, Editor), Vol. 11, Academic Press Inc., New York, pp. 3–15, 1967.
29. Moore, S., J. Biol. Chem. *243*, 6281–6283, 1968.
30. Samejima, K., Dairman, W., and Udenfriend, S., Anal. Biochem. *42*, 222–236, 1971.
31. Samejima, K., Dairman, W., Stone, J., and Udenfriend, S., Anal. Biochem. *42*, 237–247, 1971.
32. Weigele, M., De Bernardo, S. L., Tengi, J. P., and Leimgruber, W., J. Amer. Chem. Soc. *94*, 5927–5928, 1972.
33. Moore, S., In Chemistry and Biology of Peptides (J. Meienhofer, Editor), Ann Arbor Science Publications, Inc., Ann Arbor, Mich., pp. 629–653, 1972.
34. Hirs, C. H. W., Moore, S., and Stein, W. H., J. Biol. Chem. *195*, 669–683, 1952.
35. Hirs, C. H. W., Moore, S., and Stein, W. H., J. Amer. Chem. Soc. *76*, 6063–6065, 1954.
36. Manning, J. M., and Moore, S., J. Biol. Chem. *243*, 5591–5597, 1968.
37. Manning, J. M., J. Amer. Chem. Soc. *92*, 7449–7454, 1970.
38. Schram, E., Dustin, J. P., Moore, S., and Bigwood, E. J., Anal. Chim. Acta *9*, 149–161, 1953.
39. Dustin, J. P., Czakowska, C., Moore, S., and Bigwood, E. J., Anal. Chim. Acta *9*, 256–262, 1953.
40. Stein, W. H., and Moore, S. J. Biol. Chem. *211*, 915–926, 1954.
41. Stein, W. H., J. Biol. Chem. *201*, 45–58, 1953.
42. Tallan, H. H., Moore S., and Stein, W. H., J. Biol. Chem. *211*, 927–939, 1954.
43. Tallan, H. H., Moore, S., and Stein, W. H., J. Biol. Chem. *206*, 825–834, 1954.
44. Tallan, H. H., Bella, S. T., Stein, W. H., and Moore, S., J. Biol. Chem. *217*, 703–708, 1955.
45. Tallan, H. H., Moore, S., and Stein, W. H., J. Biol. Chem. *219*, 257–264, 1956
46. Tallan, H. H., Moore, S., and Stein, W. H., J. Biol. Chem. *230*, 707–716, 1958.
47. Hirs, C. H. W., Moore, S., and Stein, W. H., J. Biol. Chem. *219*, 623–642, 1956.
48. Sanger, F., Science *129*, 1340–1344, 1959.
49. Edman, P., Acta Chem. Scand. *10*, 761–768, 1956.
50. Hirs, C. H. W., J. Biol. Chem. *235*, 625–632, 1960.
51. Hirs, C. H. W., Moore, S., and Stein, W. H., J. Biol. Chem. *235*, 633–647, 1960.
52. Spackman, D. H., Stein, W. H., and Moore, S., J. Biol. Chem. *235*, 648–659, 1960.
53. Smyth, D. G., Stein, W. H., and Moore, S., J. Biol. Chem. *237*, 1845–1850, 1962.
54. Smyth, D. G., Stein, W. H., and Moore, S., J. Biol. Chem. *238*, 227–234, 1963.
55. Anfinsen, C. B., Redfield, R. R., Choate, W. L., Page, J., and Carroll, W. R., J. Biol. Chem. *207*, 201–210, 1954.
56. Redfield, R. R., and Anfinsen, C. B., J. Biol. Chem. *221*, 385–404, 1956.
57. Potts, J. T., Berger, A., Cooke, J., and Anfinsen, C. B., J. Biol. Chem. *237*, 1851–1855, 1962.
58. Linderstrøm-Lang, K., Lane Medical Lectures, Stanford University Publications, Vol. 6, 1–115, 1952.
59. Gundlach, H. G., Stein, W. H., and Moore, S., J. Biol. Chem. *234*, 1754–1760, 1959.
60. Crestfield, A. M., Stein, W. H., and Moore, S., J. Biol. Chem. *238*, 2413–2420; *238*, 2421–2428, 1963.
61. Heinrikson, R. L., J. Biol. Chem. *241*, 1393–1405. 1966.
62. Hirs, C. H. W., Halmann, M., and Kycia, J. H., In T. W. Goodwin and I. Lindberg, (Editors), Biological Structure and Function, Vol. 1, Academic Press, Inc., New York, p. 41, 1961.
63. Perutz, M. F., Science *140*, 863–869, 1963.
64. Kendrew, J. C., Science *139*, 1259–1266, 1963.
65. Kartha, G., Bello, J., and Harker, D., Nature *213*, 862–865, 1967.
66. Wyckoff, H. W., Hardman, K. D., Allewell, N. M., Inagami, T., Johnson, L. N., and Richards, F. M., J. Biol. Chem. *242*, 3984–3988, 1967.
67. Richards, F. M., C. R. Trav. Lab. Carlsberg *29*, 322–328, 329–346, 1955.

68. Richards, F. M., and Vithayathil, P. J., J. Biol. Chem. *234*, 1459–1465, 1959.
69. Brown, D. M., and Todd, A. R., In The Nucleic Acids: Chemistry and Biology (E. Chargaff and J. N. Davidson, Editors), Vol. 1, Academic Press, Inc., New York, p. 409, 1955.
70. Richards, F. M., and Wyckoff, H. G., In The Enzymes (P. D. Boyer, Editor), Third Ed., Vol. 4, Academic Press, Inc., New York, pp. 647–806, 1971.
71. Gutte, B., and Merrifield, R. B., J. Amer. Chem. Soc. *91*, 501–502, 1969.
72. Gutte, B., and Merrifield, R. B., J. Biol. Chem. *246*, 1922–1941, 1971.
73. Hirschmann, R., Nutt, R. F., Veber, D. F., Vitali, R. A., Varga, S. L., Jacob, T.A., Holly, F. W., and Denkewalter, R. G., J. Amer. Chem. Soc. *91*, 507–508, 1969.
74. White, F. H., J. Biol. Chem. *235*, 383–389, 1960.
75. Goldberger, R. F., Epstein. C. J., and Anfinsen, C. B., J. Biol. Chem. *328*, 628–635, 1963.
76. Anfinsen, C. B., Les Prix Nobel en 1972, Stockholm, 1973.
77. Anfinsen, C. B., J. Biol. Chem. *221*, 405–412, 1956.
78. Lin, M. C., J. Biol. Chem. *245*, 6726–6731, 1970.
79. Lin, M. C., Gutte, B., Moore, S., and Merrifield, R. B., J. Biol. Chem. *245*, 5169–5170, 1970.
80. Lin, M. C., Gutte, B., Caldi. D. G., Moore, S., and Merrifield, R. B., J. Biol. Chem. *247*, 4768–4774, 1972.
81. Anfinsen, C. B., Aqvist, S. E. G., Cooke, J. P., and Jonsson, B., J. Biol. Chem. *234*, 1118–1123, 1959.
82. Beintema, J. J., and Gruber, M., Biochim. Biophys. Acta *147*, 612–614, 1967.
83. Jackson, R. L., and Hirs, C. H. W., J. Biol. Chem. *245*, 637–653, 1970.
84. Phelan, J. J., and Hirs, C. H. W., J. Biol. Chem. *245*, 654–661, 1970.
85. Laskowski, M., Sr., In The Enzymes (P. D. Boyer, Editor), 3. Ed., Vol. 4, Academic Press, Inc., New York, pp. 289–311, 1971.
86. Avery, O. T., MacLeod, C. M., and McCarty, M., J. Exp. Med. *79*, 137–157, 1944.
87. McCarty, M., and Avery, O. T., J. Exp. Med. *83*, 89–96, 1946.
88. McCarty, M., J. Gen. Physiol. *29*, 123–139, 1946.
89. Kunitz, M., J. Gen. Physiol. *33*, 349–362, 1950.
90. Lindberg, U., Biochemistry *6*, 335–342, 1967.
91. Price, P. A., Liu, T.-Y., Stein, W. H., and Moore, S., J. Biol. Chem. *244*, 917–923, 1969.
92. Salnikow, J., Moore, S., and Stein, W. H., J. Biol: Chem. *245*, 5685–5690, 1970.
93. Salnikow, J., Liao, T.-H., Moore, S., and Stein, W. H., J. Biol. Chem., *248*, 1480—1488, 1973.
94. Liao, T.-H., Salnikow, J., Moore, S., and Stein, W. H., J. Biol. Chem., *248* 1489—1495, 1973.
95. Gross, E., and Witkop, B., J. Biol. Chem. *257*, 1856–1860, 1962.
96. Catley, B. J., Moore, S., and Stein, W. H., J. Biol. Chem. *244*, 933–936, 1969.
97. Hugli, T. E., and Stein, W. H., J. Biol. Chem. *246*, 7191–7200, 1971.
98. Riordan, J. F., Sokolovsky, M., and Vallee, B. L., J. Amer. Chem. Soc. *88*, 4104–4105, 1966.
99. Price, P. A., Moore, S., and Stein, W. H., J. Biol. Chem. *244*, 924–928, 1969.
100. Salnikow, J., and Murphy, D., J. Biol. Chem., *248*, 1499—1501, 1973.
101. Price, P. A., Stein, W. H., and Moore, S., J. Biol. Chem. *244*, 929–932, 1969.
102. Takahashi, K., Stein, W. H., and Moore, S., J. Biol. Chem. *242*, 4682–4690, 1967.
103. Liu, T.-Y., Stein, W. H., Moore, S., and Elliott, S. D., J. Biol. Chem. *240*, 1143–1149, 1965.
104. Liu, T.-Y., J. Biol. Chem. *242*, 4029–4032, 1967.
105. Rajagopalan, T. G., Stein, W. H., and Moore, S., J. Biol. Chem. *241*, 4295–3297, 1966.
106. Stark, G. R., Stein, W. H., and Moore, S., J. Biol. Chem. *235*, 3177–3181, 1960.
107. Wöhler, F., Pogg. Ann. *12*, 253–256, 1828.
108. Cerami, A., and Manning, J. M., Proc. Nat. Acad. Sci. USA *68*, 1180–1183, 1971.

Chemistry 1973

ERNST OTTO FISCHER and GEOFFREY WILKINSON

for their pioneering work, performed independently, on the chemistry of the organometallic, so called sandwich compounds

THE NOBEL PRIZE FOR CHEMISTRY

Speech by Professor INGVAR LINDQVIST of the Royal Academy of Sciences

Your Majesty, Your Royal Highnesses, Ladies and Gentlemen,

Throughout history, there has been a tendency for people to take a stereotyped view of their fellow men in other occupations or with different backgrounds. I think that we all would agree that these stereotypes are harmful, and yet every day we can see evidence for the hold they have, even on intelligent people. Popularly held ideas about what a chemist does and how he works are perhaps more harmless than many such stereotypes, but for the chemist himself, the misconceptions can sometimes be a little irritating. In the first place, chemists are assumed always to work on technological developments rather than the discovery of new concepts. At the same time, the whole field of chemistry, including practical applications, rests firmly on the foundations laid down during hundreds of years of the impartial quest for truth. Chemists are also thought of as reasoning by the use of intellectual processes, without the use of much fantasy, at least as far as their research is concerned. But the history of chemistry is full of colourful, imaginative personalities, and what is more, the important fact is that they would never have been able to make their discoveries without tempering the use of logical reasoning and knowledge with fantasy and intuition. Koestler has on some occasion claimed that scientific activity is closely related to the process of artistic creation, and there is certainly at least a grain of truth in this assertion.

It seems to be appropriate to put forward these views when speaking of this year's Nobel laureates in chemistry, professors Fischer and Wilkinson. The facts were available for all the chemists of the world to see, but the right interpretation was lacking. Once the correct hypothesis was arrived at by fantasy or intuition, it readily lent itself to simple processes of logical deduction. I am of course referring to the way in which they together with the former Nobel laureate Woodward reached the conclusion that certain compounds could not be understood without the introduction of a new concept, namely that of the sandwich compound. This expression applies to the structure of these compounds, which can be thought of as a metal atom—the filling—sandwiched between two flat molecules. Now the science of chemistry involves, of course, more than flashes of visionary inspiration, and both Fischer and Wilkinson did not hesitate to confirm and develop the concept of the sandwich compound by an intensive experimental effort. This they did by the successful synthesis of a large number of compounds which were analogous to the initially discovered ferrocene (named by Woodward in analogy with benzene), but with other metals than iron, and with other flat molecules than the cyclopentadienyl group found in ferrocene. Amongst other things,

Fischer managed to surprise chemists by preparing a sandwich of chromium between two benzene molecules. The culinary exploits were pursued further with the progress to open sandwiches, having a flat molecule on one side of the metal atom, and with only small molecules such as carbonyl, methyl or ethyl groups on the other side. Perhaps the most interesting development was when Wilkinson managed to prepare a sandwich compound with a direct chemical bond between the metal atom rhenium and hydrogen. This was a completely new type of chemical bond at the time.

The discovery and confirmation of the new bonding and structural principles applying to sandwich compounds is the notable achievement of the two Nobel prize winners. It is a fact that every discovery within the realm of fundamental chemical research has potential applications. The very circumstance of the stability of sandwich compounds has led to involved theoretical discussions, which have also played their part in important advances in theoretical and applied research on catalysis. We cannot yet predict the practical importance that sandwich compounds will have in the future, nor whether they will finally be shown to occur in biological systems. One thing that is quite sure, however, is that all workers in the field of sandwich compounds in the future will be familiar with the names of Fischer and Wilkinson.

Professor Fischer,

Die Entdeckungen vollständig neuer Prinzipien der chemischen Bindung und Struktur sind immer grosse Augenblicke in der Geschichte der Chemie gewesen. Sie haben zu einer solchen Entdeckung in hervorragender Weise beigetragen. Ich überbringe Ihnen die wärmsten Glückwünsche der Königlichen Schwedischen Akademie der Wissenschaften.

Professor Wilkinson,

The discovery of quite new types of chemical bonds and structures has always been considered as milestones in the history of chemistry. You have contributed to such a discovery in a decisive way. May I convey to you the warmest congratulations of the Royal Swedish Academy of Sciences.

Professor Fischer, Professor Wilkinson. May I request you to receive the Nobel Prize for Chemistry from the hands of His Majesty the King.

ERNST OTTO FISCHER

Translation from the German text

I was born in Solln, near Munich, on 10 November 1918 as the third child of the Professor of Physics at the Technical College of Munich, Dr. Karl T. Fischer (died 1953), and his wife, Valentine, *née* Danzer (died 1935). After completing four years at elementary school I went on to grammar school in 1929, from which I graduated in 1937 with my *Abitur*. Following a subsequent period of "work service" and shortly before the end of my two years' compulsory military service, the Second World War broke out. I served in Poland, France and Russia. In the winter of 1941/2 I began to study Chemistry at the Technical College in Munich during a period of study leave. I was released by the Americans in the autumn of 1945, and resumed my study of Chemistry in Munich after the reopening of the Technical College in 1946. I graduated in 1949. I took up a position as scientific assistant to Professor Walter Hieber in the Inorganic Chemistry Department, and under his guidance I dedicated myself to working on my doctoral thesis, "The Mechanisms of Carbon Monoxide Reactions of Nickel II Salts in the Presence of Dithionites and Sulfoxylates". After receiving my doctorate in 1952, I was invited by Professor Hieber to continue my activities at the college and consequently chose to specialise in the study of transition metal and organometallic chemistry. I wrote my university teaching thesis on "The Metal Complexes of Cyclopentadienes and Indenes". I was appointed a lecturer at the Technical College in 1955 and in 1956 I completed a scientific sojourn of many months in the United States. In 1957 I was appointed Professor at the University of Munich. After turning down an offer of the Chair of Inorganic Chemistry at the University of Jena I was appointed Senior Professor at the University of Munich in 1959 . In 1957 I was awarded the Chemistry Prize by the Göttingen Academy of Sciences. The Society of German Chemists awarded me the Alfred Stock Memorial Prize in 1959. In 1960 I refused an appointment as Senior Professor in the Department of Inorganic Chemistry at the University of Marburg. In 1964 I took the Chair of Inorganic Chemistry at the Technical College of Munich, which had been vacated by Professor Hieber. In the same year I was elected a member of the Mathematics/Natural Science section of the Bavarian Academy of Sciences; in 1969 I was appointed a member of the German Academy of Scientists Leopoldina. In 1972 I was given an honorary doctorate by the Faculty of Chemistry and Pharmacy of the University of Munich.

Lectures on my fields, particularly those on metallic complexes of cyclopentadienes and indenes, metal-π-complexes of six-ringed aromatics, mono-, di- and oligo-olefins and most recently metalcarbonyl carbene and carbyne complexes, led me on lecture tours of the United States, Australia, Venezuela, Brazil, Israel and Lebanon, as well as numerous European countries, including the former Soviet Union. In 1969 I was Firestone Lecturer at the University of Wisconsin, Madison, Wisconsin, USA; in 1971 Visiting Professor at the University of Florida, Gainesville, USA, as well as the first Inorganic Chemistry Pacific West Coast Lecturer. In the spring of 1973 I held lectures as the Arthur D. Little Visiting Professor at the Massachusetts Institute of Technology, Cambridge, Massachusetts, USA; and that was followed by a period when I was Visiting Distinguished Lecturer at the University of Rochester, Rochester, New York, USA.

ON THE ROAD TO CARBENE AND CARBYNE COMPLEXES

Nobel Lecture, 11 December 1973
by
ERNST OTTO FISCHER
Inorganic Chemistry Laboratory, Technical University, Munich,
Federal Republic of Germany

Translation from the German text

INTRODUCTION

In the year 1960, I had the honour of giving a talk at this university[*] about sandwich complexes on which we were working at that time. I think I do not have to repeat the results of those investigations today. I would like to talk instead about a field of research in which we have been intensely interested in recent years: namely, the field of carbene complexes and, more recently, carbyne complexes. If we substitute one of the hydrogen atoms in a hydrocarbon of the alkane type — for example, ethane — by a metal atom, which can of course bind many more ligands, we arrive at an organometallic compound in which the organic radical is bound to the metal atom by a σ-bond (Fig. 1a). The earliest compounds of this kind were prepared more than a hundred years ago; the first was cacodyl, prepared by R. Bunsen (1), and then zinc dialkyls were prepared by E. Frankland (2). Later V. Grignard was able to synthesise alkyl magnesium halides by treating magnesium with alkyl halides (3). Grignard was awarded the Nobel Prize in 1912 for this effort. We may further recall the organo-aluminium compounds (4) of K. Ziegler which form the basis for the low pressure polymerisation, for example of ethylene. Ziegler and G. Natta were together honoured with the Nobel Prize in 1963 for their work on organometallic compounds.

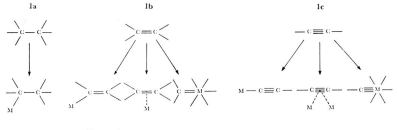

M = metal or metal complex

Fig. 1. Derivation of organometallic compounds from hydrocarbon derivatives.

If we then go over to a system with two carbon atoms connected by a double bond, i.e. a molecule of the alkene type, the roads leading to organometallic

[*]Royal Technical University, Stockholm

compounds branch out (Fig. 1b). In the first place, on substituting a radical by a metal atom, we get, as before, the σ compounds, for example, the vinyl lithium derivatives. In the second type, only the π electrons of the double bond are used for binding the organic molecule to the metal atom. In this way, we obtain π complexes (5,6) (Fig. 1b), the first representative of this being Zeise's salt $K[PtCl_3(C_2H_4)]$, which was prepared as early as 1827 (7). Such metal π complexes of olefins appear especially with transition metals. Main group elements are less capable of forming such a bond. The sandwich complexes (8,9) in which the bond between the metal and ligand takes place not only through two π electrons alone but also through a delocalised cyclic π electron system, may also be included in this type of compound. As an example of this we may mention dibenzene chromium(0) (10), in which the chromium atom lies between two parallel benzene rings that face each other in a congruent fashion.

We get a third type of compound by formally separating the double bond and by fixing one of the halves to a transition metal radical. This idea is realised in the transition metal carbene complexes, in which carbenes CRR' that have a short life in the free state are stabilised by being bound to the metal. The first part of my lecture will be devoted to complexes of this type.

Finally, if we consider molecules with a carbon-carbon triple bond, of the kind that is present in alkynes, we find here also three paths towards metal derivatives (Fig. 1c). As in the earlier cases, we can build up σ compounds and then utilise the two π bonds to synthesise, for example, complexes in which the two metal-ligand bonds are situated more or less perpendicular to each other. Finally, if we imagine the triple bond to be separated and one half to be substituted by a metal complex radical, we arrive at carbyne complexes. I shall deal with these complexes in the second part of my lecture.

TRANSITION METAL CARBENE COMPLEXES
PREPARATION OF THE EARLIEST CARBENE COMPLEXES

A. Maasböl and I reported some stable carbene complexes for the first time in the year 1964 in a short communication (12). We had treated hexacarbonyl tungsten with phenyl (and methyl) lithium in ether with the intention of adding the carbanion to the carbon ion of a CO ligand which has been positivised with respect to the oxygen ion. By this reaction we did in fact get lithium acyl pentacarbonyl tungstates, which could then be converted into pentacarbonyl [hydroxy(organyl)carbene]tungsten(0) complexes by acidifying in aqueous solution (Eq. 1).

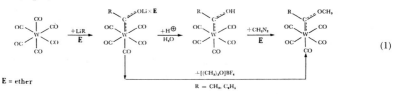

E = ether

(1)

However, we very soon found that these complexes are not very stable. They tend to cleave the carbene ligand with a simultaneous hydrogen displacement/shift. We then get aldehydes — as also independently found by Japanese researchers (13). Only recently we have managed to prepare these hydroxy carbene complexes in an analytically pure form (14). However, even without isolating them, these complexes could be converted into the much more stable methoxy carbene compounds by treating them at an early stage with diazomethane (12).

We soon found a more elegant method of getting the methoxy carbene compounds by directly alkylating (15) the lithium acyl carbonyl metallates with trialkyl oxonium tetrafluoroborates (16) (Eq. 1). H. Meerwein had also synthesised these compounds in a similar manner. This method of preparation combines several advantages: the method is clear and easy to carry out and gives very high yields. It also makes it possible to prepare a wide spectrum of carbene complexes. Thus, for example, instead of phenyl lithium, many other organolithium compounds (17–23) can be used. Similarly, instead of the hexacarbonyl tungsten used first, one can use hexacarbonyl chromium (17), hexacarbonyl molybdenum (17), the di-metal decacarbonyls of manganese (24, 25), technetium (25) and rhenium (25), pentacarbonyl iron (26) and tetracarbonyl nickel (27). However, the corresponding carbene complexes become increasingly unstable in the above sequence. Finally, substituted metal carbonyl (27–30) can also be subjected to carbanion addition followed by alkylation. The carbene complexes are generally quite stable, diamagnetic, easily soluble in organic solvents, and sublimable. Before going into the details about their reactions, I would like to deal briefly with the binding conditions of the carbene ligand/metal bond.

OUR UNDERSTANDING OF THE NATURE OF THE BOND AND SPECTROSCOPIC RESULTS

SPECULATIONS ABOUT THE BONDS AND SPECTROSCOPIC RESULTS

The first X-ray structure determination (31), carried out by O. S. Mills in collaboration with us, on pentacarbonyl [methoxy(phenyl)carbene] chromium (0) confirmed our notion about the bonds postulated earlier. Our notion was that the carbon atom of carbene is present in the sp^2-hybridised form. It should therefore have a vacant p-orbital and should hence show an electron deficit.

This strong electron deficiency is compensated mainly through a $p\pi$-$p\pi$ bond between one of the free electron pairs of the oxygen atom of the methoxy group and the unoccupied p-state of the carbene carbon atom. A $d\pi$-$p\pi$ reverse bonding takes place — to a smaller yet appreciable extent — from an occupied central metal orbital with a suitable symmetry to this vacant p-orbital of the carbene carbon. This can be deduced from the distances between the carbene carbon atom and the oxygen atom on the one hand, and between the carbene carbon atom and the central chromium atom on the other hand.

The $C_{carbene}$-O distance, which was found to be 1.33 Å, lies between the numerical values for a single bond (1.43 Å in diethyl ether) and a double bond (1.23 Å in acetone). While the Cr-C_{CO} distance in the carbene complex is 1.87 Å on average, the Cr-$C_{carbene}$ distance was found to be 2.04 Å. For a pure chromium-carbon σ bond, one would expect the distance to be 2.21 Å according to the arguments of F. A. Cotton (32). According to these arguments, the bond order for the Cr-$C_{carbene}$ bond is much smaller than for the chromium-C_{CO} bond in the same complex, but greater than the bond order in a single bond. That the phenyl group does not form any pπ-pπ bond with the carbene carbon — at least in the lattice — is clear from the intense twisting of the plane formed by the atoms Cr, C and O and the twisting of the phenyl ring. We can see at the same time that the double bond character of the $C_{carbene}$-O bond is so strong that cis and trans isomers with respect to the CO bond can easily occur (Fig. 2). In the case of pentacarbonyl [methoxy(phenyl)carbene]chromium(0) we only find molecules of the trans type in the lattice, but at low temperatures the cis isomer can also be detected by ^{1}H-NMR spectroscopy (33, 34).

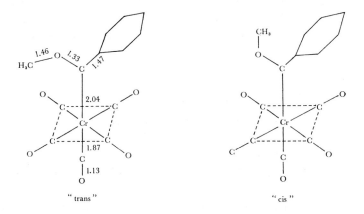

Fig. 2. Structure of pentacarbonyl[methoxy(phenyl)carbene]chromium(0).

Further important insights into the bond situations in carbene complexes are provided by the v_{CO} spectra (20, 35–37). As we know, the carbon monoxide ligands in metal carbonyls can be regarded as weak donor systems. They provide electron density from the free electron pair of the carbon to unoccupied orbitals of the metal atom. This would lead to a formally negative charging of the metal. This negative charge is broken down to a great extent by a reverse transport of charge density from the metal to the carbon monoxide through dπ-pπ back bonding; that is, the carbon has an acceptor function in addition to its donor function. This donor/acceptor ratio of the CO ligands of a complex represents a very sensitive probe for the electronic properties of the other ligands bonded to the metal. This ratio can be qualitatively estimated by determining the CO stretching vibrations (v_{CO}).

Let us now compare carbon monoxide and methoxy (phenyl) carbene as ligands in complexes of the type $(CO)_5CrL$ (L=CO or $[C(OCH_3)C_6H_5]$) with regard to their ν_{CO} absorptions: while the total symmetrical Raman-active ν_{CO} stretching vibration in $Cr(CO)_6$ appears at 2108 cm^{-1} (A_{1g}) (38), we find that the CO absorption of the CO group in the trans-position with respect to the carbene ligand is shifted drastically towards lower wave numbers, namely 1953 cm^{-1} (A_1) (17); that is, the carbene ligand has a much greater σ donor/π acceptor ratio than CO. In other words the carbene ligand is on the whole positively polarised, and the $Cr(CO)_5$ part is negatively polarised. The dipole moments of the complexes are therefore also relatively high, about 5 Debye.

In what follows, I would not like to go into the details of the spectroscopic studies as such. It may, however, be pointed out briefly that ^{13}C-NMR measurements are extremely useful for research in this area of chemistry. In one such early study on pentacarbonyl [methoxy(phenyl)carbene]chromium(0), C. G. Kreiter (39) could show that the carbene carbon atom is highly positivised. The chemical shift that was found, namely 351.42 ppm, is in fact within the range of values of carbonium ions in organic chemistry. This modern method has thus confirmed our original proposal once again.

With its high "positive" charge character, the carbene carbon acts as an electrophilic centre. This is of great importance when one is looking at the reactions of these compounds. We shall come back to these reactions in the course of this lecture.

SYNTHESES OF OTHER CARBENE COMPLEXES

Our first paper on metal carbene complexes was published in 1964. This area of work has expanded rapidly since then. A number of fairly extensive review articles (40–43) on the chemistry of carbene complexes have now appeared. I shall therefore pick out only a few examples, particularly of those involving interesting syntheses.

In 1968, K. Öfele treated 1.1dichloro 2.3diphenyl cyclopropene(2) with disodium pentacarbonyl chromate in our laboratory and obtained pentacarbonyl (2.3diphenyl cyclopropenylidene) chromium(0) with the elimination of sodium chloride (44). This carbene compound is stable up to 200° C and has the characteristic property that the carbene ligand no longer has any hetero atom. The electronic saturation of the carbene carbon takes place in this reaction through the three-ring π system (Eq. 2).

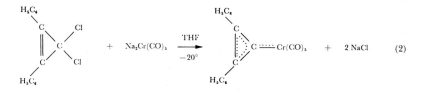

X-ray structural analysis (45) showed that the three C-C distances in the ligand are not identical: the distance between the two phenyl substituted C atoms is somewhat shorter than the other two. The carbene carbon-chromium distance is 2.05 Å and lies in the range of the values found for our carbene complexes, i.e. this complex must be a genuine carbene complex.

Another fine method of synthesis was published by R. L. Richards *et al.* in 1969 (46). They found that in the reaction of alcohols with certain isonitrile complexes, for example the isonitrile platinum complex, addition of the alkoxy group at the carbon atom and the hydrogen at the nitrogen atom of the isonitrile ligand takes place, resulting in the corresponding carbene complexes, for example (Eq. 3).

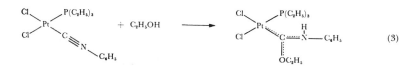

$$\tag{3}$$

This method of synthesis has since led to several such compounds. We notice here the similarity between the chemistry of isonitrile complexes and carbon monoxide complexes.

In 1971 we succeeded, for the first time once again, in transferring a carbene ligand from one complex to another (26, 47). For example, if we irradiate a solution of cyclopentadienyl(carbonyl)[methoxy(phenyl)carbene] nitrosylmolybdenum(0) in the presence of an excess of pentacarbonyl iron, we get tetracarbonyl[methoxy(phenyl)carbene]iron(0) with the simultaneous formation of cyclopentadienyl(dicarbonyl)nitrosylmolybdenum (Eq. 4).

$$\tag{4}$$

Finally, another method of synthesis was developed recently (1971) by M. F. Lappert *et al.* (48). They treated an electron-rich olefinic system, such as N.N.N'.N'-tetraphenyl bis dihydroimidazolylidene, with a suitable complex. In this way, they were able to split the double bond and fix the carbene halves at the metal, for example (Eq. 5).

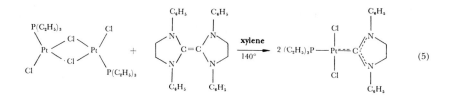

$$\tag{5}$$

This was a brief survey of other methods for the synthesis of carbene complexes found independently by other workers.

POSSIBLE REACTIONS OF CARBENE COMPLEXES

In what follows, I shall confine myself to carbene complexes of our type and I shall show, with some examples from recent times, the kinds of reactions that we could observe with these complexes.

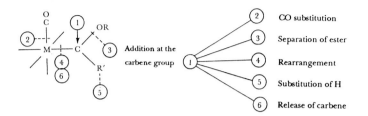

Fig. 3. Possible reactions of alkoxy carbene complexes.

We have already pointed out that the carbene carbon atom is an electrophilic centre and should therefore have a high nucleophilic reactivity. Hence, according to our present understanding, in most reactions a nucleophile would be added to the carbene carbon atom in the primary reaction stage. In some cases, for example in the case of some phosphines (49) and tertiary amines (50), such addition products can be isolated in an analytically pure form under certain conditions (① in Fig. 3). Another possible reaction is that the nucleophilic agent substitutes a carbon monoxide group in the complex while preserving the carbene ligand (② in Fig. 3). The carbene complex can also be regarded in a very formal way from the point of view of an ester-like system ($X = C \langle^{OCH_3}_R$ where $X = M (CO)_5$ instead of $X = O$) since the oxygen atom as well as the metal atom in the $M(CO)_5$ radical have two electrons less than the number of electrons required for attaining the rare gas configuration. Therefore, it is not surprising that the OR radical can be substituted by amino, thio and seleno groups (③ in Fig. 3). This leads us to the synthesis of amino (organyl) carbene complexes (36, 51–54), thio (organyl) carbene complexes (51, 55) and seleno (organyl) carbene complexes. The synthesis of the last two series of complexes requires some experimental skill.

We can also observe reactions which lead to a more stable arrangement of the entire system through a primary addition followed by a rearrangement (④ in Fig. 3). It can be noticed further that the hydrogen atoms of alkyl groups in the α position with respect to the carbene carbon atom are so acidic — because of the electron pull of the $M(CO)_5$ radical — that their acidity is similar to that of the hydrogen atoms in nitromethane (⑤ in Fig. 3). Finally the separation of the carbene ligand from the metal complex opens up new synthetic routes in organic chemistry (⑥ in Fig. 3).

ADDITION AND CO SUBSTITUTION

If we treat trialkyl phosphines with pentacarbonyl [alkoxy(organyl)carbene] complexes of chromium(0) and tungsten(0), for example, in hexane at temperatures below −30°C, the corresponding phosphorylide complexes (which are addition compounds) can be isolated in an analytically pure form and studied (49). The erstwhile carbene carbon is now sp³ hybridised and now has only a σ-bond at the central metal. In the case of triaryl and mixed alkyl aryl phosphines, the addition-dissociation equilibrium (57) (Fig. 4) lies to a great extent on the side of the starting materials, so that the ylide complexes can be detected only by spectroscopy. Figure 4 shows the reaction scheme for pentacarbonyl[methoxy(methyl)carbene]chromium(0) and tertiary phosphines.

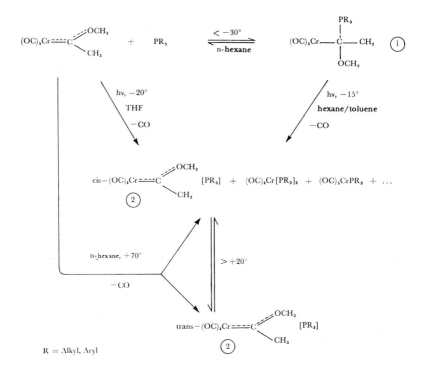

Fig. 4. Reaction of pentacarbonyl[methoxy(methyl)carbene]chromium(0) with tertiary phosphines.

On irradiating solutions of these ylide complexes in hexane-toluene mixtures at −15°C, we get the cis-tetracarbonyl[alkoxy(organyl)carbene] phosphine complexes (58) with elimination of a CO ligand from the cis position. That is, the phosphine which is at first added to the carbene carbon of the starting complex, substitutes a CO ligand at the metal atom and the carbene grouping is formed once again. To a smaller extent, the carbene ligand is also substituted by phosphine. We arrive at the same products if we employ only the equilibrium mixtures of pentacarbonyl carbene complexes and phosphines under slightly modified conditions (at −20°C in THF) instead of the isolated ylide complexes (58). But, if the reaction is carried out thermally — at 70°C in hexane — mixtures of the cis and trans isomers are formed instead of the pure cis tetracarbonyl carbene phosphine complexes (59, 60). We were able to isolate the cis and trans isomers in pure form (60). On heating solutions of either of these components (separately) isomerisation takes place until an equilibrium is attained (61). We were especially interested in the mechanism and found that the isomerisation reaction (62) follows first order kinetics, that the reaction rate is not influenced by the presence of the free ligands, phosphine and carbon monoxide, and that the isomerisation rate is greater in tetracarbonyl[methoxy(methyl)carbene]triethyl phosphine chromium(0) than in the corresponding tricyclohexyl phosphine complex. Now, how do we visualise this process of isomerism? The findings suggest an intramolecular mechanism in which none of the bonds of the six monodentate ligands with the metal are broken or formed afresh during the change from the cis isomer to the trans isomer and *vice versa*. Instead there could be a twist in the two planes formed by any three ligands, by 120° in opposite directions (twist mechanism) (Fig. 5).

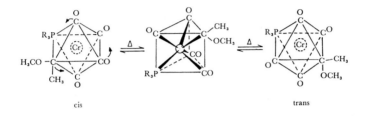

cis   trans

Fig. 5. Hypothesis regarding the isomerisation of tetracarbonyl[methoxy(methyl)carbene] phosphine chromium(0).

Since this process passes through a trigonal-prismatic transition state with an increased steric inhibition, it is also quite understandable that the compound with the highly cumbersome tricyclohexyl phosphine ligand isomerises at a slower rate than the corresponding triethyl phosphine complex.

TRANSITION METAL/CARBENE COMPLEX RADICALS AS PROTECTIVE AMINO GROUPS FOR AMINO ACIDS AND PEPTIDES

If we treat the alkoxy carbene complexes with primary or secondary amines, instead of phosphines, we observe a new type of reaction which is similar to the reactions of esters. We have been greatly interested in this type of reaction in recent times and it has shown us a new approach to the chemistry of peptides — quite a surprising approach for a chemist dealing with complexes. We could show that the alkoxy group of alkoxy (organyl) carbene complexes can be substituted not only by mono or dialkyl amino radicals but also by free amino groups of amino acid esters and peptide esters (63, 64). The principle of this reaction is shown in Eq. 6.

Even at 20°C, not only simple amino acid esters but also polyfunctional amino acid esters react with alkoxy (organyl) carbene complexes in ether solution without protection of the third function. The organometallic radical thus represents a new and interesting protecting group, especially because it can be easily separated again by treatment with trifluoro acetic acid. In some cases, this separation can be done even with acetic acid, in a milder reaction.

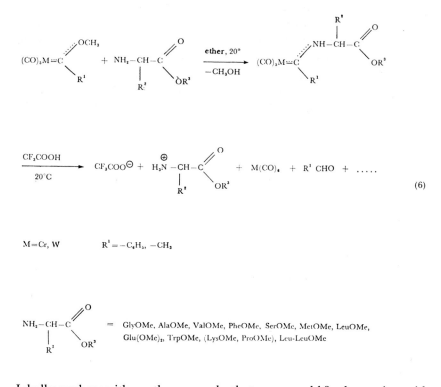

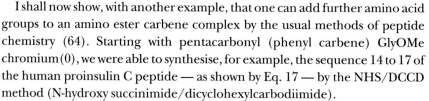

I shall now show, with another example, that one can add further amino acid groups to an amino ester carbene complex by the usual methods of peptide chemistry (64). Starting with pentacarbonyl (phenyl carbene) GlyOMe chromium(0), we were able to synthesise, for example, the sequence 14 to 17 of the human proinsulin C peptide — as shown by Eq. 17 — by the NHS/DCCD method (N-hydroxy succinimide/dicyclohexylcarbodiimide).

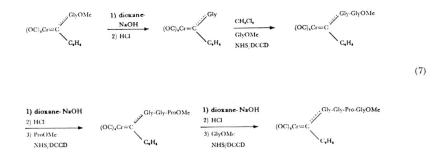

$$(7)$$

We are now working with E. Wünsch in this area and we believe that the utilisation of such carbene complexes could offer a number of new routes and advantages to the peptide chemist:

(1) Such amino acid or peptide derivatives are yellow and hence easy to distinguish, for example in chromatographic methods.

(2) This protecting group can be separated under milder conditions; the reaction products that are formed in addition to the amino acid or peptide esters — mainly aldehydes and metal hexacarbonyls — are volatile and can therefore be easily separated.

(3) Most carbene complexes of amino acid esters and many dipeptide esters are volatile and can be studied by mass spectroscopy.

(4) We have here a method by which heavy metal atoms like tungsten can be incorporated in peptides and free amino groups can thus be marked.

ADDITION AND REARRANGEMENT REACTIONS

Here are two recent examples to illustrate this type of reaction of carbene complexes. Pentacarbonyl[methyl(thiomethyl)carbene]chromium(0) and tungsten(0) react at low temperatures with hydrogen bromide to form pentacarbonyl[(1-bromoethyl)methylsulfide] complexes (65) (Eq. 8).

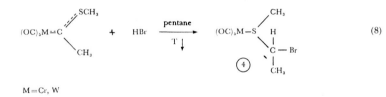

$$(8)$$

M = Cr, W

In this process, the original carbene carbon loses its bond with the transition metal and sulphur occupies this position. The second example shows that such a reaction need not always result in an uncharged system. That is, instead of thiocarbene complexes, if we use aminocarbene complexes in the reaction with hydrogen halides, salt-like compounds can be isolated (66). After the reaction we find the halogen at the metal, the hydrogen at the eliminated carbene ligand, and we get imonium halogeno pentacarbonyl metallates.

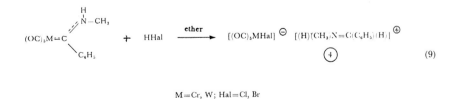

$$(9)$$

M = Cr, W; Hal = Cl, Br

We thus have a method for synthesising such cations and this method allows great variations in this type of compound, which are not easily accessible.

HYDROGEN SUBSTITUTION AT THE α-C ATOM

C. G. Kreiter (67) was the first to show the acidity of hydrogen atoms bound to the α-C atom of alkoxy (alkyl) carbene complexes by ^{1}H-NMR spectroscopic studies. Solutions of pentacarbonyl[methoxy(methyl)carbene]chromium(0) in CH_3OD, in the presence of catalytic quantities of sodium methylate, exchange all the hydrogen atoms at the methyl group bound to the carbene carbon with deuterium (Eq. 10).

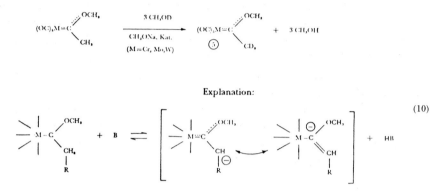

Explanation:

(10)

The base is thus evidently in a position to form an anion through reversible elimination of a proton in the α position with respect to the carbene carbon. This reaction can also be utilised to introduce new groups into the carbene ligand at this position (40, 68).

By its nature, this acidity is closely connected with the strong positively charged character of the carbene carbon atom. Let me once again therefore take up at this point the unusual ^{13}C-NMR spectroscopic shifts of this atom, with the examples of some characteristic chromium(0) complexes.

$$(OC)_5 \ CrC[N(CH_3)_2] \ C_6H_5 \qquad 270,6 \ ppm^{b)}$$
$$(OC)_5 \ CrC(OCH_3) \ C_6H_5 \qquad 351,4 \ ppm^{a)}$$
$$(OC)_5 \ CrC(C_6H_5) \ C_6H_5 \qquad 399,4 \ ppm^{b)}$$

a) in C_6D_6

b) in CD_3COCD_3

Fig. 6. ^{13}C-NMR shifts of $C_{carbene}$ atoms of some carbene complexes (values of δ relative to int. TMS).

If we start with the methoxy (phenyl) carbene complex, for which a shift of 351.4 ppm was measured (39), and replace the methoxy group with the dimethyl amino group which has a better stabilising effect, the value of δ drops as expected to 270.6 ppm (66). On the other hand, in the diphenyl carbene complex, δ has a value of 399.4 ppm (69). We have been working on this compound recently and we have in this compound an extremely highly positivised carbon atom, even when composed to the values found for organic carbonium ions. The two phenyl groups are therefore now hardly capable of compensating for the electron deficit at the carbene carbon. This chromium carbene complex is much more unstable than the homologous tungsten compound reported recently by C. P. Casey (70).

ELIMINATION OF THE CARBENE LIGAND

Reaction with acids

I do not want to try the patience of the organic chemists present here with too many details about the chemistry of complex compounds. I shall therefore mention here some applications of carbene complexes that may be of interest to organic chemists. I think the epoch of "inorganic" and "organic" chemistry as watertight compartments is now over. We must now consider all possible avenues that nature offers us.

The road to organic chemistry will be open when it becomes possible to separate the carbene ligand from the metal under conditions that are not too drastic. This is now possible with hydrogen halides in methylene chloride at a temperature as low as −78° (71) (Eq. 11).

$$(OC)_5W=C\overset{OCH_3}{\underset{C_6H_5}{\big\langle}} \quad + \quad HX \quad \xrightarrow[-78°C]{CH_2Cl_2} \quad (OC)_5XWH \quad + \quad \left\{ C\overset{OCH_3}{\underset{C_6H_5}{\big\langle}} \right\}$$

(11)

$$X = J, Br, Cl$$

This reaction leads to pentacarbonyl halogeno tungsten hydrides. Unlike the corresponding anions, these compounds were not known earlier, as far as we know. The neutral hydride complexes are very unstable and are almost completely dissociated in aqueous solution into hydronium cations and pentacarbonyl halogen tungstate anions. The anions can be precipitated in the form of tetramethyl ammonium salts (71), for example (Eq. 12).

$$(OC)_5IWH \quad + \quad [N(CH_3)_4]Br \quad \xrightarrow{H_2O} \quad [(OC)_5WI]^{\ominus} \; [N(CH_3)_4]^{\oplus} \downarrow \quad + \quad HBr$$

(12)

We found some indications concerning the fate of the cleaved carbene ligand from other studies. That is, if we treat tetracarbonyl[methoxy(organyl)carbene] triphenyl phosphine chromium(0) with benzoic acid or acetic acid in boiling ether, we can isolate the α-methoxy organyl esters of these acids (72) (Eq. 13).

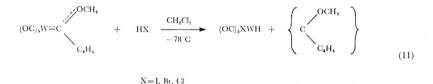

(13)

$$R,R' = CH_3, C_6H_5$$

This secondary reaction amounts to a formal insertion of the carbene radical into the OH group of the carboxylic acid. Triphenyl phosphine, which is added to the reaction mixture, merely serves to improve the separation of the organometallic radical in the form of the poorly soluble tetracarbonyl bis (triphenyl phosphine) chromium(0) complex. The reaction with HCl proceeds similarly; but the insertion products formed, namely, α-halogen organyl (methyl) ethers, react immediately with the phosphine present to form the corresponding phosphonium salts (72).

The following question of course arises in this connection: What happens to the eliminated carbene ligand when no suitable reaction partner is available? The reaction conditions employed are of decisive importance in answering the question.

Reaction with pyridine

Right at the beginning of our studies on carbene complexes, we observed that the carbene radical can be easily broken off from the metal with the help of pyridine, and that the metal fragment can be separated in the form of carbonyl pyridine chromium complexes (73). In the case of alkoxy (alkyl) carbene complexes, a hydrogen atom of the eliminated carbene radical shifts, with the help of the base, towards the original carbene carbon atom, with the formation of enol ether (73,74) (Eq. 14).

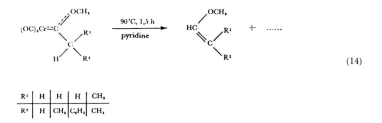

$$ \tag{14} $$

R¹	H	H	H	CH₃
R²	H	CH₃	C₂H₅	CH₃

Thermal decomposition

To check whether the base has an effect on the (secondary) reaction of the carbene ligand, we decomposed pentacarbonyl[methoxy(methyl)carbene] chromium(0) purely thermally at 150°C in decalin. Under these conditions we observed the exclusive formation of the dimer; to be precise, as a mixture of the cis and trans isomers (74). The reaction must therefore be as follows (Eq. 15).

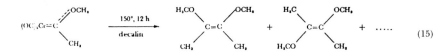

$$ \tag{15} $$

Since the shifting of a hydrogen atom is not possible in the case of the methoxy (phenyl) carbene ligand, only dimerisation takes place in the reactions with bases as well as with thermal decomposition.

Reactions with elements of the sixth group in the periodic table

We are of course especially interested in those reactions which lead to products that are not easily accessible by the conventional methods of organic chemistry and which can be prepared easily with our complexes. We found such an example in the reaction of pentacarbonyl[methoxy(aryl)carbene]chromium(0) complexes with oxygen, sulphur and selenium (76). By this reaction we can easily get the corresponding methyl esters, thio-O-methyl esters and seleno-O-methyl esters. The latter two types of compounds seem important to us from the synthetic point of view (Eq. 16).

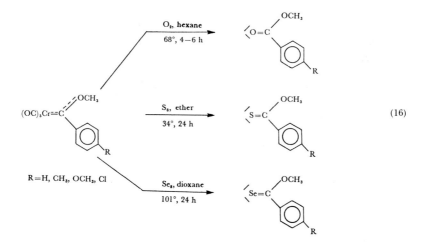

$$(16)$$

Reactions with vinyl ethers and N-vinyl pyrrolidones

At quite an early stage of our studies on carbene complexes, we had argued that these compounds would deserve their name only if they underwent reactions typical of carbenes.

In this connection the organic chemist will immediately recall the formation of cyclopropane derivatives from olefins and carbenes. We very soon found that this reaction was also possible with our complexes and those $C = C$ double bonds which are deficient in electrons and are either polarised or easily polarisable (77–81). As an example of this, I would like to mention the reaction of pentacarbonyl[methoxy(phenyl)carbene]chromium(0), molybdenum(0) and tungsten(0) with ethyl vinyl ether (79). However, we get the corresponding cyclopropane derivatives only if we cleave the carbene ligand in an autoclave at 50°C under CO atmosphere at a pressure of 170 atm. (Eq. 17).

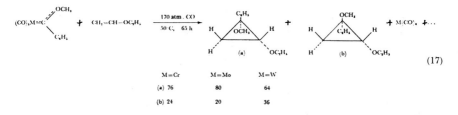

	M=Cr	M=Mo	M=W
(a)	76	80	64
(b)	24	20	36

(17)

As expected, we find two isomers [(a) and (b) in Eq. 17]. Under the same reaction conditions, the ratio of the two isomers depends on the choice of the central metal. This seems to us to be a rather definite pointer that this reaction takes place not through a "free" methoxy (phenyl) carbene, but that, to the contrary, the metal atom is involved at the decisive stage of the reaction.

When we try to carry out this reaction, under similar conditions, with vinyl pyrrolidone (2) as the olefin component, we get, instead of the expected cyclopropane derivative, quite surprisingly, 1-[4-methoxy-4-phenyl-butene(1)-on(3)yl] pyrrolidone(2) (82) (Eq. 18).

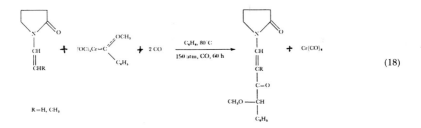

(18)

How do we explain the formation of this unexpected product, in which we find carbon monoxide also added to the carbene ligand and pyrrolidone? (Fig. 7).

We now believe that the carbene ligand first reac's with carbon monoxide to form methoxy(phenyl)ketene. This forms a cyclobutanone derivative with the polarised olefin, which is converted into the product found — via a ring opening.

This hypothesis has been reinforced by the subsequent finding that on using N-(β-methyl vinyl)pyrrolidone(2) instead of N-vinyl-pyrrolidone(2), we could isolate the postulated four-ring system in addition to the open-chain end product (82).

Our original idea of using carbon monoxide only for cleaving the carbene ligand thus led us to an unexpected result and showed at the same time that the reactivity of carbon monoxide with respect to organic systems should not be ignored.

In an attempt to obtain the cyclopropane derivatives which we wanted, we reacted the same starting materials thermally in benzene in the absence of CO. But in this attempt also, we did not get the desired compounds,

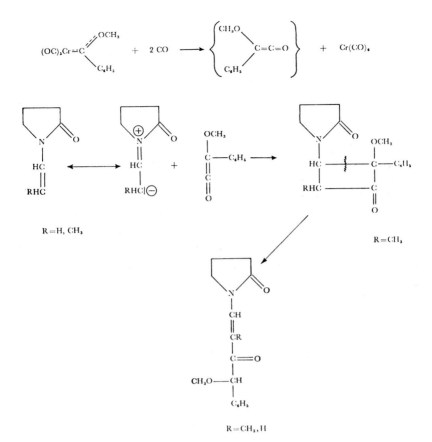

Fig. 7. Hypothesis regarding the course of the reaction of pentacarbonyl[methoxy(phenyl)carbene] chromium(0) with N-vinyl-pyrrolidone(2) under 150 atm. CO pressure.

but, surprisingly, the corresponding substituted α-methoxy styrenes (83) (Eq. 19).

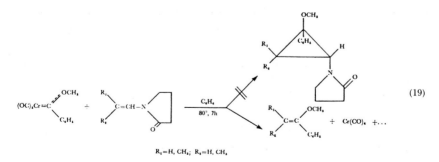

$$(19)$$

A possible course of the reaction might be as follows:

N-vinyl-pyrrolidone(2) also has a nucleophilic centre at the oxygen atom. This nucleophilic centre could attack the electrophilic carbene carbon and separate the carbene ligand from the metal. The intermediate product thus formed — irrespective of whether it has an open chain form or a six-membered ring — then undergoes a cleavage similar to the heterolytic fragmentation reported by C. A. Grob (84) (Fig. 8).

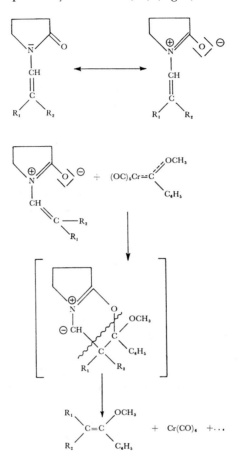

Fig. 8. Hypothesis regarding the course of the reaction of pentacarbonyl[methoxy(phenyl)carbene] chromium(0) with N-vinyl-pyrrolidone(2) and β-substituted N-vinyl-pyrrolidones under normal pressure.

Reactions with electrophilic carbenes

As shown right at the beginning of this lecture, the carbene ligand in carbene complexes of our type provides a "nucleophilic" behaviour with respect to the metal fragment. One of our pet ideas was thus to combine the carbene ligand with an electrophilic carbene. We therefore treated pentacarbonyl[methoxy(phenyl)carbene]chromium(0) with phenyl (trichloromethyl) mercury (85). Compounds of this kind have been studied in detail by D. Seyferth and they are recognised as starting materials for dihalo carbenes (86). The carbene complex and the carbenoid compound could be made to react at 80°C in benzene to form β, β-dichloro-α-methoxy styrene (85) (Eq. 20).

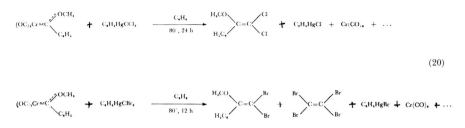

$$(20)$$

This combination reaction is very sensitive to temperature conditions. Even greater complications arose on using phenyl (tribromomethyl) mercury, with the formation of mixtures of olefins.

With this small selection from our recent research results, I think I have been able to show you the wide variety of possible reactions offered by the chemistry of transition metal carbene complexes.

I would now like to report to you our findings in another related area on which we have been working very intensively recently: the chemistry of transition metal carbine complexes.

TRANSITION METAL CARBYNE COMPLEXES

To explore all the possible reactions of transition metal carbene complexes, we had attempted some years ago to make our complexes react with electrophilic reaction partners in addition to nucleophilic reaction partners. Our idea was to exchange the methoxy group of methoxy (organyl) carbene complexes by a halogen with the help of borontrihalides and thus to arrive at halogeno (organyl) carbene complexes. We did observe a fast reaction but found only decomposition products. But recently, in collaboration with G. Kreis, we carried out this reaction at very low temperatures and could isolate well-defined compounds which were, however, thermally quite unstable (87). Their composition was equivalent to the sum of a metal tetracarbonyl fragment, a halogen and the carbene ligand minus the methoxy group (Eq. 21).

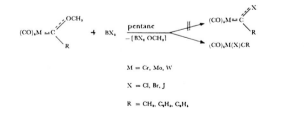

$$(21)$$

$$M = Cr, Mo, W$$

$$X = Cl, Br, J$$

$$R = CH_3, C_2H_5, C_6H_5$$

The IR spectra indicated the presence of disubstituted hexacarbonyls with two different ligands in the trans position (trans $(CO)_4MR^1R^2$). Moreover, the cryoscopic determination of molecular weight showed the presence of a monomer complex. Together with other spectroscopic findings, especially from ^{13}C and ^{1}H-NMR studies, this could be interpreted only if we assumed that, besides the four CO ligands, a halogen and a CR group bonded to the metal had to be present (Fig. 9).

Fig. 9. Structure and model of bonds for $(CO)_4(X)MCR$.

We would like to propose the name "carbyne complexes" for this new type of compound, for two reasons: (1) on the analogy of "carbene complexes" and (2) on the analogy of the term "alkyne", because on the basis of the diamagnetism of these compounds, we must postulate a formal metal-carbon triple bond.

X-RAY STRUCTURE ANALYSES

Such a triple bond should result in a very short distance between the metal and the carbyne carbon. To answer this question and to confirm the proposed structure, X-ray structural analyses have been carried out in our institute by G. Huttner *et al.* on three carbyne complexes so far (88).

The first such study was done on trans-(iodo)tetracarbonyl(phenyl-carbyne)tungsten(0) (87, 88) (Fig. 10).

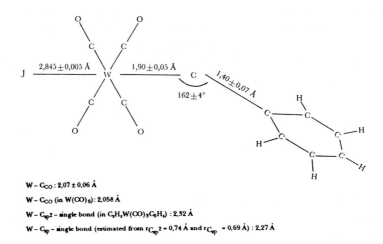

W – C_{CO} : 2,07 ± 0,06 Å

W – C_{CO} (in W(CO)$_6$): 2,058 Å

W – C_{sp^2} – single bond (in $C_5H_5W(CO)_3C_6H_5$) : 2,32 Å

W – C_{sp} – single bond (estimated from $r_{C_{sp^2}} = 0,74$ Å and $r_{C_{sp}} = 0,69$ Å) : 2,27 Å

Fig. 10. Molecular structure of trans-(iodo)tetracarbonyl(phenylcarbyne)tungsten(0).

This study essentially confirmed our ideas and gave an extremely short tungsten-carbon distance of 1.90 Å. Instead of the linear arrangement of metal, $C_{carbyne}$ and $C_{1,4\,(phenyl)}$ atoms, however, we found a clear bending of about 162°. Since we could not explain at this stage whether this bending was due to electronic or lattice effects, we immediately undertook the study of another complex. Figure 11 shows the result: the structure of trans-(iodo)-tetracarbonyl(methyl carbyne)chromium(0) (88).

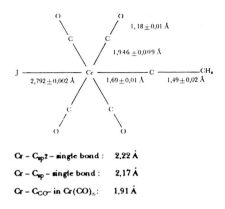

Cr – C$_{sp}$2 – single bond : 2,22 Å

Cr – C$_{sp}$ – single bond : 2,17 Å

Cr – C$_{CO}$ in Cr(CO)$_6$: 1,91 Å

Fig. 11. Molecular structure of trans-(iodo)tetracarbonyl(methylcarbyne)chromium(0).

In this compound we found not only the expected linear arrangement of the chromium, carbon and methyl group, but also the shortest distance between chromium and carbon found so far, namely 1.69 Å. This value is appreciably shorter than the Cr-C$_{CO}$ distance in the same complex (1.946 Å or in hexacarbonyl chromium (1.91 Å).

Subsequently we were interested in the question whether second substituents in the starting carbene complex can influence the orientation of the halogen in the resulting carbyne complex. To answer this question, we first treated cis-tetracarbonyl[methoxy(methyl)carbene]trimethyl phosphine, arsine and stibine chromium(0) with borontrihalides (89) (Eq. 22).

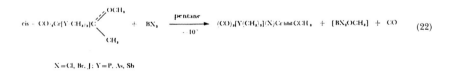

X = Cl, Br, J; Y = P, As, Sb

The reaction proceeded as smoothly as before, but in the compounds with the composition $(CO)_3[Y(CH_3)_3](X)Cr \equiv CCH_3$ (X = Cl, Br, I and Y = P, As, Sb) that were formed, the mutual spatial arrangement of the ligands could not at first be clearly determined. An X-ray structural analysis was therefore carried out on a representative compound of this type (88–90) (Fig. 12).

For (bromo)tricarbonyl(methyl carbyne)trimethylphosphinechromium(0), we found a meridional arrangement of the three substituents and, again, a trans-arrangement of the halogen and carbyne ligand. We are at present studying how a carbene complex with an initial trans-configuration behaves in the reaction with borontrihalides (89).

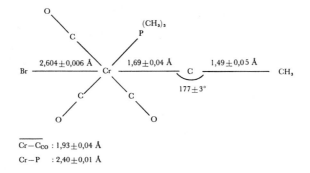

$$\overline{Cr-C_{CO}} : 1,93\pm0,04 \text{ Å}$$
$$Cr-P \quad : 2,40\pm0,01 \text{ Å}$$

Fig. 12. Molecular structure of mer-(bromo)tricarbonyl(methylcarbine)trimethylphosphine chromium(0).

REACTIONS OF OTHER PENTACARBONYL CARBENE COMPLEXES WITH BORONTRIHALIDES

We thought it would also be interesting to study the effects of changes in the organic radical of the carbyne ligand on the stability and behaviour of these compounds. For this purpose, we treated a number of phenyl-substituted pentacarbonyl[methoxy(aryl)carbene]tungsten(0) complexes with borontribromide (91) (Eq. 23).

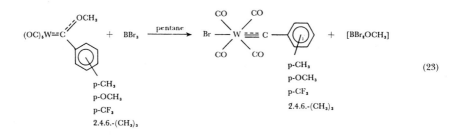

(23)

^{13}C-NMR spectroscopy seemed to us to be a suitable tool for studying electronic changes in this case. Figure 13 compares the chemical shifts of the carbyne carbon atoms of the resulting trans-(bromo)tetracarbonyl(aryl carbyne)tungsten(0) complexes (87, 92).

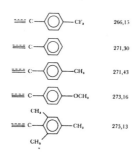

Fig. 13. ^{13}C-NMR shifts of $C_{carbyne}$ atoms of some trans-Br(CO)$_4$W-Ar complexes (values of δ, CD$_2$Cl$_2$ relative to int. TMS).

Quite unexpectedly, we find in this series that the p-CF₃ derivative has the lowest value of δ, i.e. the strongest screening for the carbyne carbon atom; we find a much weaker screening for the 2.4.6 trimethyl phenyl compound. We need more data to interpret this result exactly and experiments for this purpose are in progress.

We could further show that not only methoxy(organyl)carbenecomplexes react with borontrihalides in the manner described above. We found that trans-(bromo)tetracarbonyl(phenyl carbyne)chromium(0) and trans-(bromo)tetracarbonyl(phenyl carbyne)tungsten(0) can also be obtained by treating pentacarbonyl[hydroxy(phenyl)carbene]chromium(0) (93) as well as pentacarbonyl(phenyl carbene)glycinemethylestertungsten(0) (64) (Eqs. 24 and 25) with borontrihalides.

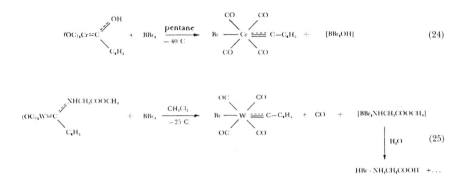

(24)

(25)

I would like to point out especially that the reaction of the aminoacidcarbene complex with borontribromide offers another convenient way of cleaving the metal pentacarbonyl (phenyl carbene)yl protecting group — under very mild conditions, namely, even at −25°C.

The fact that experimental results cannot always be generalised is shown by the reaction of cis-(bromo)tetracarbonyl[hydroxy(methyl)carbene]manganese with borontribromide. The reaction does not lead to the analogous carbyne complex, but to a product in which the hydrogen atom of the hydroxy group is replaced by the BBr₂ radical (92) (Eq. 26).

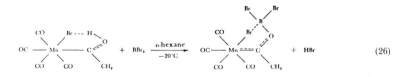

(26)

The situation here does not seem to be conducive to the formation of a carbyne complex because of the "fixation" of the OH group by the formation of a bridge with the bromide ligand in the cis position. Another interesting question was how pentacarbonyl[ethoxy(diethylamino)carbene]tungsten(0) would react with borontrihalide, because — as we have seen earlier — in principle the alkoxy group as well as the amino group can be cleaved. The experiment led to the exclusive formation of trans-(bromo)tetracarbonyl(diethyl amino carbyne)tungsten(0) (95) — a compound that can be handled relatively easily. Its stability is probably due to the interaction of the metal/carbon bond with the free electron pair of nitrogen. This explanation is also supported by ^{13}C-NMR measurements (Eq. 27).

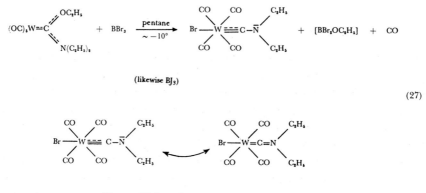

(27)

$\delta C_{carbine} = 235,6$ ppm (CD$_2$Cl$_2$, int. TMS)

REACTIONS OF PENTACARBONYL CARBENE COMPLEXES WITH ALUMINIUM AND GALLIUM HALIDES

We could extend the scope of our method of synthesis by using aluminium trichloride, aluminium tribromide and gallium trichloride instead of borontrihalides (96) (Eq. 28). With these compounds we also obtained carbyne complexes in good yields.

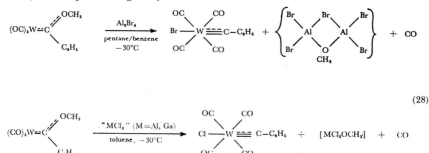

(28)

REACTIONS OF LITHIUM BENZOYL PENTACARBONYL TUNGSTATES WITH TRIPHENYLPHOSPHINE DIBROMIDE

A new method of synthesis — new in principle — was discovered from the reaction of lithium benzoyl pentacarbonyl tungstate with triphenyl phosphine dibromide, at low temperatures in ether (97) (Eq. 29).

(29)

The first step is presumably the establishment of a $C_{carbene}$-O-P bond with the formation of lithium bromide. The intermediate product thus formed could then be stabilised by reaction between the second bromine atom and the metal, elimination of a CO ligand and the cleavage of triphenyl phosphine oxide under thermodynamically favourable conditions to form the carbyne complex.

REACTIVITY OF THE CARBYNE LIGAND

With the carbyne complexes also, we did not wish to confine ourselves to the preparation and spectroscopic study of new representatives of this type of compound. We therefore began, in the meantime, to study their reactivity. We first looked for a possible way of comparing such a metal-carbon triple bond with a carbon-carbon triple bond. We found this in the reaction of dimethyl amine with trans-(bromo)tetracarbonyl(phenyl acetylenyl carbyne) tungsten(0) (98). This latter compound can be obtained by treating pentacarbonyl [ethoxy(phenylacetylenyl)carbene]tungsten(0) (21) with borontribromide. We found that at −40°C in ether only addition to the organic triple bond takes place while the carbyne-metal bond remains unchanged (99) (Eq. 30).

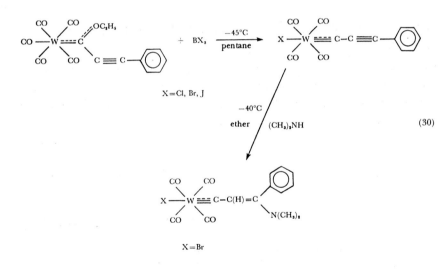

(30)

At the same time we also tried to study how the carbyne ligand behaves when it is cleaved from the metal. As with the carbene complexes, a dimerisation takes place in the absence of a suitable reaction partner, but alkines are formed in this case (100). The conditions for the decomposition are very mild.

Tolane or dimethyl acetylene can be obtained in this way in non-polar solvents even at 30°C. We get the same result on heating the solid methyl carbyne complex to 50°C (Eq. 31).

We think that the path is now open for utilising carbyne complexes to synthesise organic compounds. As far as we know, no systematic source of "carbyne complexes" is now available for the purpose of synthesis. Because the

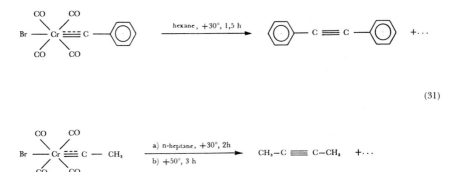

(31)

cleavage conditions are mild, there is a wide scope here for interesting applications.

That brings me to the end of my lecture. I hope I have shown you that organometallic chemistry offers many more interesting possibilities.

ACKNOWLEDGEMENTS

The research results that I have had the honour to report are largely the work of my colleagues.

I would like to thank the following for their contributions: Miss K. Weiss, Dr. K. H. Dötz, Dr. H. Fischer, Dr. F. R. Kreissl, Dr. S. Riedmüller, Dr. K. Schmid, Dr. A. de Renzi, Mr. B. Dorrer, Mr. W. Kalbfus, Mr. H. J. Kalder, Mr. G. Kreis, Mr. E. W. Meineke, Mr. D. Plabst, Mr. K. Richter, Mr. U. Schubert, Mr. A. Schwanzer, Mr. T. Selmayr, Mr. S. Walz and Mr. W. Held. Thanks are also due to my colleagues at our institute and their colleagues: Dr. J. Müller for analysing the mass spectra, Dr. C. G. Kreiter for the ^{13}C-NMR arrangements and Dr. G. Huttner, Mr. W. Gartzke and Mr. H. Lorenz for X-ray structure analysis.

REFERENCES

1. Bunsen, R., Lieb. Ann. Chem. *42*, 41 (1842).
2. Frankland, E., Lieb. Ann. Chem. *71*, 171, 213 (1849), 95, 36 (1855).
3. Grignard, V., Compt. Rend. *130*, 1322 (1900).
4. Ziegler, K. (Nobel-Vortrag from 12.12.1963); Angew. Chem. *76*, 545, (1964) and references therein.
5. Fischer, E. O. and Werner, H., Metall. π-Komplexe mit di- and oligoolefinischen Liganden, Verlag Chemie, Weinheim 1963.
6. Herberhold, M., Metal-π-Complexes, Vol. II. Complexes with Monoolefinic Ligands, Elsevier Publ. Comp., Amsterdam 1972.
7. Zeise, W. C., Pogg. Annalen *9*, 632 (1827).
8. a) Fischer, E. O. and Fritz, H. P., Adv. Inorg. Chem. Radiochem. *1*, 55 (1959), Angew. Chem. *73*, 353 (1961).
   b) Fischer, E. O., Angew. Chem. *67*, 475 (1955).
9. Wilkinson, G. and Cotton, F. A., Progr. Inorg. Chem. *1*, 1 (1959).
10. Fischer, E. O. and Hafner, W., Z. Naturforsch. *10b*, 665 (1955).
11. Hübel, W., Derivatives from Metal Carbonyls and Acetylenic Compounds in Wender, I. and Pino, P., Organic Syntheses via Metal Carbonyls, Interscience Publ. 1968.
12. Fischer, E. O. and Maasböl, A., Angew. Chem. *76*, 645 (1964); Angew. Chem. Int. Ed. *3*, 580 (1964).

13. Ryang, M., Rhee, I. and Tsutsumi, S., Bull. Chem. Soc. Japan *37*, 341 (1964).
14. Fischer, E. O., Kreis, G. and Kreißl, F., J. organomet. Chem. *56*, C37 (1973).
15. Aumann, R. and Fischer, E. O., Chem. Ber. *101*, 954 (1968).
16. Meerwein, H., Hinz, G., Hofmann, P., Kroning, E. and Pfeil, E., J. prakt. Chem. (2) *147*, 257 (1937); Meerwein, H., Battenberg, E., Gold, H., Pfeil, E. and Willfang, G., J. prakt. Chem. (2) *154*, 83 (1940).
17. Fischer, E. O. and Maasböl, A., Chem. Ber. *100*, 2445 (1967).
18. Fischer, E. O. und Kollmeier, H. J., Angew. Chem. *82*, 325 (1970); Angew. Chem. Int. Ed. *9*, 309 (1970).
19. Fischer, E. O., Winkler, E., Kreiter, C. G., Huttner, G. and Krieg, B., Angew. Chem. *83*, 1021 (1971); Angew. Chem. Int. Ed. *10*, 922 (1971).
20. Fischer, E. O., Kreiter, C. G., Kollmeier, H. J., Müller, J. and Fischer, R. D., J. organomet. Chem. *28*, 237 (1971).
21. Fischer, E. O. and Kreißl, F. R., J. organomet. Chem. *35*, C47 (1972).
22. Fischer, E. O., Kreißl, F. R., Kreiter, C. G. and Meineke, E. W., Chem. Ber. *105*, 2558 (1972).
23. Wilson, J. W. and Fischer, E. O., J. organomet. Chem. *57*, C63 (1973).
24. Fischer, E. O. and Offhaus, E., Chem. Ber. *102*, 2449 (1969).
25. Fischer, E. O., Offhaus, E., Müller, J. and Nöthe, D., Chem. Ber. *105*, 3027 (1972).
26. Fischer, E. O., Beck, H.-J., Kreiter, C. G., Lynch, J., Müller, J. and Winkler, E., Chem. Ber. *105*, 162 (1972).
27. Fischer, E. O., Kreißl, F. R., Winkler, E. and Kreiter, C. G., Chem. Ber. *105*, 588 (1972).
28. Fischer, E. O. and Riedel, A., Chem. Ber. *101*, 156 (1968).
29. Fischer, E. O. and Aumann, R., Chem. Ber. *102*, 1495 (1969).
30. Fischer, E. O. and Beck, H.-J., Chem. Ber. *104*, 3101 (1971).
31. Mills, O. S. and Redhouse, A. D., J. Chem. Soc. *A* 642 (1968).
32. Cotton, F. A. and Richardson, D. C., Inorg. Chem. *5*, 1851 (1966).
33. Moser, E. and Fischer, E. O., J. organomet. Chem. *13*, 209 (1968).
34. Kreiter, C. G. and Fischer, E. O., Angew. Chem. *81*, 780 (1969); Angew. Chem. Int. Ed. *8*, 761 (1969).
35. Kreiter, C. G. and Fischer, E. O., Chem. Ber. *103*, 1561 (1970).
36. Connor, J. A. and Fischer, E. O., J. Chem. Soc. *A* 578 (1969).
37. Fischer, E. O. and Kollmeier, H. J., Chem. Ber. *104*, 1339 (1971).
38. Hawkins, N. J., Mattraw, H. C., Sabol, W. W. and Carpenter, D. R., J. Chem. Phys. *23*, 2422 (1955).
39. Kreiter, C. G. and Formácek, V., Angew. Chem. *84*, 155 (1972); Angew. Chem. Int. Ed. *11*, 141 (1972).
40. Fischer, E. O., Pure appl. Chem. *24*, 407 (1970); *30*, 353 (1972).
41. Cardin, D. J., Cetinkaya, B. and Lappert, M. F., Chem. Rev. *72*, 545 (1972).
42. Cotton, F. A. and Lukehart, C. M., Progr. Inorg. Chem. *16*, 487 (1972).
43. Cardin, D. J., Cetinkaya, B., Doyle, M. J. and Lappert, M. F., Chem. Soc. Rev. *2*, 99 (1973).
44. Öfele, K., Angew. Chem. *80*, 1032 (1968); Angew. Chem. Int. Ed. *7*, 950 (1968).
45. Huttner, G., Schelle, S. and Mills, O. S., Angew. Chem. *81*, 536 (1969); Angew. Chem. Int. Ed. *8*, 515 (1969).
46. Badley, E. M., Chatt, J., Richards, R. L. and Sim, G. A., Chem. Commun. *1969*, 1322.
47. Fischer, E. O. and Beck, H.-J., Angew. Chem. *82*, 44 (1970); Angew. Chem. Int. Ed. *9*, 72 (1970).
48. Cardin, D. J., Cetinkaya, B., Lappert, M. F., Manojlovic-Muir, Lj. and Muir, K. W., Chem. Commun. *1971*, 400.
49. Kreißl, F. R., Fischer, E. O., Kreiter, C. G. and Fischer, H., Chem. Ber. *106*, 1262 (1973).
50. Kreißl, F. R. and Fischer, E. O., Chem. Ber. *107*, 183 (1974).
51. Klabunde, U. and Fischer, E. O., J. Am. Chem. Soc. *89*, 7141 (1967).
52. Fischer, E. O., Heckl, B. and Werner, H., J. organomet. Chem. *28*, 359 (1971).
53. Fischer, E. O. and Leupold, M., Chem. Ber. *105*, 599 (1972).
54. Fischer, E. O. and Fontana, S., J. organomet. Chem. *40*, 367 (1972).
55. Fischer, E. O., Leupold, M., Kreiter, C. G. and Müller, J., Chem. Ber. *105*, 150 (1972).
56. Fischer, E. O., Kreis, G., Kreißl, F. R., Kreiter, C. G. and Müller, J., Chem. Ber. *106*, 3910 (1973).
57. Fischer, H., Fischer, E. O., Kreiter, C. G. and Werner H., Chem. Ber. *107*, 2459 (1974).

58. Fischer, H., Fischer, E. O. and Kreißl, F. R., J. organomet. Chem. *64*, C41 (1974).
59. Werner, H. and Rascher, H., Inorg. Chim. Acta *2*, 181 (1968).
60. Fischer, E. O. and Fischer, H., Chem. Ber., *107*, 657 (1974).
61. Fischer, H. and Fischer, E. O., Chem. Ber., *107*, 673 (1974).
62. Fischer, H., Fischer, E. O. and Werner, H., J. organomet. Chem. *73*, 331 (1974).
63. Weiß, K. and Fischer, E. O., Chem. Ber. *106*, 1277 (1973).
64. Weiß, K. and Fischer, E. O., unpublished.
65. Fischer, E. O. and Kreis, G., Chem. Ber. *106*, 2310 (1973).
66. Fischer, E. O., Schmid, K. R., Kalbfus, W. and Kreiter, C. G., Chem. Ber. *106*, 3893 (1973).
67. Kreiter, C. G., Angew. Chem. *80*, 402 (1968); Angew. Chem. Int. Ed. *7*, 390 (1968).
68. Casey, C. P., Boggs, R. A. and Anderson, R. L., J. Am. Chem. Soc. *94*, 8947 (1972).
69. Fischer, E. O., Held, W., Riedmüller, S. and Köhler, F., unpublished.
70. Casey, C. P. and Burkhardt, T. J., J. Am. Chem. Soc. *95*, 5833 (1973).
71. Fischer, E. O., Walz, S. and Kreis, G., unpublished.
72. Schubert, U. and Fischer, E. O., Chem. Ber. *106*, 3882 (1973).
73. Fischer, E. O. and Maasböl, A., J. organomet. Chem. *12*, P15 (1968).
74. Fischer, E. O. and Plabst, D., Chem. Ber. *107*, 3326 (1974).
75. Fischer, E. O., Heckl, B., Dötz, K. H., Müller, J. and Werner, H., J. organomet. Chem. *16*, P29 (1969).
76. Fischer, E. O. and Riedmüller, S., Chem. Ber. *107*, 915 (1974).
77. Fischer, E. O. and Dötz, K. H., Chem. Ber. *103*, 1273 (1970).
78. Dötz, K. H. and Fischer, E. O., Chem. Ber. *105*, 1356 (1972).
79. Fischer, E. O. and Dötz, K. H., Chem. Ber. *105*, 3966 (1972).
80. Cooke, M. D. and Fischer, E. O., J. organomet. Chem. *56*, 279 (1973).
81. See also: Fischer, E. O., Weiß, K. and Burger, K., Chem. Ber. *106*, 1581 (1973).
82. Dorrer, B. and Fischer, E. O., Chem. Ber. *107*, 2683 (1974).
83. Fischer, E. O. and Dorrer, B., Chem. Ber. *107*, 374 (1974).
84. Grob, C. A., Angew. Chem. *81*, 543 (1969); Angew. Chem. Int. Ed. *8*, 535 (1969).
85. de Renzi, A. and Fischer, E. O., Inorg. chim. Acta, *8*, 185 (1974).
86. Seyferth, D., Burlitch, J. M., Minasz, R. J., Yick-Pui Mui, J., Simmons, H. D., Jr., Treiber, A. J. H. and Dowd, S. R., J. Am. Chem. Soc. *87*, 4259 (1965).
87. Fischer, E. O., Kreis, G., Kreiter, C. G., Müller, J., Huttner, G. and Lorenz, H., Angew. Chem. *85*, 618 (1973); Angew. Chem. Int. Ed. *12*, 564 (1973).
88. Huttner, G. Lorenz, H. and Gartzke, W., Angew. Chem. *86*, 667 (1974).
89. Fischer, E. O. and Richter, K., unpublished.
90. Huttner, G. and Lorenz, H., Chem. Ber. *107*, 996 (1974).
91. Fischer, E. O. and Schwanzer, A., unpublished.
92. Fischer, E. O., Schwanzer, A. and Kreiter, C. G., unpublished.
93. Fischer, E. O. and Kreis, G., unpublished.
94. Fischer, E. O. and Meineke, E. W., unpublished.
95. Fischer, E. O., Kreis, G., Kreißl, F. R., Kalbfus, W. and Winkler, E., J. organomet. Chem., *65*, C53 (1974).
96. Fischer, E. O. and Walz, S., unpublished.
97. Fischer, H. and Fischer, E. O., J. organomet. Chem. *69*, C1 (1974).
98. Fischer, E. O., Kreis, G., Kalder, H. J. and Kreißl, F. R., unpublished.
99. Fischer, E. O. and Kalder, H. J., unpublished.
100. Fischer, E. O. and Plabst, D., unpublished.

Geoffrey Wilkinson

GEOFFREY WILKINSON

I was born in Springside, a village close to Todmorden in west Yorkshire on July 14th, 1921. The house where I was born and indeed most of the village has been demolished by the local council as being unfit for habitation. My father, and his father, also Geoffrey, were both master house painters and decorators, the latter, youngest of twelve children having migrated from Boroughbridge in Yorkshire about 1880. My mother's family were originally of hill farming stock but many of my relations were weavers in the local cotton mills and indeed my mother went into the mill at an early age. My first introduction to chemistry came at a quite early age through my mother's elder brother. A well known organist and choirmaster he had married into a family that owned a small chemical company making Epsom and Glauber's salt for the pharmaceutical industry. I used to play around in their small laboratory as well as go with my uncle on visits to various chemical companies.

The oldest of three children, I was educated in the local council primary school and after winning a County Scholarship in 1932, went to Todmorden Secondary School. This small school has had an unusual record of scholarly achievement, including two Nobel Laureates within 25 years. I actually had the same Physics teacher as Sir John Cockroft, but physics was never my favourite subject.

In 1939 I obtained a Royal Scholarship for study at the Imperial College of Science and Technology where I graduated in 1941. As it was wartime, I was directed to stay on and did some research under the supervision of my predecessor, Professor H. V. A. Briscoe. In late 1942, Professor F. A. Paneth was recruiting young chemists for the nuclear energy project which I joined. I was sent out to Canada in January 1943 and remained in Montreal and later Chalk River until I could leave in 1946. Having been attracted by the prospect of California, I wrote to, and was accepted by Professor Glenn T. Seaborg. For the next four years in Berkeley I was engaged mostly on nuclear taxonomy and made many new neutron deficient isotopes using the cyclotrons of the Radiation Laboratory.

On a visit in 1949 to England, Briscoe advised me that I was unlikely to get an academic position in England in nuclear chemistry so that when I went as Research Associate to the Massachusetts Institute of Technology in 1950, I began to return to my first interest as a student—transition metal complexes such as carbonyls and olefin complexes.

In 1951 I was offered an Assistant Professorship at Harvard University, largely because of my nuclear background. I was at Harvard from September

1951 until I returned to England in December, 1955, with a sabbatical break of nine months in Copenhagen in Professor Jannik Bjerrum's laboratory as a John Simon Guggenheim Fellow. At Harvard, I still did some nuclear work on excitation functions for protons on cobalt but I had already begun to work on olefin complexes so that I was primed for the appearance of the celebrated Kealy and Pauson note on dicyclopentadienyliron in Nature in early 1952.

In June 1955, I was appointed to the chair of Inorganic Chemistry at Imperial College in the University of London, which, at that time was the only established chair in the subject in the United Kingdom and took up the position in January 1956. I have been at the College ever since and have worked, with a relatively few students and postdoctoral fellows, almost entirely on the complexes of transition metals. I have been much interested in the complex chemistry of ruthenium, rhodium and rhenium, in compounds of unsaturated hydrocarbons and with metal to hydrogen bonds. The latter led to work on homogeneous catalytic reactions such as hydrogenation and hydro-formylation of olefins.

In 1952 I married Lise Sølver, only daughter of Professor and Mrs. Svend Aage Schou, lately Rector of Denmark's Pharmaceutical High School and we have two daughters.

AWARDS

1965—American Chemical Society Award in Inorganic Chemistry—Fellow of the Royal Society of London
1968—Lavoisier Medal, French Chemical Society—Foreign Member, Royal Danish Academy of Science
1970—Foreign Member, American Academy of Arts & Sciences
1971—Chemical Society Award in Transition Metal Chemistry
1974—Consejero de Honor, Spanish Council for Scientific Research

THE LONG SEARCH FOR STABLE TRANSITION METAL ALKYLS

Nobel Lecture, December 11, 1973

by

GEOFFREY WILKINSON

Imperial College of Science & Technology, London, England

Chemical compounds in which there is a single bond between a saturated carbon atom and a transition metal atom are of unusual importance. Quite aside from the significance and role in Nature of the cobalt to carbon bonds in the vitamin B_{12} system and possible metal to carbon bonds in other biological systems, we need only consider that during the time taken to deliver this lecture, many thousands, if not tens of thousands of tons of chemical compounds are being transformed or synthesised industrially in processes which at some stage involve a transition metal to carbon bond. The nonchemist will probably be most familiar with polyethylene or polypropylene in the form of domestic utensils, packaging materials, children's toys and so on. These materials are made by Ziegler-Natta* or Philipps' catalysis using titanium and chromium respectively. However, transition metal compounds are used as catalysts in the synthesis of synthetic rubbers and other polymers, and of a variety of simple compunds used as industrial solvents or intermediates. For example alcohols are made from olefins, carbon monoxide and hydrogen by use of cobalt or rhodium catalysts, acetic acid is made by carbonylation of methanol using rhodium catalysts and acrylonitrile is dimerised to adiponitrile (for nylon) by nickel catalysts. We should also not forget that the huge quantities of petroleum hydrocarbons processed by the oil and petrochemical industry are re-formed over platinum, platinum-rhenium or platinum-germanium supported on alumina.

In all of these processes, every single molecule, at some point in the catalytic cycle, is involved in the formation of a metal to carbon single bond. Of course, catalytic processes require that the metal to carbon bond be unstable —or, more accurately, that it be labile and able readily to undergo chemical reactions such as

$$M-CR_3 + CO \rightarrow M-\overset{\displaystyle O}{\overset{\displaystyle \|}{C}}-CR_3$$

$$M-CR_3 + H_2 \rightarrow MH + HCR_3$$

Today however, I am concerned not with catalysis but with the synthesis of simple, stable metal compounds that have single bonds to saturated carbon.

* Nobel Laureates in Chemistry, 1963.

The synthesis of stable metal alkyls, and indeed, the nature of the transition metal to carbon bond, is a problem that has been with us for a long time.

The first attempts to make such compounds were shortly after Frankland's epoch-making discovery of diethylzinc. Thus in 1859 Buckton wrote, "a rich harvest can scarcely fail to be reaped from submitting to the action of diethylzinc the metallic compounds of the other groups" (1). However, he failed with the transition metals, silver, copper and platinum, as indeed did other workers in the 1800's.

We must not forget however, that a transition metal organo compound, though of a quite different type had been made much earlier by the Danish pharmacist Zeise. By the interaction of alchohol and chloroplatinic acid he had made the salt $K[C_2H_4PtCl]$ known now as Zeise's salt. There followed much discussion of the exact constitution of this substance but it was to be over 125 years before the structure was determined and the way in which ethylene is bound to the metal fully understood.

The first stable metal alkyls to be isolated were those of platinum (2) and gold (3) by William Pope and his co-workers Peachy and Gibson at the Municipal School of Technology in Manchester. The structure of the trimethylplatinum compound $[(CH_3)_3PtCl]_4$ was not to be determined by X-ray diffraction for another 40 years. At the meeting of the Chemical Society in London on March 21st, 1907, at which the discovery was announced, the Chairman, Sir Henry Roscoe, complimented the authors on opening out an entirely new branch of investigation "which might indeed be said to be a wonderful find". This promise took a very long time indeed to be fulfilled. Even in 1955 it could be written (4) "It will be apparant from this overall picture of alkyls and aryls of the transition metals that the often heard generalization that they are much less stable and accessible than those of non-transition metals is quite true".

The main reason for the failure to isolate stable compounds, despite evidence that alkyls or aryls were present in solution at low temperatures, was that the interaction of transition metal halides with Grignard reagents or lithium alkyls usually gave either coupled products or products from decompositon of coupled species. Typically, ferric chloride was used to make coupled products, e.g., diphenyl.

$$C_6H_5MgBr \xrightarrow{\quad FeCl_3 \quad}$$

It was this coupling that Kealy and Pauson (5) were trying to utilize, to make dihydrofulvalene, which led to their synthesis of dicyclopentadienyl iron

$$C_5H_5MgBr \xrightarrow{\quad FeCl_3 \quad} (C_5H_5)_2Fe \quad not \quad C_5H_5{-}C_5H_5$$

It might be interesting to note in passing that my own conclusion about the

structure of this iron compound was based on two points. Firstly, my knowledge of the instability of transition metal alkyls and aryls, secondly my intuition concerning the nature, uncertain at that time, of the binding of ethylene
in Zeise's salt and of butadiene in Reihlen's compound, $C_4H_6Fe(CO)_3$. I
was convinced that ethylene was bound "sideways" (I) and that butadiene in
its *cis* form could act as a chelate (II), both double bonds being bound to the
metal.

(I)   (II)

These structures were later confirmed by others. Thus I wanted to involve
both double bonds of the cyclopentadienyl ring in the structure originally
written by Kealy and Pauson and Miller, Tebboth and Tremaine (III), in
a butadiene-like fashion (IV) resulting in a

"sandwich" structure.

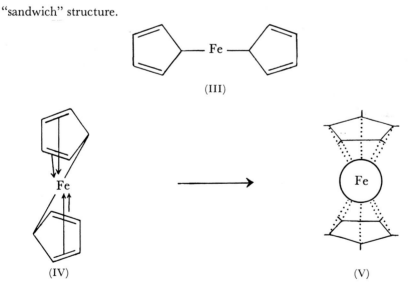

(IV)   (V)

Having read Linus Pauling's* famous book, *The Nature of the Chemical Bond,* and heard about resonance, this meant that I could write various
resonance forms of IV, which directly led to the idea that all the carbon
atoms were equivalent as in V, that is to the well known structure of the
molecule now known as ferrocene*.

* Nobel Laureate in Chemistry 1954.
* This name was coined by Dr. Mark C. Whiting in March or April, 1952.

During the next few years a number of stable compunds in which alkyl or aryl groups were present were synthesised. The first was a phenyltitanium alkoxide $C_6H_5Ti(OC_3H_7)_3$ (6) and others soon followed. However, in essentially all of the compounds a special type of ligand was present. These ligands were what are referred to as π-acid or π-bonding ligands. Examples are π-C_5H_5, CO, PEt_3, etc. Representative types of these alkyls are (VI—IX).

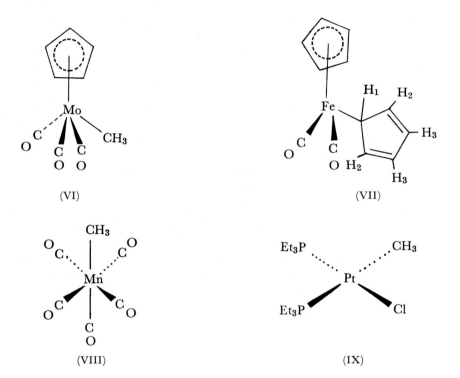

The compund (VII) turned out to be especially interesting. We observed only two signals in the proton magnetic resonance spectrum whereas the σ-C_5H_5 group alone should have had a complex spectrum; the infrared spectrum however was consistant with (VII). Because of the analogy with photography where, if one takes a picture of a moving wheel with a short exposure time (infrared), the picture is sharp, whereas if one uses a longer time (n.m.r.) it is blurred, I had to draw the conclusion that the σ-C_5H_5 was actually quite slowly rotating via a 1 : 2 shift. This was the first recognition of what are now known as fluxional molecules, this particular type being called "ring whizzers".

So the view developed that in order to have stable compunds with alkyl or aryl groups bound to a metal, some "stabilizing" group also had to be present. As recently as 1968, it was written, (7). "By any criterion, simple transition metal alkyls are unstable" and "In contrast to the simple alkyls, some metal complexes bearing other ligands in addition to alkyl or aryl groups are strikingly stable".

However, it was not too well recognised that the presence of such "stabilizing" ligands, is no guarantee of stability because other factors are involved. Indeed, the activity of many metal complexes with π-bonding ligands present in catalytic reactions depends on the lability of the metal of carbon bonds.

Despite all the intense study of metal-carbon bonds and the arguments about their stability, it is remarkable that only few bond energies are known. The available thermodynamic data show that M-C bonds are not exceptionally weak and the bond energies are quite comparable to those of non-transition metal to carbon bonds. It could be argued that since the compounds for which bond energies have been determined have "stabilizing" π-bonding ligands present the M-C bond energies are abnormally high, but this view cannot be sustained. There is no reason to assume this to be so, nor is there any reason to assume that carbon would differ appreciably in its capability to bond to transition metals compared to other first row elements such as oxygen and nitrogen or to the halogens. For both oxygen and nitrogen, metal compounds in high oxidation states are well known, e.g., the alkoxides such as $V(OR)_4$ and dialkylamides such as $W(NR_2)_6$. So, accepting that there is no thermodynamic reason for the instability of simple alkyls, the conclusion was that they are kinetically unstable.

There are several ways by which a metal alkyl can decompose, but for transition metal compounds one of the best established is the so-called hydride transfer-alkene elimination reaction. Here, a hydrogen atom is transferred from the second or β-carbon of the alkyl chain to the metal. The intermediate hydrido-alkene complex can then lose alkene and the resulting metal hydride decompose further e.g., to metal and hydrogen.

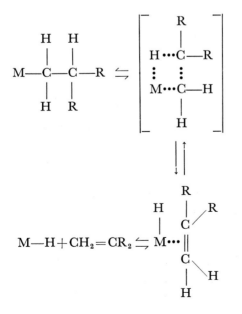

The reverse reaction, namely the generation of a metal alkyl from a metal hydride plus alkene is the key reaction in many catalytic cycles involving alkenes, hydrogen and metal species. It is involved for example in homogeneous hydrogenation of unsaturated organic compounds and in the hydroformylation reaction in which aldehydes or alcohols are synthesised from alkenes, carbon monoxide and hydrogen.

Two relevant studies on this decomposition reaction are illustrative. Firstly, a comparison (8) of the relative stabilities of dialkylmangenese compounds made *in situ* shows that those alkyls, e.g., $Mn(CH_3)_2$ or $Mn(CH_2C_6H_5)_2$, that cannot readily undergo this reaction are most stable. Secondly, a comparison (9) of the decomposition products of two similar alkyls, one that has a β-hydrogen, $Bu_3PCuCH_2CH_2CH_2CH_3$, and one that has not, Bu_3PCuCH_2C $(CH_3)_2C_6H_5$, shows that the former decomposes via H-transfer-alkene elimination, whereas the second decomposes by homolytic fission and a free radical pathway.

It will be observed that the H-transfer scheme involves a change in coordination number of the metal. Thus the alkyl group occupies only one coordination site, whereas in the intermediate hydrido-alkene, two sites are involved one for M-H and one for olefin coordination. Thus, one way in which an alkyl could be stabilized against decomposition is by coordinative saturation of the metal. If the sites required for the reaction to proceed are occupied by firmly bound ligands, then there is no possible pathway for decomposition. The stability of the alkyls with π-bonded ligands referred to above are prime examples of this situation. Other examples are the alkyls of the substitution-inert octahedral metal ions, Cr^{III}, Co^{III} and Rh^{III} such as the Werner*-type complex ion (**X**).

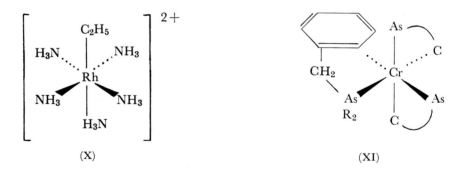

(**X**)   (**XI**)

A different type is on the chelated, coordinatively saturated alkyl (**X**) which is thermally stable to 350° C.

There seemed, to be another way, however, by which stable alkyls could be obtained, namely by making the H-transfer reaction impossible. Thus if the β-carbon atom were to be replaced by silicon or some other element that could *not* form a double bond to carbon the formation of alkene becomes impossible, even if there were a hydrogen atom on silicon; alternatively to have no hydrogen on the β atom. We illustrated this concept (10) by use of the trimethylsilylmethyl—$CH_2Si(CH_3)_3$ group as the alkyl and were able to

---

* Nobel Laureate in Chemistry 1913.

isolate for the first time a number of kinetically stable alkyls such as $Cr[CH_2Si(CH_3)_3]_4$.

A number of other groups, including of course the carbon analogue, neopentyl—$CH_2C(CH_3)_3$, fit the required criteria of having no β-hydrogen, and at the present alkyls are known with the following groups:

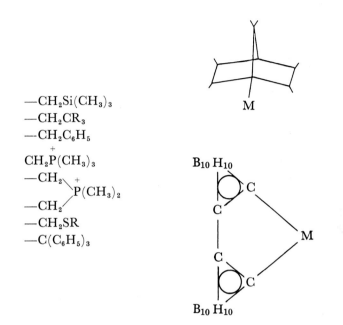

—$CH_2Si(CH_3)_3$
—CH_2CR_3
—$CH_2C_6H_5$
$CH_2\overset{+}{P}(CH_3)_3$
—CH_2
$\overset{+}{P}(CH_3)_2$
—CH_2
—CH_2SR
—$C(C_6H_5)_3$

It may be noted that some of these alkyls, notably those of titanium, have been recognised as Ziegler-Natta type catalysts for the polymerisation of alkenes (11).

The simplest of all alkyl groups is, of course, the methyl group. Although tetramethyltitanium has been known for some years, it decomposes above — 70° very readily. However, by blocking the remaining two vacant coordination sites of the tetrahedral alkyl, e.g., by dipyridyl, the thermal stability can be substantially increased. Despite its instability, vibrational spectroscopic studies (12) of $Ti(CH_3)_4$ suggest that the Ti-C bond strength, as measured by its force constant, is quite comparable to those of the tetramethyls of Si, Ge, Sn and Pb. By contrast, the least stable of these, tetramethyl lead, can be distilled at its boiling point (110°) without decomposition. The instability of $Ti(CH_3)_4$ is thus clearly kinetic, there being readily accessible pathways for decomposition possible because of its coordinative unsaturation. Although the chloride methyls of niobium and tantalum, $(CH_3)_3MCl_2$, have been known for some time and are reasonably stable, and we succeeded in characterising a rather unstable tetramethyl of chromium, it appeared that if a coordinatively saturated methyl could be synthesized, there was good reason to expect it to be stable. The obvious candidate was tungsten for which the hexa alkoxides and dialkylamides as well as the halides, are known. We succeeded in synthesizing hexamethyltungsten by the classical reaction of methyllithium with tungsten hexachloride (13). The reaction sequence is a very complex one and in order to obtain any $W(CH_3)_6$ it is necessary to use only half of the theoretical quantity of methyllithium, probably to avoid the formation of a methyl anion of the type known for other metals, e.g., $Li_4Mo_2(CH_3)_8$. The final step probably also involves the disproportionation of a reduced species:

$$W^V \rightarrow W^{IV} + W^{VI}$$

Once obtained, hexamethyltungsten is reasonably stable and can readily be characterised spectroscopically. It also has a number of interesting chemical reactions. It soon became evident that the compound, though octahedral is not coordination saturated and that some of the reactions are very facile because coordinative unsaturation allows initial coordination of the reagent. Using tertiary phosphines, we were able to isolate 7-coordinate adducts, $W(CH_3)_6PR_3$. One of the unusual reactions was that with nitric oxide which quantitatively gives a compound $(CH_3)_4W[ON(CH_3)NO]_2$ which contains two N-nitroso-N-methyl hydroxylaminato rings. These are probably generated by initial coordination of nitric oxide followed by methyl transfer as in the sequence:

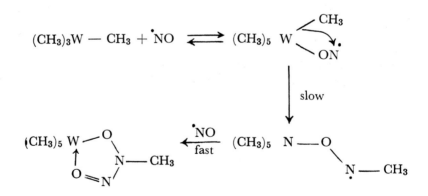

This sequence can happen only twice as the coordination number then reaches eight, which appears to be the maximum for tungsten in the VI oxidation state. Although the synthesis of other methyls is in principle possible, there is more art to it than science. The only other methyl in a very high oxidation state we have yet made is oxotetramethylrhenium(IV), $ReO(CH_3)_4$, but oxo or similar methyls of tungsten, molybdenum, osmium etc., may well be stable if suitable synthetic methods are found.*

So finally, in conclusion we can say that the effort of well over 100 years to synthesize stable transition metal alkyls has finally succeeded. The long established view that the transition metal to carbon bond is weak is now un-

* *Note added in proof.* Hexamethylrhenium (K. Mertis and G. Wilkinson) and pentamethyltantalum (R. Schrock, DuPont, Wilmington, private communication) have recently been synthesized.

tenable and must be discarded. We can expect other types of transition metal alkyls to be made in due course and can hope that in addition to their own intrinsic interest some of them may find uses in catalytic or other syntheses. The use of titanium and zirconium alkyls in alkene polymerisation and the use of alumina treated with hexamethyltungsten for alkene metathesis (14) give good grounds for optimism.

REFERENCES

1. Buckton, G. B., *Proc. Roy. Soc.,* 1859, *9,* 309.
2. Pope, W. J. and Peachy, S. J., *Proc. Chem. Soc.,* 1907, *23,* 86; *J. Chem. Soc., 1909,* 371.
3. Pope, W. J. and Gibson, C. S., *Trans. Chem. Soc.,* 1907, *91,* 2061.
4. Cotton, F. A., *Chem. Rev.,* 1955, *55,* 551.
5. Kealy, T. J. and Pauson, P. L., *Nature,* 1952, *168,* 1039; for the independent earlier discovery of $C_{10}H_{10}Fe$ see Miller, S. A., Tebboth, J. A. and Tremaine, J. F., *J. Chem. Soc., 1952,* 632.
6. Herman, D. F. and Nelson, W. K., *J. Amer. Chem. Soc.,* 1953, *75,* 3877.
7. Parshall, G. W. and Mrowca, J. J., *Adv. Organometal. Chem.,* 1968, *7,* 157.
8. Tamura, M. and Kochi, J., *J. Organometal. Chem.,* 1971, 29, 11.
9. Whitesides, G. M., *et al., J. Amer. Chem. Soc.,* 1972, *94,* 232.
10. Yagupsky, G., Mowat, W., Shortland, A. & Wilkinson, G., *Chem. Comm., 1970,* 1369; *J. C. S. Dalton, 1972,* 533.
11. see e.g., Ballard, D. G. H., *23rd. Internat. Congress Pure Appl. Chem.,* Butterworths, 1971.
12. Eysel, H. H., Siebert, H., Groh, G. and Berthold, H. J. *Spectrochim. Acta,* 1970, *26A,* 1595.
13. Shortland, A. J. and Wilkinson, G., *J. C. S. Dalton, 1973,* 872.
14. Mowat, W., Smith, I., and Whan, D. A., *Chem. Comm., 1974,* 34.

Chemistry 1974

PAUL J. FLORY

for his fundamental achievements, both theoretical and experimental, in the physical chemistry of the macromolecules

THE NOBEL PRIZE FOR CHEMISTRY

Speech by Professor STIG CLAESSON of the Royal Academy of Sciences
Translation from the Swedish text

Your Majesty, Your Royal Highnesses, Ladies and Gentlemen,
This year's Nobel prize in chemistry has been awarded to Professor Paul Flory
for his fundamental contributions to the physical chemistry of macromolecules.

Macromolecules include biologically important materials such as cellulose,
albumins and nucleic acids, and all of our plastics and synthetic fibers.

Macromolecules are often referred to as chain molecules and can be
compared to a pearl necklace. They consist of long chains of atoms which, when
magnified one hundred million times, appear as a pearl necklace. The pearls
represent the atoms in the chain. One should realize that this chain is much
longer than the necklaces being worn here this evening. To obtain a representative model of a macromolecule all of the necklaces here in this hall should
be connected together in a single long chain.

One can readily appreciate that the development of a theory for these
molecules presented considerable difficulties. The forms of the chain itself,
whether extended or coiled, represents a property difficult to rationalize.

A statistical description is of necessity required, and Professor Flory has
made major contributions to the development of such a theory. The problem
is more difficult, however. How can one compare different molecules in different solvents?

When chain molecules are dissolved in different solvents they become
coiled to different degrees, depending on the interaction between repulsive and
attractive forces in the solution. In a good solvent the chain molecules are
extended. In a poor solvent, in contrast, the chain molecules assume a highly
coiled form.

Professor Flory showed that if one takes a solution of extended chain molecules in a good solvent, and slowly cools the solution, then the molecules
become progressively more coiled until they are no longer soluble.

Thus, there must be an intermediate temperature where the attractive and
repulsive forces are balanced. At this temperature the molecules assume a kind
of standard condition that can be used, generally, to characterize their
properties.

This temperature Professor Flory named the theta temperature. A corresponding temperature exists for real gases at which they follow the ideal gas
law. This temperature is called the Boyle temperature after Robert Boyle who
discovered the gas laws. By analogy, the theta temperature for macromolecules
is often referred to as the Flory temperature.

Profssor Flory showed also that it was possible to define a constant for chain molecules, now called Flory's universal constant, which can be compared in significance to the gas constant.

When one, in retrospect, reads about an important scientific discovery, one often feels that the work was remarkably simple. This actually indicates, however, that it was brilliant insight in a new and until then unexplored research area. This is highly characteristic of Professor Flory's scientific discoveries, not only those concerned with the Flory temperature and Flory's universal constant but also many of his other important research studies. Further examples are found in his investigation of the relationship between the reaction mechanism and the length of the chains formed when chain molecules are prepared synthetically, as well as his important contribution to the theory of crystallization and rubber elasticity. These achievements have been of major importance for technological developments in the plastics industry.

In recent years Professor Flory has investigated, both theoretically and experimentally, the relation between rotational characteristics of the chain links and the form of the chain molecules. This is of fundamental significance for the understanding of both biological macromolecules and synthetic chain molecules.

During the time Professor Flory has been active as a scientist, macromolecular chemistry has been transformed from primitive semi-empirical observations into a highly developed science. This evolution has come about through major contributions by research groups from both universities and many of the world's largest industrial laboratories. Professor Flory has remained a leading researcher in the area during this entire period, giving further evidence of his unique position as a scientist.

Professor Flory,

I have tried to describe briefly the fundamental importance of your many contributions to macromolecular chemistry and in particular those concepts introduced by you and now referred to as the Flory-temperature and the Flory universal constant.

On behalf of the Royal Academy of Sciences I wish to convey to you our warmest congratulations and I now ask you to receive your prize from the hands of His Majesty the King.

PAUL J. FLORY

I was born on 19 June, 1910, in Sterling, Illinois, of Huguenot-German parentage, mine being the sixth generation native to America. My father was Ezra Flory, a clergyman-educator; my mother, nee Martha Brumbaugh, had been a schoolteacher. Both were descended from generations of farmers in the New World. They were the first of their families of record to have attended college.

My interest in science, and in chemistry in particular, was kindled by a remarkable teacher, Carl W. Holl, Professor of Chemistry at Manchester College, a liberal arts college in Indiana, where I graduated in 1931. With his encouragement, I entered the Graduate School of The Ohio State University where my interests turned to physical chemistry. Research for my dissertation was in the field of photochemistry and spectroscopy. It was carried out under the guidance of the late Professor Herrick L. Johnston whose boundless zeal for scientific research made a lasting impression on his students.

Upon completion of my Ph.D. in 1934, I joined the Central Research Department of the DuPont Company. There it was my good fortune to be assigned to the small group headed by Dr. Wallace H. Carothers, inventor of nylon and neoprene, and a scientist of extraordinary breadth and originality. It was through the association with him that I first became interested in exploration of the fundamentals of polymerization and polymeric substances. His conviction that polymers are valid objects of scientific inquiry proved contagious. The time was propitious, for the hypothesis that polymers are in fact covalently linked macromolecules had been established by the works of Staudinger and of Carothers only a few years earlier.

A year after the untimely death of Carothers, in 1937, I joined the Basic Science Research Laboratory of the University of Cincinnati for a period of two years. With the outbreak of World War II and the urgency of research and development on synthetic rubber, supply of which was imperiled, I returned to industry, first at the Esso (now Exxon) Laboratories of the Standard Oil Development Company (1940—43) and later at the Research Laboratory of the Goodyear Tire and Rubber Company (1943—48). Provision of opportunities for continuation of basic research by these two industrial laboratories to the limit that the severe pressures of the times would allow, and their liberal policies on publication, permitted continuation of the beginnings of a scientific career which might otherwise have been stifled by the exigencies of those difficult years.

In the Spring of 1948 it was my privilege to hold the George Fisher Baker Non-Resident Lectureship in Chemistry at Cornell University. The invitation on behalf of the Department of Chemistry had been tendered by the late Professor Peter J. W. Debye, then Chairman of that Department. The experience of this lectureship and the stimulating asociations with the Cornell faculty led me to accept, without hesitation, their offer of a professorship commencing in the Autumn of 1948. There followed a most productive and satisfying period of research and teaching. "Principles of Polymer Chemistry," published by the Cornell University Press in 1953, was an outgrowth of the Baker Lectures.

It was during the Baker Lectureship that I perceived a way to treat the effect of excluded volume on the configuration of polymer chains. I had long suspected that the effect would be non-asymptotic with the length of the chain; that is, that the perturbation of the configuration by the exclusion of one segment of the chain from the space occupied by another would increase without limit as the chain is lengthened. The treatment of the effect by resort to a relatively simple "smoothed density" model confirmed this expectation and provided an expression relating the perturbation of the configuration to the chain length and the effective volume of a chain segment. It became apparent that the physical properties of dilute solutions of macromolecules could not be properly treated and comprehended without taking account of the perturbation of the macromolecule by these intramolecular interactions. The hydrodynamic theories of dilute polymer solutions developed a year or two earlier by Kirkwood and by Debye were therefore reinterpreted in light of the excluded volume effect. Agreement with a broad range of experimental information on viscosities, diffusion coefficients and sedimentation velocities was demonstrated soon thereafter.

Out of these developments came the formulation of the hydrodynamic constant called Φ, and the recognition of the Theta point at which excluded volume interactions are neutralized. Criteria for experimental identification of the Theta point are easily applied. Ideal behavior of polymers, natural and synthetic, under Theta conditions has subsequently received abundant confirmation in many laboratories. These findings are most gratifying. More importantly, they provide the essential basis for rational interpretation of physical measurements on dilute polymer solutions, and hence for the quantitative characterization of macromolecules.

In 1957 my family and I moved to Pittsburgh where I undertook to establish a broad program of basic research at the Mellon Institute. The opportunity to achieve this objective having been subsequently withdrawn, I accepted a professorship in the Department of Chemistry at Stanford University in 1961. In 1966, I was appointed to the J. G. Jackson—C. J. Wood Professorship in Chemistry at Stanford.

The change in situation upon moving to Stanford afforded the opportunity to recast my research efforts in new directions. Two areas have dominated the interests of my coworkers and myself since 1961. The one concerns the spatial configuration of chain molecules and the treatment of their configuration-dependent properties by rigorous mathematical methods; the other constitutes a new approach to an old subject, namely, the thermodynamics of solutions.

Our investigations in the former area have proceeded from foundations laid by Professor M. V. Volkenstein and his collaborators in the Soviet Union, and were supplemented by major contributions of the late Professor Kazuo Nagai in Japan. Theory and methods in their present state of development permit realistic, quantitative correlations of the properties of chain molecules with their chemical constitution and structure. They have been applied to a wide variety of macromolecules, both natural and synthetic, including polypeptides and polynucleotides in the former category. The success of these efforts has been due in no small measure to the outstanding students and research fellows who have collaborated with me at Stanford during the past thirteen years. A book entitled "Statistical Mechanics of Chain Molecules," published in 1969, summarizes the development of the theory and its applications up to that date.

Mrs. Flory, the former Emily Catherine Tabor, and I were married in 1936. We have three children: Susan, wife of Professor George S. Springer of the Department of Mechanical Engineering at the University of Michigan; Melinda, wife of Professor Donald E. Groom of the Department of Physics at the University of Utah; and Dr. Paul John Flory, Jr., currently a post-doctoral Research Associate at the Medical Nobel Institute in Stockholm. We have four grandchildren: Elizabeth Springer, Mary Springer, Susanna Groom and Jeremy Groom.

Honors and Awards
Joseph Sullivant Medal, The Ohio State University, 1945.
Baekeland Award, New Jersey Section, American Chemical Society, 1947.
Sc.D. (Honorary), Manchester College (Indiana), 1950.
Colwyn Medal, Institution of Rubber Industry, Great Britain, 1954
Nichols Medal, New York Section, American Chemical Society, 1962.
High-Polymer Physics Prize, American Physical Society, 1962.
Laurea Honoris Causa, Politecnico di Milano, 1964.
International Award in Plastics Science and Engineering, 25th Anniversary, Society of Plastics Engineers, 1967.
Charles Goodyear Medal, American Chemical Society, 1968.
Peter Debye Award in Physical Chemistry, American Chemical Society, 1969.
D.Sc., Honoris Causa, University of Manchester, England, 1969.
Sc.D. (Honorary), The Ohio State University, 1970.
Charles Frederick Chandler Medal, Columbia University, 1970.
First Award for Excellence-Chemistry, The Carborundum Company, 1971.
Cresson Medal, The Franklin Institute, 1971.
John G. Kirkwood Medal, Yale University, 1971.
J. Willard Gibbs Medal, Chicago Section, American Chemical Society, 1973.
Priestley Medal, American Chemical Society, 1974.
Nobel Prize in Chemistry, 1974.

Paul J. Flory died in 1985.

SPATIAL CONFIGURATION OF MACROMOLE-CULAR CHAINS

Nobel Lecture, December 11, 1974

by

PAUL J. FLORY

Department of Chemistry

Stanford University, Stanford, California

The science of macromolecules has developed from primitive beginnings to a flourishing field of investigative activities within the comparatively brief span of some forty years. A wealth of knowledge has been acquired and new points of view have illumined various branches of the subject. These advances are the fruits of efforts of many dedicated investigators working in laboratories spread around the world. In a very real sense, I am before you on this occasion as their representative.

In these circumstances, the presentation of a lecture of a scope commensurate with the supreme honor the Royal Swedish Academy of Sciences has bestowed in granting me the Nobel Prize for Chemistry is an insuperable task. Rather than attempt to cover the field comprehensively in keeping with the generous citation by the Royal Academy of Sciences, I have chosen to dwell on a single theme. This theme is central to the growth of ideas and concepts concerning macromolecules and their properties. Implemented by methods that have emerged in recent years, researches along lines I shall attempt to highlight in this lecture give promise of far-reaching advances in our understanding of macromolecular substances—materials that are invaluable to mankind.

These polymeric substances are distinguished at the molecular level from other materials by the concatenation of atoms or groups to form chains, often of great length. That chemical structures of this design should occur is implicit in the multivalency manifested by certain atoms, notably carbon, silicon, oxygen, nitrogen, sulfur and phosphorus, and in the capacity of these atoms to enter into sequential combinations. The concept of a chain molecule consisting of atoms covalently linked is as old as modern chemistry. It dates from the origins of the graphic formula introduced by Couper in 1858 and advanced by Kekulé, Loschmidt and others shortly thereafter. Nothing in chemical theory, either then apparent or later revealed, sets a limit on the number of atoms that may be thus joined together. The rules of chemical valency, even in their most primitive form, anticipate the occurrence of macromolecular structures.

The importance of macromolecular substances, or polymers, is matched by their ubiquity. Examples too numerous to mention abound in biological systems. They comprise the structural materials of both plants and animals. Macromolecules elaborated through processes of evolution perform intricate regulatory and reproductive functions in living cells. Synthetic polymers in great variety are familiar in articles of commerce. The prevailing structural motif is the linear chain of serially connected atoms, groups or structural units. Departures from strict linearity may sometimes occur through the agency of occasional branched units that impart a ramified pattern to the over-all structure. Linearity is predominant in most macromolecular substances, however.

It is noteworthy that the chemical bonds in macromolecules differ in no discernible respect from those in "monomeric" compounds of low molecular weight. The same rules of valency apply; the lengths of the bonds, e.g., C—C, C—H, C—O, etc., are the same as the corresponding bonds in monomeric molecules within limits of experimental measurement. This seemingly trivial observation has two important implications: first, the chemistry of macromolecules is coextensive with that of low molecular substances; second, the chemical basis for the special properties of polymers that equip them for so many applications and functions, both in nature and in the artifacts of man, is not therefore to be sought in peculiarities of chemical bonding but rather in their macromolecular constitution, specifically, in the attributes of long molecular chains.

Comprehension of the spatial relationships between the atoms of a molecule is a universal prerequisite for bridging the connection between the graphic formula and the properties of the substance so constituted. Structural chemistry has provided a wealth of information on bond lengths and bond angles. By means of this information the graphic formula, primarily a topological device, has been superseded by the structural formula and by the space model that affords a quantitative representation of the molecule in three dimensions. The stage was thus set for the consideration of rotations about chemical bonds, i.e., for conformational analysis of conventional organic compounds, especially cyclic ones. A proper account of bond rotations obviously is essential for a definitive analysis of the spatial geometry of a molecule whose structure permits such rotations.

The configuration of a linear macromolecule in space involves circumstances of much greater complexity. A portion of such a molecule is shown schematically in Figure 1. Consecutive bonds comprising the chain skeleton are joined at angles θ fixed within narrow limits. Rotations φ may occur about these skeletal bonds. Each such rotation is subject, however, to a potential determined by the character of the bond itself and by hindrances imposed by steric interactions between pendant atoms and groups. The number and variety of configurations (or conformations in the language of organic chemistry) that may be generated by execution of rotations about each of the skeletal bonds of a long chain, comprising thousands of bonds in a typical polymer, is prodigious beyond comprehension. When the macromolecule is free of constraints, e.g., when in dilute solution, all of these configurations are accessible. Analysis of the manner in which such a molecule may arrange itself in space finds close analogies elsewhere in science, e.g., in the familiar problem of random walk, in diffusion, in the mathematical treatment of systems in one dimension, and in the behavior of real gases.

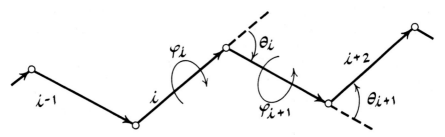

Fig. 1. Representation of the skeletal bonds of a section of a chain molecule showing supplements θ of bond angles, and torsional rotations φ for bonds i, i+1, etc.

Inquiry into the spatial configuration of these long-chain molecules, fascinating in itself, derives compelling motivation from its close relevancy to the properties imparted by such molecules to the materials comprising them. Indeed, most of the properties that distinguish polymers from other substances are intimately related to the spatial configurations of their molecules, these configurations being available in profusion as noted. The phenomenon of rubber-like elasticity, the hydrodynamic and thermodynamic properties of polymer solutions, and various optical properties are but a few that reflect the spatial character of the random macromolecule. The subject is the nexus between chemical constitution and physical and chemical properties of polymeric substances, both biological and synthetic.

The importance of gaining a grasp of the spatial character of polymeric chains became evident immediately upon the establishment, *ca.* 1930, of the hypothesis that they are covalently linked molecules rather than aggregates of smaller molecules, an achievement due in large measure to the compelling evidence adduced and forcefully presented by H. Staudinger, Nobelist for 1953. In 1932 K. H. Meyer[1] adumbrated the theory of rubber-like elasticity by calling attention to the capacity of randomly coiled polymer chains to accommodate large deformations owing to the variety of configurations accessible to them.

W. Kuhn[2] and E. Guth and H. Mark[3] made the first attempts at mathematical description of the spatial configurations of random chains. The complexities of bond geometry and of bond rotations, poorly understood at the time, were circumvented by taking refuge in the analogy to unrestricted random flights, the theory of which had been fully developed by Lord Rayleigh. The skeletal bonds of the molecular chain were thus likened to the steps in a random walk in three dimensions, the steps being uncorrelated one to another. Restrictions imposed by bond angles and hindrances to rotation were dismissed on the grounds that they should not affect the form of the results.

For a random flight chain consisting of n bonds each of length *l*, the mean-square of the distance r between the ends of the chain is given by the familiar relation

$$\langle r^2 \rangle = nl^2 \tag{1}$$

The angle brackets denote the average taken over all configurations. Kuhn[4] argued that the consequences of fixed bond angles and hindrances to rotation could be accommodated by letting several bonds of the chain molecule be represented by one longer "equivalent" bond, or step, of the random flight. This would require n to be diminished and l to be increased in Eq. 1. Equivalently, one may preserve the identification of n and l with the actual molecular quantities and replace Eq. (1) with

$$\langle r^2 \rangle = C\, nl^2, \tag{2}$$

where C is a constant for polymers of a given homologous series, i.e., for polymers differing in length but composed of identical monomeric units. The proportionality between $\langle r^2 \rangle$ and chain length expressed in Eq. (2) may be shown to hold for any random chain of finite flexibility, regardless of the structure, provided that the chain is of sufficient length and that it is unperturbed by external forces or by *effects due to excluded volume (cf. seq.)*.

The result expressed in Eq. (2) is of the utmost importance. Closely associated with it is the assertion that the density distribution $W(\mathbf{r})$ of values of the end-to-end vector $\mathbf{r}$ must be Gaussian for chains of sufficient length, irrespective of their chemical structure, provided only that the structure admits of some degree of flexibility. Hence, for large n the distribution of values of r is determined by the single parameter $\langle r^2 \rangle$ that defines the Gaussian distribution.

Much of polymer theory has been propounded on the basis of the Kuhn "equivalent" random flight chain, with adjustment of n and l, or of C, as required to match experimental determination of $\langle r^2 \rangle$ or of other configuration-dependent quantities. The validity of this model therefore invites critical examination. Its *intrinsic artificiality* is its foremost deficiency. Actual bond lengths, bond angles and rotational hindrances cannot be incorporated in this model. Hence, contact is broken at the outset with the features of chemical constitution that distinguish macromolecular chains of one kind from those of another. The model is therefore incapable of accounting for the vast differences in properties exhibited by the great variety of polymeric substances.

The random flight chain is patently unsuited for the treatment of constitutive properties that are configuration-dependent, e.g., dipole moments, optical polarizabilities and dichroism. Inasmuch as the contribution to one of these properties from a structural unit of the chain is a vector or tensor, it cannot be referenced to an equivalent bond that is a mere line. Moreover, the equivalent bond cannot be embedded unambiguously in the real structure.

Methods have recently been devised for treating macromolecular chains in a realistic manner. They take full account of the structural geometry of the given chain and, in excellent approximation, of the potentials affecting bond rotations as well. Before discussing these method, however, I must direct your attention to another aspect of the subject. I refer to the notorious effect of volume exclusion in a polymer chain.

At the hazard of seeming trite, I should begin by pointing out that the chain molecule is forbidden to adopt a configuration in which two of its parts, or segments, occupy the same space. The fact is indisputable; its consequences are less obvious. It will be apparent, however, that volume exclusion vitiates

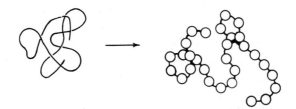

Fig. 2. The effect of excluded volume. The configuration on the left represents the random coil in absence of volume exclusion, the chain being equivalent to a line in space. In the sketch on the right, the units of the chain occupy finite domains from which other units are excluded, with the result that the average size of the configuration is increased.

the analogy between the trajectory of a particle executing a random flight and the molecular chain, a material body. The particle may cross its own path at will, but self intersections of the polymer chain are forbidden.

The effect of excluded volume must be dealt with regardless of the model chosen for representation of the chain. In practice, elimination of the effect of volume exclusion is a prerequisite to the analysis of experimental results, as I will explain in more detail later.

The closely related problems of random flights with disallowance of self intersections and of volume exclusion within long-chain molecules have attracted the attention of many theorists. A variety of mathematical techniques have been applied to the treatment of these problems, and a profusion of theories have been put forward, some with a high order of sophistication. Extensive numerical computations of random walks on lattices of various sorts also have been carried out. Convergence of results obtained by the many investigators captivated by the subject over the past quarter century seems at last to be discernible. I shall confine myself to a brief sketch of an early, comparatively simple approach to the solution of this problem.[5] The results it yields contrast with its simplicity.

Returning to the analogy of the trajectory traced by a particle undergoing a sequence of finite displacements, we consider only those trajectories that are free of intersections as being acceptable for the chain molecule. Directions of successive steps may or may not be correlated, i.e., restrictions on bond angles and rotational hindrances may or may not be operative; this is immaterial with respect to the matter immediately at hand. Obviously, the set of eligible configurations will occupy a larger domain, on the average, than those having one or more self intersections. Hence, volume exclusion must cause $\langle r^2 \rangle$ to increase. The associated expansion of the spatial configuration is illustrated in Fig. 2. Other configuration-dependent quantities may be affected as well.

This much is readily evident. Assessment of the magnitude of the perturbation of the configuration and its dependence on chain length require a more penetrating examination.

The problem has two interrelated parts: (i) the mutual exclusion of the space occupied by segments comprising the chain tends to disperse them over a larger volume, and (ii) the concomitant alteration of the chain configuration opposes expansion of the chain. Volume exclusion (i) is commonplace. It is prevalent in conventional dilute solutions and in real gases, molecules of which mutually exclude one another. In the polymer chain the same rules of exclusion apply, but treatment of the problem is complicated by its association with (ii).

Pursuing the analogies to dilute solutions and gases, we adopt a "smoothed density" or "mean field" model. The segments of the chain, x in number, are considered to pervade a volume V, the connections between them being ignored insofar as part (i) is concerned. The segment need not be defined explicitly; it may be identified with a repeating unit or some other approximately isometric portion of the chain. In any case, x will be proportional to the number n of bonds; in general $x \neq n$, however. For simplicity, we may consider the segment density ϱ to be uniform throughout the volume V; that is, $\varrho = x/V$ within V and $\varrho = 0$ outside of V. This volume should be proportional to $\langle r^2 \rangle^{3/2}$, where $\langle r^2 \rangle$ is the mean-square separation of the ends of the chain averaged over those configurations *not disallowed by excluded volume interactions*. Accordingly, we let

$$V = A \langle r^2 \rangle^{3/2}, \tag{3}$$

where A is a numerical factor expected to be of the order of magnitude of unity.

It is necessary to digress at this point for the purpose of drawing a distinction between $\langle r^2 \rangle$ for the chain perturbed by the effects of excluded volume and $\langle r^2 \rangle_0$ for the unperturbed chain in the absence of such effects. If α denotes the factor by which a linear dimension of the configuration is altered, then

$$\langle r^2 \rangle = \alpha^2 \langle r^2 \rangle_0 \tag{4}$$

Equation (2), having been derived without regard for excluded volume interactions, should be replaced by

$$\langle r^2 \rangle_0 = C \, nl^2, \tag{2'}$$

where C reaches a constant value with increase in n for any series of finitely flexible chains.

The smoothed density within the domain of a linear macromolecule having a molecular weight of 100,000 or greater (i.e., n > 1000) is low, only on the order of one percent or less of the space being occupied by chain segments. For a random dispersion of the segments over the volume V, encounters in which segments overlap are rare in the sense that few of them are thus involved. However, the expectation that such a dispersion is entirely free of overlaps between any pair of segments is very small for a long chain. The attrition of configurations due to excluded volume is therefore severe.

In light of the low segment density, it suffices to consider only binary encounters. Hence, if β is the volume excluded by a segment, the probability that an arbitrary distribution of their centers within the volume V is free of conflicts between any pair of segments is

$$P_{(i)} \approx \prod_{i=1}^{x} (1-i\beta/V) \approx \exp(-\beta x^2/2V). \tag{5}$$

Introduction of Eq. (3) and (4) gives

$$P_{(i)} = \exp(-\beta x^2/2A\langle r^2\rangle_0^{3/2}a^3) \tag{6}$$

or, in terms of the conventional parameter z defined by

$$z = (3/2\pi)^{3/2} (\langle r^2\rangle_0/x)^{-3/2}x^{1/2}\beta, \tag{7}$$

$$P_{(i)} = \exp[-2^{1/2}(\pi/3)^{3/2}A^{-1}za^{-3}]. \tag{8}$$

Since $\langle r^2\rangle_0$ is proportional to x for long chains (see Eq. (2′)), z depends on the square-root of the chain length for a given series of polymer homologs.

We require also the possibility $P_{(ii)}$ of a set of configurations having the average density corresponding to the dilation a^3 relative to the probability of a set of configurations for which the density of segments corresponds to $a^3 = 1$. For the former, the mean-squared separation of the ends of the chain is $\langle r^2\rangle$; for the latter it is $\langle r^2\rangle_0$. The distribution of chain vectors $\mathbf{r}$ for the unperturbed chain is approximately Gaussian as noted above. That is to say, the probability that $\mathbf{r}$ falls in the range $\mathbf{r}$ to $\mathbf{r}+d\mathbf{r}$ is

$$W(\mathbf{r})d\mathbf{r} = \text{Const } \exp(-3r^2/2\langle r^2\rangle_0)d\mathbf{r}, \tag{9}$$

where $d\mathbf{r}$ denotes the element of volume. The required factor is the ratio of the probabilities for the dilated and the undilated sets of configurations. These probabilities, obtained by taking the products of $W(\mathbf{r})d\mathbf{r}$ over the respective sets of configurations, are expressed by $W(\mathbf{r})$ according to Eq. (9) with r^2 therein replaced by the respective mean values, $\langle r^2\rangle$ and $\langle r^2\rangle_0$, for the perturbed and unperturbed sets. Bearing in mind that the volume element $d\mathbf{r}$ is dilated as well, we thus obtain

$$P_{(ii)} = [(d\mathbf{r})/(d\mathbf{r})_0]\exp[-3(\langle r^2\rangle-\langle r^2\rangle_0)/2\langle r^2\rangle_0]$$
$$= a^3\exp[-(3/2) (a^2-1)]. \tag{10}$$

The combined probability of the state defined by the dilation a^3 is

$$P_{(i)}P_{(ii)} = a^3\exp[-2^{1/2}(\pi/3)^{3/2}A^{-1}za^{-3}-(3/2) (a^2-1)]. \tag{11}$$

Solution for the value of a that maximizes this expression gives

$$a^5-a^3 = 2^{1/2}(\pi/3)^{3/2}A^{-1}z. \tag{12}$$

Recalling that z is proportional to $x^{1/2}\beta$ according to Eq. (7), one may express this result alternatively as follows

$$a^5-a^3 = Bx^{1/2}\beta, \tag{12′}$$

where $B = (\langle r^2\rangle_0/x)^{-3/2}(2A)^{-1}$ is a constant for a given series of polymer homologs.

In the full treatment[5,6] of the problem along the lines sketched briefly above, the continuous variation of the mean segment density with distance from the center of the molecule is taken into account, and the appropriate sums are executed over all configurations of the chain. The squared radius of gyration s^2, i.e., the mean-square of the distances of the segments from their center of gravity, is preferable to r^2 as a parameter with which to characterize the spatial distribution.[7] Treatments carried out with these refinements affirm the essential validity of the result expressed by Eq. (12) or (12'). They show conclusively[7,8] that the form of the result should hold in the limit of large values of $\beta x^{1/2}$, i.e., for large excluded volume and/or high chain length, and hence for $a \gg 1$. In this limit, $(a^5 - a^3)/z = 1.67$ according to H. Fujita and T. Norisuye.[8] For $a < {\sim}1.4$, however, this ratio decreases, reaching a value of 1.276 at $a = 1$.[8,9]

The general utility of the foregoing result derived from the most elementary considerations is thus substantiated by elaboration and refinement of the analysis, the quantitative inaccuracy of Eqs. (12) and (12') in the range $1.0 < a \leqslant 1.4$ notwithstanding. The relationship between a and the parameter z prescribed by these equations, especially as refined by Fujita and Norisuye,[8] appears to be well supported by experiment.[10,11]

The principal conclusions to be drawn from the foregoing results are the following: the expansion of the configuration due to volume exclusion increases with chain length *without limit* for $\beta > 0$; for very large values of $\beta x^{1/2}$ relative to $(\langle r^2 \rangle_0/x)^{3/2}$ it should increase as the 1/10 power of the chain length. The sustained increase of the perturbation with chain length reflects the fact that interactions between segments that are remote in sequence along the chain are dominant in affecting the dimensions of the chain. It is on this account that the excluded volume effect is often referred to as a long-range inter-action.[9-12]

The problem has been treated by a variety of other procedures.[9-12] Notable amongst these treatments is the self-consistent field theory of S. F. Edwards.[12] The asymptotic dependence of a on the one-tenth power of the chain length, and hence the dependence of $\langle r^2 \rangle$ on $n^{6/5}$ for large values of the parameter z, has been confirmed.[12]

The dilute solution is the milieu chosen for most physicochemical ex-periments conducted for the purpose of characterizing polymers. The effect of excluded volume is reflected in the properties of the polymer molecule thus determined, and must be taken into account if the measurements are to be properly interpreted. The viscosity of a dilute polymer solution is illustrative. Its usefulness for the characterization of polymers gained recognition largely through the work of Staudinger and his collaborators.

Results are usually expressed as the intrinsic viscosity $[\eta]$ defined as the ratio of the increase in the relative viscosity η_{rel} by the polymeric solute to its concentration c in the limit of infinite dilution. That is,

$$[\eta] = \lim_{c \to 0} [(\eta_{rel} - 1)/c]$$

the concentration c being expressed in weight per unit volume. The increment in viscosity due to a polymer molecule is proportional to its hydrodynamic volume, which in turn should be proportional to $\langle r^2 \rangle^{3/2}$ for a typical polymer chain. Hence, $\eta_{rel}-1$ should be proportional to the product of $\langle r^2 \rangle^{3/2}$ and the number density of solute molecules given by c/M where M is the molecular weight. It follows that

$$[\eta] = \Phi \langle r^2 \rangle^{3/2}/M, \tag{13}$$

where Φ is a constant of proportionality.[6,13] Substitution from Eq. (4) and rearrangement of the result gives

$$[\eta] = \Phi(\langle r^2 \rangle_0/M)^{3/2}M^{1/2}a^3 \tag{13'}$$

The ratio $\langle r^2 \rangle_0/M$ should be constant for a series of homologs of varying molecular weight, provided of course that the molecular weight, and hence the chain length, is sufficiently large.

If the excluded volume effect could be ignored, the intrinsic viscosity should vary proportionally to $M^{1/2}$. Since, however, a increases with M, a stronger dependence on M generally is observed. Often the dependence of $[\eta]$ on molecular weight can be represented in satisfactory approximation by the empirical relation

$$[\eta] = KM^a \tag{14}$$

where $0.5 \leqslant a \leqslant 0.8$. Typical results are shown by the upper sets of data in Figs. 3 and 4 for polystyrene dissolved in benzene[14] and for poly(methyl methacrylate) in methyl ethyl ketone,[15] respectively. Values of a^3 are in the range 1.4 to 5. At the asymptote for chains of great length and large excluded volume β, the exponent a should reach 0.80 according to the treatment given above. Although this limit is seldom reached within the accessible range of molecular weights, the effects of excluded volume can be substantial. They must be taken into account in the interpretation of hydrodynamic measurements.[13,16] Otherwise, the dependences of the intrinsic viscosity and the translational friction coefficient on molecular chain length are quite incomprehensible.

Measurement of light scattering as a function of angle, a method introduced by the late P. Debye, affords a convenient means for determining the mean-square radius of gyration. Small-angle scattering of x-rays (and lately of neutrons) offers an alternative for securing the same information. From the radius of gyration one may obtain the parameter $\langle r^2 \rangle$ upon which attention is focused here. The results are affected, of course, by the perturbation due to excluded volume. Inasmuch as the perturbation is dependent on the solvent and temperature, the results directly obtained by these methods are not intrinsically characteristic of the macromolecule. Values obtained for $\langle r^2 \rangle$ from the intrinsic viscosity by use of Eq. (13), or by other methods, must also be construed to be jointly dependent on the macromolecule and its environment.

If the factor a were known, the necessary correction could be introduced readily to obtain the more substantive quantities, such as $\langle r^2 \rangle_0$ and $\langle s^2 \rangle_0$ that characterize the macromolecule itself and are generally quite independent of the solvent. Evaluation of a according to Eq. (11) and (12) would require the excluded volume β. This parameter depends on the solvent in a manner that eludes prediction. Fairly extensive experimental measurements are required for its estimation, or for otherwise making correction for the expansion a.

All these difficulties are circumvented if measurements on the polymer solution are conducted under conditions such that the effects of excluded volume are suppressed. The resistance of atoms to superposition cannot, of course, be set aside. But the consequences thereof can be neutralized. We have only to recall that the effects of excluded volume in a gas comprising real molecules of finite size are exactly compensated by intermolecular attractions at the Boyle temperature (up to moderately high gas densities). At this temperature the real gas masquerades as an ideal one.

For the macromolecule in solution, realization of the analogous condition requires a relatively poor solvent in which the polymer segments prefer self-contacts over contacts with the solvent. The incidence of self-contacts may then be adjusted by manipulating the temperature and/or the solvent composition until the required balance is established. Carrying the analogy to a real gas a step further, we require the excluded volume integral for the interaction between a pair of segments to vanish; that is, we require that $\beta=0$. This is the necessary and sufficient condition.[5,6,13]

As already noted, estimation of the value of β is difficult; the prediction of conditions under which β shall precisely vanish would be even more precarious. However, the "Theta point," so-called, at which this condition is met is readily identified with high accuracy by any of several experimental procedures. An excluded volume of zero connotes a second virial coefficient of zero, and hence conformance of the osmotic pressure to the celebrated law of J. H. van't Hoff. The Theta point may be located directly from osmotic pressure determinations, from light scattering measured as a function of concentration, or from determination of the precipitation point as a function of molecular weight.[6,13]

The efficacy of this procedure, validated a number of years ago with the collaboration of T. G. Fox, W. R. Krigbaum, and others,[13,17,18] is illustrated in Figs. 3 and 4 by the lower plots of data representing intrinsic viscosities measured under ideal, or Theta conditions.[6] The slopes of the lines drawn through the lower sets of points are exactly $1/2$, as required by Eq. (13′) when $\beta = 0$ and hence $a=1$. The excellent agreement here illustrated has been abundantly confirmed for linear macromolecules of the widest variety, ranging from polyisobutylene and polyethylene to polyribonucleotides.[19] At the Theta point the mean-square chain vector $\langle r^2 \rangle_0$ and the mean-square radius of gyration $\langle s^2 \rangle_0$ invariably are found to be proportional to chain length.

A highly effective strategy for characterization of macromolecules emerges from these findings. By conducting experiments at the Theta point, the disconcerting (albeit interesting!) effects of excluded volume on experimentally measured quantities may be eliminated. Parameters (e.g., $\langle r^2 \rangle_0$ and $\langle s^2 \rangle_0$) are thus obtained that are characteristic of the molecular chain. They are found to be virtually independent of the nature of the "Theta solvent" selected. Having eliminated the effects of long range interactions, one may turn

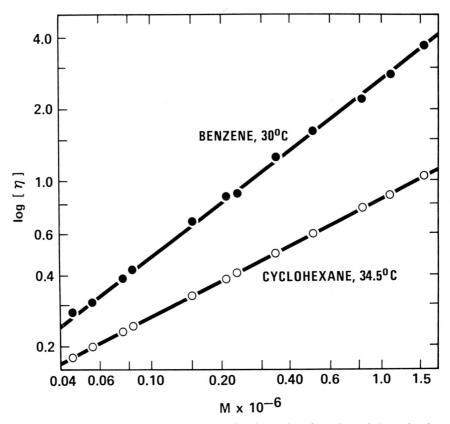

Fig. 3. Intrinsic viscosities of polystyrene fractions plotted against their molecular. weights on logarithmic scales in accordance with Eq. (14). The upper set of data was determined in benzene, a good solvent for this polymer. The lower set of data was determined in cyclohexane at the Theta point. The slopes of the lines are $a = 0.75$ and 0.50, respectively. From the results of Altares, Wyman and Allen.[14]

attention to the role of short range features: structural geometry, bond rotation potentials, and steric interactions between near-neighboring groups. It is here that the influences of chemical architecture are laid bare. If the marked differences in properties that distinguish the great variety of polymeric substances, both natural and synthetic, are to be rationally understood in fundamental, molecular terms, this must be the focus of future research.

Rigorous theoretical methods have recently become available for dealing realistically with short-range features peculiar to a given structure. Most of the remainder of this lecture is devoted to a brief overview of these methods. Although the field is comparatively new and its exploration has only begun, space will not permit a digest of the results already obtained.

The broad objective of the methods to which we now turn attention is to treat the structure and conformations accessible to the chain molecule in such a manner as will enable one to calculate configuration-dependent quantities and to average them over all conformations, or spatial configurations, of the unperturbed chain. The properties under consideration are constitutive; they

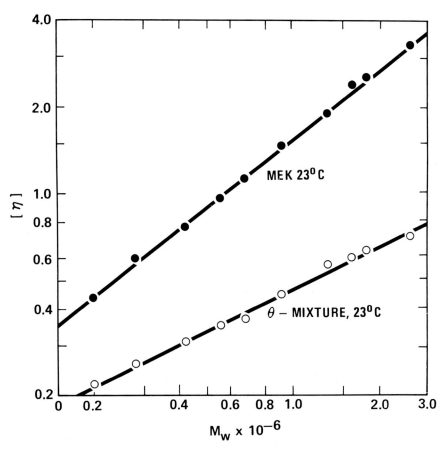

Fig. 4. Intrinsic viscosities of fractions of poly(methyl methacrylate) according to Chinai and Samuels[15] plotted as in Fig. 3. The upper set of points was measured in methyl ethyl ketone, a good solvent. The lower set was determined in a mixture of methyl ethyl ketone and isopropanol at the Theta point. Slopes are $a = 0.79$ and 0.50, respectively.

represent sums of contributions from the individual units, or chemical groupings, comprising the chain. In addition to $\langle r^2 \rangle_0$ and $\langle s^2 \rangle_0$, they include: mean-square dipole moments; the optical anisotropies underlying strain birefringence, depolarized light scattering and electric birefringence; dichroism; and the higher moments, both scalar and tensor, of the chain vector **r**. Classical statistical mechanics provides the basis for evaluating the configurational averages of these quantities. Since bond lengths and bond angles ordinarily may be regarded as fixed, the bond rotations φ are the variables over which averaging must be carried out. The procedure rests on the *rotational isomeric state scheme*, the foundations for which were set forth in large measure by M. V. Volkenstein[20] and his colleagues[21] in Leningrad in the late 1950's and early 1960's. It is best explained by examples.

Consider rotation about an internal bond of an n-alkane chain. As is now well established,[22,23] the three staggered conformations, trans(t), gauche-plus(g⁺) and its mirror image, gauche-minus(g⁻), are of lower energy than

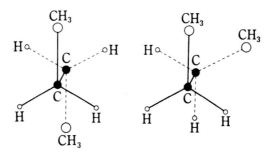

Fig. 5. Two of the staggered conformations for n-butane: trans on the left and gauche-minus on the right.

the eclipsed forms. The t and g⁻ conformations of n–butane are shown in Fig. 5. The energies of the eclipsed conformations separating t from g⁺ and t from g⁻ are about 3.5 kcal. mol⁻¹ above the energy of the trans conformation. Hence, in good approximation, it is justified to consider each bond to occur in one of three *rotational isomeric states* centered near (but not necessarily precisely at) the energy minima associated with the three staggered conformations.[20-24] The gauche minima lie at an energy of about 500 cal. mol⁻¹ above trans. Each of the former is therefore disfavored compared to the latter by a "statistical weight" factor we choose to call $\sigma \approx \exp(-E_g/RT)$, where E_g is about 500 cal. mol⁻¹; thus, $\sigma \approx 0.5$ at T = 400 K.

A complication arises from the fact that the potentials affecting bond rotations usually are neighbor dependent; i.e., the potential affecting φ_i depends on the rotations φ_{i-1} and φ_{i+1}. Bond rotations cannot, therefore, be treated independently.[20,21,24,25] The source of this interdependence in the case of an n-alkane chain is illustrated in Fig. 6 showing a pair of consecutive bonds in three of their nine conformations. In the conformations tt, tg⁺, g⁺t, tg⁻ and g⁻t, the two methylene groups pendant to this pair of bonds are well separated. For gauche rotations g⁺g⁺ and g⁻g⁻ of the same hand (Fig. 6b), these groups are proximate but not appreciably overlapped. Semi-empirical calculations[21,24,26,27] show the intramolecular energy for these two equivalent conformations to be very nearly equal to the sum (*ca.* 1000 cal. mol⁻¹) for two well-separated gauche bonds; i.e., the interdependence of the pair of rotations is negligible. In the remaining conformations, g⁺g⁻ and g⁻g⁺, the steric overlap is severe (Fig. 6c). It may be alleviated somewhat by compromising rotations, but the excess energy associated therewith is nevertheless about 2.0 kcal. mol⁻¹. Hence, a statistical weight factor $\omega \approx \exp(-2000/RT)$ is required for each such pair.[24,26,28] Inspection of models in detail shows that interactions dependent upon rotations about three, four of five consecutive bonds are disallowed by interferences of shorter range and hence may be ignored.[24] It suffices therefore to consider first neighbors only.

The occurrence of interactions that depend on pairs of skeletal bonds is the rule in chain molecules. In some of them, notably in vinyl polymers, such interactions may affect most of the conformations. Hence, interdependence of rotations usually plays a major role in determining the spatial configuration

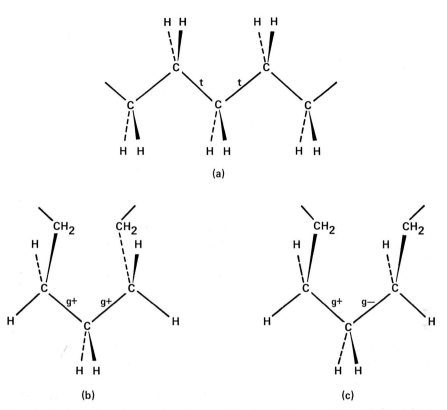

Fig. 6. Conformations for a pair of consecutive bonds in an n-alkane chain: (a), tt, (b), g^+g^+; (c), g^+g^-. Wedged bonds project forward from the plane of the central bonds, dashed bonds project behind this plane.

of the chain. The rotational isometric state approximation, whereby the continuous variation of each φ is replaced by discrete states, provides the key to mathematical solution of the problem posed by rotational interdependence.[20,21,24,25]

It is necessary therefore to consider the bonds pairwise consecutively, and to formulate a set of statistical weights for bond i that take account of the state of bond i−1. These statistical weights are conveniently presented in the form of an array, or matrix, as follows:

$$\mathbf{U}_i = \begin{bmatrix} u^{tt} & u^{tg^+} & u^{tg^-} \\ u^{g^+t} & u^{g^+g^+} & u^{g^+g^-} \\ u^{g^-t} & u^{g^-g^+} & u^{g^-g^-} \end{bmatrix}_i , \tag{15}$$

where the rows are indexed in the order t, g^+, g^- to the state of bond i−1, and the columns are indexed to the state of bond i in the same order. According to the analysis of the alkane chain conformations presented briefly above, $\mathbf{U}_i$ takes the form[24,26,28]

$$\mathbf{U_i} = \begin{bmatrix} 1 & \sigma & \sigma \\ 1 & \sigma & \sigma\omega \\ 1 & \sigma\omega & \sigma \end{bmatrix}_i \tag{16}$$

for any bond $1 < i < n$.

A conformation of the chain is specified in the rotational isometric state approximation by stipulation of the states for all internal bonds 2 to $n-1$ inclusive; e.g., by $g^+ttg^-g^-$, etc. Owing to the three-fold symmetry of the terminal methyl groups of the alkane chain, rotations about the terminal bonds are inconsequential and hence are ignored. The statistical weight for the specified conformation of the chain is obtained by selecting the appropriate factor for each bond from the array (15) according to the state of this bond and of its predecessor, and taking the product of such factors for all bonds 2 to $n-1$. In the example above this product is $u^{g^+} u^{g^+t} u^{tt} u^{tg^-} u^{g^-g^-}$, etc. It will be obvious that the first superscripted index in one of the factors u must repeat the second index of its predecessor since these indices refer to the same bond.

The configuration partition function, representing the sum of all such factors, one for each conformation of the chain as represented by the scheme of rotational isomeric states, is

$$Z = \sum_{\text{all states}} u_2 u_3 \ldots u_1 \ldots u_{n-1}, \tag{17}$$

where the subscripts are serial indexes. Each u_i must be assigned as specified above. The sum, which extends over all ordered combinations of rotational states, may be generated identically as the product of the arrays $\mathbf{U_i}$ treated as matrices. That is, according to the rules of matrix multiplication

$$Z = \prod_{i=1}^{n} \mathbf{U_i}, \tag{18}$$

where $\mathbf{U_1} = $ row $(1, 0, 0)$ and $\mathbf{U_n} = $ column $(1, 1, 1)$. Matrix multiplication generates products precisely of the character to which attention is directed at the close of the preceding paragraph. Serial multiplication of the statistical weight matrices generates this product for each and every conformation of the chain, and Eq. (18) with the operators $\mathbf{U_1}$ and $\mathbf{U_n}$ appended gives their sum.

The foregoing procedure for evaluation of Z is a minor variant of the method of H. A. Kramers and G. H. Wannier [29] for treating a hypothetical one-dimensional ferromagnet or lattice. A number of interesting characteristics of the chain molecule can be deduced from the partition function by application of familiar techniques of statistical mechanics. I shall resist the temptation to elaborate these beyond mentioning two properties of the molecule that may be derived directly from the partition function, namely, the incidences of the various rotational states and combinations thereof, and the equilibrium constants between isomeric structures of the chain in the presence of catalysts effectuating their inter-conversion. Vinyl polymers having the structure depicted in Fig. 7 with $R' \neq R$ afford examples wherein the study of equilibria

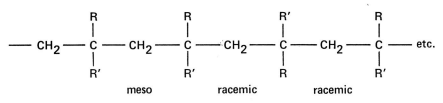

Fig. 7. A vinyl polymer chain shown in projection in its planar (fully extended) conformation. If the substituents R and R' differ (e.g., if $R = C_6H_5$ and $R' = H$ as in polystyrene), diastereomeric dyads must be distinguished as indicated for the stereochemical structure shown.

between various diastereomeric forms arising from the local chirality of individual skeletal bonds has been especially fruitful.[30]

Consider the evaluation of a configuration-dependent property for a given configuration, or conformation, of the chain. Since the configuration is seldom "given", the problem as stated is artificial. Its solution, however, is a necessary precursor to the ultimate goal, which is to obtain the average of the property over all configurations. A property or characteristic of the chain that will serve for illustration is the end-to-end vector $\mathbf{r}$. Suppose we wish to express this vector with reference to the first two bonds of the chain. For definiteness, let a Cartesian coordinate system be affixed to these two bonds with its X_1-axis along the first bond and its Y_1-axis in the plane of bonds 1 and 2, as shown in Fig. 8.

The vector $\mathbf{r}$ is just the sum $\sum\limits_{i=1}^{n} \mathbf{l}_i$ of all of the bond vectors $\mathbf{l}_i$, each expressed in this reference frame.

In order to facilitate the task of transforming every bond vector to the reference frame affiliated with the first bond, it is helpful to define a reference frame for each skeletal bond of the chain. For example, one may place the axis X_i along bond i, the Y_i-axis in the plane of bonds $i-1$ and i, and choose the Z_i-axis to complete a right-handed Cartesian system. Let $\mathbf{T}_i$ symbolize the transformation that, by premultiplication, converts the representation of a vector in reference frame $i+1$ to its representation in the preceding reference frame i. Then bond i referred to the initial reference frame is given by

$$\mathbf{T}_1\mathbf{T}_2 \ldots \mathbf{T}_{i-1}\mathbf{l}_i,$$

where $\mathbf{l}_i$ is presented in reference frame i. The required sum is just

$$\mathbf{r} = \sum_{i=1}^{n} \mathbf{T}_1 \ldots \mathbf{T}_{i-1}\mathbf{l}_i. \tag{19}$$

This sum of products can be generated according to a simple algorithm. We first define a "generator" matrix $\mathbf{A}_i$ as follows[31,32]

$$\mathbf{A}_i = \begin{bmatrix} \mathbf{T}_i & \mathbf{l}_i \\ \mathbf{0} & 1 \end{bmatrix}, \qquad 1 < i < n, \tag{20}$$

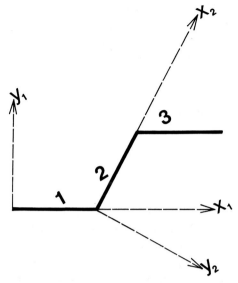

Fig. 8. Specification of the coordinate axes affixed to each of the first two bonds of the chain: X_1Y_1 for bond 1 and X_2Y_2 for bond 2.

together with the two terminal matrices

$$\mathbf{A}_1 = [\mathbf{T}_1 \quad \mathbf{1}_1], \tag{21}$$

$$\mathbf{A}_n = \begin{bmatrix} \mathbf{1}_n \\ 1 \end{bmatrix}. \tag{22}$$

In these equations $\mathbf{T}_i$ is the matrix representation of the transformation specified above and $\mathbf{0}$ is the null matrix of order 1×3. The desired vector $\mathbf{r}$ is generated identically by taking the serial product of the $\mathbf{A}$'s; i.e.,

$$\mathbf{r} = \prod_{i=1}^{n} \mathbf{A}_i, \tag{23}$$

as may easily be verified from the elementary rules of matrix multiplication. Each generator matrix $\mathbf{A}_i$ depends on the length of bond i and, through $\mathbf{T}_i$, on both the angle θ_i between bonds i and i+1 and on the angle of rotation φ_i about bond i (see Fig. 1).

In order to obtain the average of $\mathbf{r}$ over all configurations of the chain, it is necessary to evaluate the sum over all products of the kind given in Eq. (23) with each of them multiplied by the appropriate statistical weight for the specified configuration of the chain; see Eq. (17). That is,

$$\langle \mathbf{r} \rangle_0 = Z^{-1} \Sigma u_2 u_3 \ldots u_{n-1} \mathbf{A}_1 \mathbf{A}_2 \ldots \mathbf{A}_n, \tag{24}$$

where the sum includes all configurations. This sum can be generated by serial multiplication of matrices defined as follows:

$$a_i = \begin{bmatrix} u^{tt}\mathbf{A}^t & u^{tg^+}\mathbf{A}^{g^+} & u^{tg^-}\mathbf{A}^{g^-} \\ u^{g^+t}\mathbf{A}^t & u^{g^+g^+}\mathbf{A}^{g^+} & u^{g^+g^-}\mathbf{A}^{g^-} \\ u^{g^-t}\mathbf{A}^t & u^{g^-g^+}\mathbf{A}^{g^+} & u^{g^-g^-}\mathbf{A}^{g^-} \end{bmatrix}_i, \qquad 1 < i < n, \tag{25}$$

$$a_1 = [\mathbf{A}_1 \quad 0 \quad 0], \tag{26}$$

$$a_n = \text{column } (\mathbf{A}_n, \mathbf{A}_n, \mathbf{A}_n). \tag{27}$$

Then[31]

$$\langle \mathbf{r} \rangle_0 = Z^{-1} \prod_{i=1}^{n} a_i. \tag{28}$$

The matrix a_i comprises the elements of $\mathbf{U}_i$ (see Eq. (15)) joined with the $\mathbf{A}$ matrix for the rotational state of bond i as prescribed by the column index. It will be apparent that serial multiplication of the a_i according to Eq. (28) generates the statistical weight factor $u_2 u_3 \ldots u_{n-1}$ for every configuration of the chain in the same way that these factors are generated by serial multiplication of the statistical weight matrices $\mathbf{U}_i$ in Eq. (18). Simultaneously, Eq. (28) generates the product of $\mathbf{A}$'s (see Eq. (23)) that produces the vector $\mathbf{r}$ for each configuration thus weighted. The resulting products of statistical weights and of $\mathbf{A}$'s are precisely the terms required by Eq. (24). The terminal factors in Eq. (28) yield their sum.

With greater mathematical concision[31,32]

$$a_i = (\mathbf{U}_i \otimes \mathbf{E}_3) \| \mathbf{A}_i \|, \qquad 1 < i < n, \tag{29}$$

$$a_1 = \mathbf{U}_1 \otimes \mathbf{A}_1, \tag{30}$$

$$a_n = \mathbf{U}_n \otimes \mathbf{A}_n, \tag{31}$$

where $\mathbf{E}_3$ is the identity matrix of order three, $\otimes$ signifies the direct product, and $\| \mathbf{A}_i \|$ denotes the diagonal array of the matrices $\mathbf{A}_i^t$, $\mathbf{A}_i^{g^+}$ and $\mathbf{A}_i^{g^-}$.

A characteristic of the chain commanding greater interest is the quantity $\langle r^2 \rangle_0$ introduced in earlier discussion. For a given configuration of the chain, r^2 is just the scalar product of $\mathbf{r}$ with itself, i.e.,

$$r^2 = \mathbf{r} \cdot \mathbf{r} = \sum_{i=1}^{n} l^2{}_i + 2 \sum \sum_{i<j} \mathbf{l}_i \cdot \mathbf{l}_j \tag{32}$$

If each bond vector $\mathbf{l}_i$ is expressed in its own reference frame i, then

$$r^2 = \sum_{1}^{n} l_i^2 + 2 \sum \sum_{i<j} \mathbf{l}_i^T \mathbf{T}_i \mathbf{T}_{i+1} \ldots \mathbf{T}_{j-1} \mathbf{l}_j, \tag{33}$$

where $\mathbf{l}_i^T$ is the transposed, or row form of vector $\mathbf{l}_i$. These sums can be evaluated by serial multiplication of the generator matrices[24,33]

$$\mathbf{G}_i = \begin{bmatrix} 1 & 2l^T\mathbf{T} & l^2 \\ 0 & \mathbf{T} & 1 \\ 0 & 0 & 1 \end{bmatrix}_i, \qquad 1 < i < n. \tag{34}$$

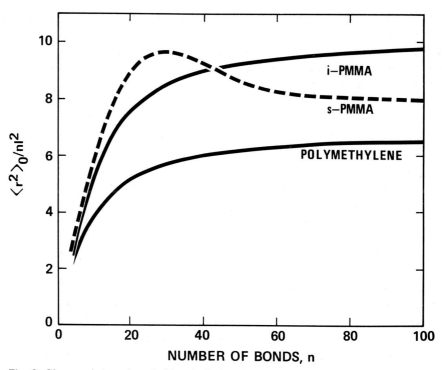

Fig. 9. Characteristic ratios $< r^2 >_0/nl^2$ plotted against the number of bonds n in the chain for polymethylene, and for isotactic and syndiotactic poly(methyl metacrylate)'s. From the calculations of Abe, Jernigan and Flory[26] and of Yoon.[34]

That is,

$$r^2 = \prod_1^n G_i \tag{35}$$

where G_1 has the form of the first row, and G_n that of final column of Eq. (34). Evaluation of $\langle r^2 \rangle_0$ proceeds exactly as set forth above for $\langle r \rangle_0$.[32,33]

The foregoing method enjoys great versatility. The chain may be of any specified length and structure. If it comprises a variety of skeletal bonds and repeat units, the factors entering into the serial products have merely to be fashioned to introduce the characteristics of the bond represented by each of the successive factors. The mathematical methods are exact; the procedure is free of approximations beyond that involved in adoption of the rotational isomeric state scheme. With judicious choice of rotational states, the error here involved is generally within the limits of accuracy of basic information on bond rotations, nonbonded interactions, etc.

Other molecular properties that may be computed by straightforward adaptation of these methods[24,32] include the higher scalar moments $\langle r^4 \rangle_0$, $\langle r^6 \rangle_0$, etc; the moment tensors formed from $\mathbf{r}$; the radius of gyration $\langle s^2 \rangle_0 = (n+1)^{-2} \sum_i \sum_j \langle r^2_{ij} \rangle$; the optical polarizability and its invariants that govern the optical anistropy as manifested in depolarized light scattering, in strain birefringence and in electric birefringence; x-ray scattering at small angles; and NMR chemical shifts.

For illustration, characteristic ratios $\langle r^2 \rangle_0 / nl^2$ are plotted in Fig. 9 against the numbers n of bonds for n-alkanes and for isotactic and syndiotactic poly-(methyl methacrylate), or PMMA. Isotactic PMMA is represented by the formula in Fig. 7 with $R = COOCH_3$ and $R' = CH_3$ and with all dyads of the meso form, i.e., with R occurring consistently above (or below) the axis of the chain. In the syndiotactic stereoisomer, the substituents R and R' alternate ₁rom one side to the other, all dyads being racemic.

For the alkane and the isotactic PMMA chains the characteristic ratios increase monotonically with chain length, approaching asymptotic values for $n \approx 100$ bonds. This behavior is typical. For syndiotactic PMMA, however, the characteristic ratio passes through a maximum at intermediate values of n, according to these computations by D. Y. Yoon.[34] This behavior can be traced[34] to the inequality of the skeletal bond angles in PMMA in conjunction with the preference for tt conformations in the syndiotactic chain.[35] The maximum exhibited in Fig. 9 for this polymer is thus a direct consequence of its constitution. This peculiarity manifests itself in the small angle scattering of x-rays and neutrons by predominantly syndiotactic PMMA of high molecular weight.[36] Scattering intensities are enhanced at angles corresponding, roughly, to distances approximating $\langle r^2 \rangle_0^{1/2}$ at the maximum in Fig. 9. This enhancement, heretofore considered anomalous, is in fact a direct consequence of the structure and configuration of syndiotactic PMMA.

It is thus apparent that subtle features of the chemical architecture of polymeric chains are manifested in their molecular properties. Treatment in terms of the artificial models much in use at present may therefore be quite misleading.

The analysis of the spatial configurations of macromolecular chains presented above is addressed primarily to an isolated molecule as it exists, for example, in a dilute solution. On theoretical grounds, the results obtained should be equally applicable to the molecules as they occur in an amorphous polymer, even in total absence of a diluent. This assertion follows unambiguously from the statistical thermodynamics of mixing of polymer chains,[5,6,37] including their mixtures with low molecular diluents. It has evoked much skepticism, however, and opinions to the contrary have been widespread. These opposing views stem primarily from qualitative arguments to the effect that difficulties inherent in the packing of long chains of consecutively connected segments to space-filling density can only be resolved either by alignment of the chains in bundle arrays, or by segregation of individual molecules in the form of compact globules. In either circumstance, the chain configuration would be altered drastically.

Whereas dense packing of polymer chains may appear to be a distressing task, a thorough examination of the problem leads to the firm conclusion that macromolecular chains whose structures offer sufficient flexibility are capable of meeting the challenge without departure or deviation from their intrinsic proclivities. In brief, the number of configurations the chains may assume is sufficiently great to guarantee numerous combinations of arrangements in which the condition of mutual exclusion of space is met throughout the system as a whole. Moreover, the task of packing chain molecules is not made easier by partial ordering of the chains or by segregating them.[6,37] Any state of organization short of complete abandonment of disorder in favor of creation of a crystalline phase offers no advantage, in a statistical-thermodynamic sense.

Theoretical arguments aside, experimental evidence is compelling in showing the chains to occur in random configurations in amorphous polymers, and further that these configurations correspond quantitatively with those of the unperturbed state discussed above.[38] The evidence comes from a variety of sources: from investigations on rubber elasticity, chemical cyclization equilibria, thermodynamics of solutions, and, most recently, from neutron scattering studies on protonated polymers in deuterated hosts (or *vice versa*).[39] The investigations last mentioned go further. They confirm the prediction made twenty-five years ago that the excluded volume perturbation should be annulled in the bulk amorphous state.[5] The excluded volume effect is therefore an aberration of the dilute solution, which, unfortunately, is the medium preferred for physicochemical characterization of macromolecules.

Knowledge gained through investigations, theoretical and experimental, on the spatial configuration and associated properties of random macromolecular chains acquires added significance and importance from its direct, quantitative applicability to the amorphous state. In a somewhat less quantitative sense, this knowledge applies to the intercrystalline regions of semicrystalline polymers as well. It is the special properties of polymeric materials in amorphous phases that render them uniquely suited to many of the functions they perform both in biological systems and in technological applications. These properties are intimately related to the nature of the spatial configurations of the constituent molecules.

Investigation of the conformations and spatial configurations of macromolecular chains is motivated therefore by considerations that go much beyond its appeal as a stimulating intellectual exercise. Acquisition of a thorough understanding of the subject must be regarded as indispensable to the comprehension of rational connections between chemical constitution and those properties that render polymers essential to living organisms and to the needs of man.

REFERENCES

1. Meyer, K. H., von Susich, G., and Valkó, F., Kolloid-Z, *59*, 208 (1932).
2. Kuhn, W., Kolloid-Z, *68*, 2 (1934).
3. Guth, E., and Mark, H., Monatsch., *65*, 93 (1934).
4. Kuhn, W., Kolloid-Z., *76*, 258 (1936); *87*, 3 (1939).
5. Flory, P. J., J. Chem. Phys., *17*, 303 (1949).
6. Flory, P. J., *Principles of Polymer Chemistry*, Cornell University Press, Ithaca, N.Y., 1953.
7. Flory, P. J., and Fisk, S., J. Chem. Phys., *44*, 2243 (1966).
8. Fujita, H., and Norisuye, T., J. Chem. Phys., *52*, 115 (1971).
9. Fixman, M., J. Chem. Phys., *23*, 1656 (1955).

10. Yamakawa, H., *Modern Theory of Polymer Solutions,* Harper and Row, New York, 1971.
11. Yamakawa, H., Pure and Appl. Chem., *31,* 179 (1972).
12. Edwards, S. F., Proc. Phys. Soc., (London), *85,* 613 (1965).
13. Fox, T. G., Jr., and Flory, P. J., J. Phys. and Coll. Chem., *53,* 197 (1949). Flory, P. J., and Fox, T. G., Jr., J. Polymer Sci., *5,* 745 (1950); J. Amer. Chem. Soc., *73,* 1904 (1951).
14. Altares, T., Wyman, D. P. and Allen, V. R., J. Polymer Sci., A, *2,* 4533 (1964).
15. Chinai, S. N., and Samuels, R. J., J. Polymer Sci., *19,* 463 (1956).
16. Mandelkern, L., and Flory, P. J., J. Chem. Phys., *20,* 212 (1952), Mandelkern, L., Krigbaum, W. R. and Flory, P. J., ibid., *20,* 1392 (1952).
17. Fox, T. G., Jr., and Flory, P. J., J. Amer. Chem. Soc., *73,* 1909, 1915 (1951).
18. Krigbaum, W. R., Mandelkern, L., and Flory, P. J., J. Polymer Sci., *9,* 381 (1952). Krigbaum, W. R. and Flory, P. J., ibid, *11,* 37 (1953).
19. Eisenberg, H., and Felsenfeld, G., J. Mol. Biol., *30,* 17 (1967). Inners, L. D., and Felsenfeld, G., ibid., *50,* 373 (1970).
20. Volkenstein, M. V., *Configurational Statistics of Polymeric Chains,* translated from the Russian ed., 1959, by S. N. and M. J. Timasheff, Interscience, New York, 1963.
21. Birshtein, T. M. and Ptitsyn, O. B., *Conformations of Macromolecules,* translated from the Russian ed., 1964, by S. N. and M. J. Timasheff, Interscience, New York, 1966.
22. Pitzer, K. S., Discussions Faraday Soc., *10,* 66 (1951).
23. Mizushima, S., Structure of Molecules and Internal Rotation, Academic Press, New York, 1954.
24. Flory, P. J., *Statistical Mechanics of Chain Molecules,* Interscience Publishers, New York, 1969.
25. Gotlib, Yu. Ya., Zh. Fiz Tekhn, *29,* 523 (1959). Birshtein, T. M., and Ptitsyn, O. B., ibid., *29,* 1048 (1959). Lifson, S., J. Chem. Phys., *30,* 964 (1959). Nagai, K., ibid., *31,* 1169 (1959), Hoeve, C. A. J., ibid., *32,* 888 (1960).
26. Abe, A., Jernigan, R. L., and Flory, P. J., J. Amer. Chem. Soc., *88,* 631 (1966).
27. Scott, R. A., and Scheraga, H. A., J. Chem. Phys., *44,* 3054 (1966).
28. Hoeve, C. A. J., J. Chem. Phys., *35,* 1266 (1961).
29. Kramers, H. A., and Wannier, G. H., Phys. Rev., *60,* 252 (1941).
30. Williams, A. D., and Flory, P. J., J. Amer. Chem. Soc., *91,* 3111, 3118 (1969). Flory, P. J., and Pickles, C. J., J. Chem. Soc., Faraday Trans. II, *69,* 632 (1973). Suter, U. W., Pucci, S., and Pino, P., J. Amer. Chem. Soc., *97* 1018 (1975).
31. Flory, P. J., Proc. Nat. Acad. Sci., *70,* 1819 (1973).
32. Flory, P. J., Macromolecules, *7,* 381 (1974).
33. Flory, P. J., and Abe, Y., J. Chem. Phys. *54,* 1351 (1971).
34. Yoon, D. Y., unpublished results, Laboratory of Macromolecular Chemistry, Stanford University.
35. Sundararajan, P. R., and Flory, P. J., J. Amer. Chem. Soc., *96,* 5025 (1974).
36. Kirste, R. G., and Kratky, O., Z. Physik, Chem. Neue Folge, *31,* 363 (1962). Kirste, R. G., Makromol. Chem., *101,* 91 (1967). Kirste, R. G., Kruse, W. A., and Ibel, K., Polymer, *16,* 120 (1975).
37. Flory, P. J., Proc. Royal Soc., A, *234,* 60 (1956). Flory, P. J., J. Polym. Sci., *49,* 105 (1961).
38. Flory, P. J., Pure & Appl. Chem., Macromolecular Chem., *8,* 1—15 (1972).
39. Kirste, R. G., Kruse, W. A., and Schelten, J., Makromol. Chem., *162,* 299 (1972). Benoit, H., Decker, D., Higgins, J. S., Picot, C., Cotton, J. P., Farnoux, B., Jannink, G., and Ober, R., Nature, Physical Sciences, *245,* 13 (1973). Ballard, D. G. H., Wignall, G. D., and Schelten, J., Eur. Polymer J., *9,* 965 (1973); ibid, *10,* 861 (1974). Fischer, E. W., Leiser, G., and Ibel, K. Polymer Letters, *13,* 39 (1975).

Chemistry 1975

JOHN WARCUP CORNFORTH

for his work on the stereochemistry of enzyme-catalyzed reactions

VLADIMIR PRELOG

for his research into the stereochemistry of organic molecules and reactions

THE NOBEL PRIZE FOR CHEMISTRY

Speech by Professor ARNE FREDGA of the Royal Academy of Sciences
Translation from the Swedish text

Your Majesties, Your Royal Highnesses, Ladies and Gentlemen,
The laureates in chemistry of this year have both studied reaction mechanisms, especially from a stereochemical, i.e. a geometrical point of view. In a chemical experiment some compounds are mixed, then something happens, and finally one can isolate one or more other compounds. What has really happened, and why, and how? The situation is as if someone had abbreviated a classical tragedy, say Hamlet, by showing only the opening scenes of the play and the final scene of the last act. The principal characters are introduced, then the scene closes and when the curtain rises again you see a number of dead bodies on the stage and a few survivors. Of course the spectators would like to know what has happened in the meantime.

What I have said is valid not least for enzymatic reactions. Many such reactions are perpetually going on in all living organisms; one could say that they really concern all of us although we don't observe them. When a chemist tries to find out what really happens, he often comes across the problem: right or left? It is the same in common life. If you leave Stockholm by Norrtull, you soon come to a place where the main road branches: the left branch leads to Oslo, the right one to Sundsvall or, if you like, to Haparanda.

Professor Cornforth has among other things studied the biological synthesis of the hydrocarbon squalene from six molecules of mevalonic acid. This hydrocarbon is necessary for the formation of steroids, which are of vital importance in many respects. The synthesis of squalene takes place in 14 steps and at each the enzyme must find the proper way. That means that there are just $2^{14} = 16384$ different routes and only one of them leads to squalene. If the enzyme should make a mistake in the first step (which it does not), the final result could be rubber or various other things but definitely not squalene. The problem at each step concerns which of two hydrogen atoms is to be eliminated, the right one or the left one. Professor Cornforth has shown which choice the enzyme makes at each of the 14 steps. For this purpose he has, with brilliant mastership, utilized the properties of the hydrogen isotopes: the ordinary hydrogen, the heavy hydrogen and the radioactive hydrogen. The last-mentioned isotope can only be used in tracer quantities, which means that only about one part per million of the participating molecules are radioactive. In a similar way, Professor Cornforth has studied several other biologically important reactions. All problems connected with the reaction mechanisms are not solved at that point, but the results constitute a very important step on the way.

Professor Prelog has worked in many fields of stereochemistry, and often the problems have been connected with the geometrical shapes of the molecules and their influence on the course of the reactions.

An impressive series of investigations deal with the "medium rings", i.e. molecules containing rings of 8 to 11 carbon atoms. Such rings are not rigid but rather limp. Parts of the ring which may seem rather distant may come into close contact with each other leading to unexpected reactions. Professor Prelog has been able to elucidate such reactions by utilizing the carbon isotopes.

Many important investigations refer to reactions between chiral molecules. The term chiral is derived from a word in ancient Greek, meaning hand. The molecules are unsymmetrical and may exist in two forms differing in the same way as a right hand and a left hand. The molecules are so small that you can't see them, but one can gain much knowledge by studying the reactions between chiral molecules of different kind.

Professor Prelog has also made important contributions to enzyme chemistry. He has studied enzymatic reactions on small molecules and in particular oxidation or reduction processes. The experiments may be more or less successful depending on how the enzyme and the other molecule fit together geometrically. By systematic experiments with various small molecules of well-defined shapes, it was possible to construct a "map" of the active part of the enzyme molecule. The results have recently been confirmed in a special case by Swedish scientists using x-ray methods.

Professor Prelog has also with ingenuity and penetration discussed and analysed the fundamental concepts of stereochemistry, not least the conditions for chirality in large and complicated molecules.

Professor Cornforth. Enzymatic reactions have always had a certain air of magic, perhaps witchcraft. Of course this is due to our imperfect knowledge of what really happens. This air of magic is, however, gradually dispersing, and your contributions, utilizing the isotopes of hydrogen, imply most striking advances. The handling of compounds with chiral methyl groups is an achievement of the highest intellectual standard.

Let me also express our admiration for the skill and perseverance with which you have pursued your work in spite of a serious physical handicap. Perhaps it had not been possible without the never-failing help and support of Mrs. Cornforth. I think she should not be forgotten on this day.

In recognition of your services to chemistry and to natural science as a whole, the Royal Academy of Sciences has decided to confer upon you the Nobel Prize. To me has been granted the privilege to convey to you the most heartly congratulations of the Academy.

Professor Prelog. Ich habe hier versucht, einen Kurzbericht über Ihre wichtigsten Leistungen in der Stereochemie zu erstatten. Das war gewiss etwas schwierig. Ihre schönen Experimentalarbeiten erstrecken sich über weite Felder der heutigen organischen Chemie. Öfters haben Sie die Fortführung Ihrer Arbeiten anderen Forschern überlassen, und viele Chemiker hohen Ranges sind zurzeit auf den Gebieten tätig, die Sie einst eröffnet haben. Sie haben auch die fundamentalen Grundlagen der Stereochemie, besonders den Chiralitätsbegriff, in tiefsinnigen Auseinandersetzungen diskutiert und klargelegt.

In Anerkennung Ihrer Verdienste um die Entwicklung der Chemie hat die Königliche Akademie der Wissenschaften entschlossen, Ihnen den Nobelpreis zu verleihen. Mir ist die Aufgabe zugefallen, Ihnen die wärmsten Glückwünsche der Akademie zu überbringen.

Professor Cornforth. In the name of the Academy I invite you to receive your prize from the hands of His Majesty the King.

Professor Prelog. Im Namen der Akademie bitte ich Sie aus den Händen Seiner Majestät des Königs den Nobelpreis in Empfang zu nehmen.

John Warcup Cornforth

JOHN WARCUP CORNFORTH

I was born on 7 September 1917 at Sydney in Australia. My father was English-born and a graduate of Oxford; my mother, born Hilda Eipper, was descended from a German minister of religion who settled in New South Wales in 1832. I was the second of four children.

Part of my childhood was spent in Sydney and part in rural New South Wales, at Armidale. When I was about ten years old the first signs of deafness (from otosclerosis) became noticeable. The total loss of hearing was a process that lasted more than a decade, but it was sufficiently gradual for me to attend Sydney Boys' High School and to profit from the teaching there. In particular a good young teacher, Leonard Basser, influenced me in the direction of chemistry; and this seemed to offer a career where deafness might not be an insuperable handicap.

I entered Sydney University at the age of 16, and though by that time unable to hear any lecture I was attracted by laboratory work in organic chemistry (which I had done in an improvised laboratory at home since the age of 14) and by the availability of the original chemical literature. In 1937 I graduated with first-class honours and a University medal. After a year of post-graduate research I won an 1851 Exhibition scholarship to work at Oxford with Robert Robinson. Two such scholarships were awarded each year, and the other was won by Rita Harradence, also of Sydney and also an organic chemist. This began an association which continues to this day. We were married in 1941, and have three children and two grandchildren.

War broke out as we journeyed to Oxford and after completing our work (on steroid synthesis) for doctorates we became part of the chemical effort on penicillin which was the major chemical project in Robinson's laboratory during the war. We made contributions, and I helped to write *The Chemistry of Penicillin* (Princeton University Press, 1949), the record of a great international effort. However, I had earlier discovered what was to prove a key reaction for the synthesis of the sterols; and after the war I returned to this pursuit. The collaboration with Robinson continued after I joined (1946) the scientific staff of the Medical Research Council and worked at its National Institute, first at Hampstead and then at Mill Hill. In the end (1951) we were able to complete, simultaneously with Woodward, the first total synthesis of the non-aromatic steroids.

At the National Institute for Medical Research I came into contact with biological scientists and formed collaborative projects with several of them. In particular George Popják and I shared an interest in cholesterol. At this time Konrad Bloch was beginning his work on the biosynthesis of the sterols and Popják and I began to concert experiments in which the disciplines of chemistry and biochemistry could be applied to this subject. We were led to devise a complete carbon-by-carbon degradation of the nineteen-carbon ring structure of cholesterol and to identify, by means of radioactive tracers, the arrangement of the acetic acid molecules from which the system is built. As knowledge of the intermediate stages became more complete our experiments could be planned to give more and more information.

In 1962 Popják and I left the service of the Medical Research Council and became co-directors of the Milstead Laboratory of Chemical Enzymology set up by Shell Research Ltd. Lord Rothschild was influential in the decision to establish this laboratory and I was his subordinate until he left Shell in 1970. At Milstead a project already conceived—the study of the stereochemistry of enzymic reactions by means of asymmetry artificially introduced by isotopic substitution—was developed. It continued after 1968, when Popják left Milstead to go to the University of California at Los Angeles, until 1975, when I left to take up my present position of Royal Society Research Professor at the University of Sussex. In 1967 I had formed a collaboration with Hermann Eggerer, then of München; and together we solved the problem of the "asymmetric methyl group", and applied the solution in some of the many ways that have proved possible.

My work has received ample recognition as it progressed: I was elected to the Royal Society in 1953; the Chemical Society has awarded me its Corday-Morgan medal (1953), Flintoff medal (1965), and Pedler (1968) and Robert Robinson (1971) lectureships; the American Chemical Society gave me its Ernest Guenther award (1968); and I received the Prix Roussel in 1972. Popják and I were jointly awarded the Biochemical Society's Ciba medal (1965); the Stouffer prize (1967); and the Royal society's Davy Medal (1968).

Throughout my scientific career my wife has been my most constant collaborator. Her experimental skill made major contributions to the work; she has eased for me beyond measure the difficulties of communication that accompany deafness; her encouragement and fortitude have been my strongest supports.

ASYMMETRY AND ENZYME ACTION

Nobel Lecture, December 12, 1975

by JOHN WARCUP CORNFORTH

University of Sussex, Falmer, Brighton, England

It must, I think, be rare to be rewarded so generously for work that was so purely a pleasure in planning and execution. I shall try, in return, to impart some of that pleasure today.

In 1948, a short but historic note by Alexander Ogston appeared in the scientific magazine *Nature,* demonstrating the importance of a particular type of stereochemical thinking in relation to biochemical processes catalysed by enzymes. Up to that time I had, as an organic chemist interested in the synthesis of natural products, the same kind of feeling for stereochemistry that a motorist night have for a system of one-way streets—a set of rules forming one more obstacle on the way to a destination. But 1948 was a year in which as well as continuing collaboration with Robert Robinson on the total synthesis of sterols, I had begun to co-operate with biological scientists at the National Institute for Medical Research; so that Ogston's note was a seed that germinated the more readily in my mind.

The essential principles of the three-dimensional structure of organic molecules had been correctly formulated by the first Nobel laureate in Chemistry, Jacobus van't Hoff, as early as 1874. In particular he (and independently Le Bel) gave a structural basis to Louis Pasteur's discovery that certain molecules can exist in two *optically active* forms that differ from each other in their effect on a beam of plane polarized light; the plane of polarization is rotated to the right when the beam passes through a solution of one form, and to the left when the other form is substituted. van't Hoff theorized that when a carbon atom in a molecule is attached to its maximum number of four other atoms, these occupy the four apices of a tetrahedron, with the carbon atom in the middle. Another way of saying the same thing is that the four atoms keep as far away from each other as they can, given that they are bound at fixed distances from the central atom. If these four atoms are all different, or at any rate if each forms part of a different group of atoms, they can occupy two distinct spatial arrangements that are *chiral:* their relationship is that of a right hand with a left hand and they are mirror images of one another (I, II).

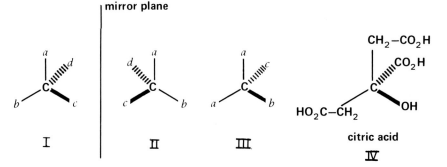

citric acid

IV

The central carbon may then be called a centre of asymmetry. On the other hand, if two of the atoms or groups are indistinguishable from each other (III), only one arrangement about the central atom is possible.

Most molecules, including enzymes, that mediate the processes of life are optically active and have centres of asymmetry, but many molecules quite important in life processes have no centre of asymmetry and one of these is citric acid (IV), which was first isolated in 1784 by the great Swedish chemist Scheele: it is a *Caabc* compound, the two acetic acid groups attached to the central carbon being identical. When processes in living cells began to be studied with the help of radioactive and stable isotopes as tracers, an apparent anomaly arose that centred round this substance.

The biochemical course of events, which was outlined quite correctly on the evidence available at the time, was studied with preparations of pigeon liver and is expressed in Scheme 1. Pyruvic acid reacts with carbon dioxide to give oxaloacetic acid, and this condenses with "active acetate" (now known to be acetyl-coenzyme A) to yield citric acid. The citric acid then undergoes oxidation with loss of carbon dioxide to 2-oxoglutaric acid, which is further oxidized to a second molecule of carbon dioxide and succinic acid. All these reactions are catalysed by enzymes (though the last step can also be done chemically) but the original stereochemical reasoning about the sequence was based on what would happen if the same reactions were carried out nonenzymatically. The reasoning went something like this: "citric acid is a symmetrical molecule of the type *Caabc;* there is no difference between the two *a* groups, which are both acetic acid residues; so that if we make one of them radioactive, for instance by using radioactive carbon dioxide in the

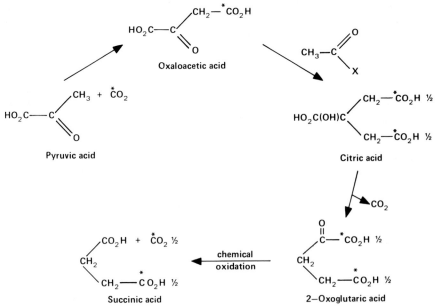

Scheme 1. Predicted distribution of isotopic label from carbon dioxide in 2-oxoglutaric acid

first step, then when the citric acid is broken down by way of 2-oxoglutaric acid to succinic acid and carbon dioxide the two residues will be affected indifferently and half the radioactivity will be in succinic acid and half will be in the second molecule of carbon dioxide liberated on oxidation".

But experiment showed that *all* the radioactivity appeared in carbon dioxide and none in succinic acid, and no explanation (except the incorrect conclusion that citric acid took no part in the biochemical sequence) could be found before Ogston's note. Although Ogston clearly grasped the principle, I am giving the explanation in rather more general terms.

Enzymes catalyse chemical reactions by binding the reactant molecules (substrates) at a specific site in the enzyme molecule. Enzymes are proteins, and proteins are made up of a large number of asymmetric units, the amino-acids. There is no element of symmetry in an enzyme, or in its specific site. Moreover, each enzyme will characteristically accept very few molecular types as substrates: small changes in shape or size from the normal substrate may result in a very much slower reaction or in none at all. Emil Fischer had this in mind when he said, as long ago as 1894, that enzyme and substrate must fit each other like lock and key.

A lock and key must fit each other; but also, if the lock has no symmetry, the key has to be oriented three-dimensionally in an unique manner for introduction into the lock; and then the key has to be turned in a particular sense to operate the lock, so that one particular side of the key executes the actual operation of moving the lock's mechanism.

So that if a molecule of oxaloacetic acid is considered as a key, then only one particular side (above or below the plane of the paper) of its ketone carbonyl group can react with the other substrate, acetyl-coenzyme A. The other side *cannot* be fitted to the enzyme without changing the whole orientation of the substrate, as if one tried to fit the wrong end of a key, or a key upside down, into a lock. This is quite different from the reactions of the same carbonyl group in free solution, where both sides are equally open to attack by a reagent.

If the oxaloacetic acid carries a radioactive label, as it does when it has been made from pyruvic acid and radioactive carbon dioxide, then the citric acid which is formed on the enzyme carries its labelled acetic acid residue in a particular orientation that is distinct from the orientation of the other acetic acid residue—the one that originates from acetyl-coenzyme A. Since the next reaction in the biochemical sequence produces a change in one of the acetic acid residues, it is obvious that *all* citric acid molecules presented to the enzyme that alters them will have the labelled residue altered, or else they will all have the unlabelled residue altered: this is a necessary consequence if the citric acid molecules must be presented to the enzyme in a particular orientation. As it happens, the labelled residue is the one altered; and the relevant stereochemistry of the process, as elucidated much later by Kenneth Hanson and Irwin Rose, is as shown (Scheme 2).

Asymmetric synthesis is not unknown to organic chemists: for example a reaction in free solution that produces a new centre of asymmetry by bringing

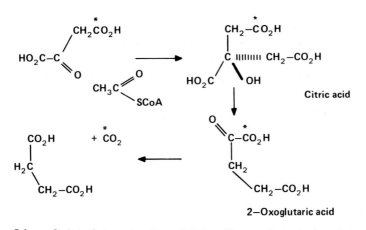

Scheme 2. Actual stereochemistry of citric acid enzymic synthesis and degradation

together a symmetrical and an unsymmetrical molecule will often produce an excess of one of the two chiral forms. But here was something of a different order: two reactions promoted with complete specificity by an asymmetric catalyst. And in both reactions the asymmetry of the catalysis is *hidden:* if it had not become possible to place the experimenter's private mark, in the shape of an isotopic label, on one of the two acetic acid residues, the Ogston effect, as it has come to be known, might have remained unsuspected for many years. In this field of work, the use of isotopes as markers is almost indispensable: replacement of an atom in a substrate molecule by one of its isotopes makes very little change in shape or chemistry; and an enzyme will always accept a labelled substrate, though it may transform it a little slower.

In 1953, Frank Westheimer, in collaboration with the biochemists Frank Loewus and Birgit Vennesland, studied yeast alcohol dehydrogenase. This enzyme catalyses the reversible transfer of a hydrogen atom between a molecule of ethanol and a molecule of a coenzyme, nicotinamide-adenine dinucleotide. The transfer neither creates nor destroys a centre of asymmetry, but nevertheless, the two hydrogen atoms on the oxygenated carbon of ethanol, and the two sides of the nicotinamide ring in the coenzyme, are stereochemically distinct in the Ogston sense. A simple test of this is to look at the rest of the molecule from the viewpoint of each hydrogen atom in turn, or from each side of the

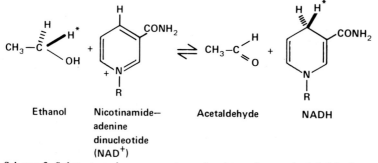

| Ethanol | Nicotinamide— adenine dinucleotide (NAD⁺) | Acetaldehyde | NADH |

Scheme 3. Substrate and coenzyme stereochemistry of yeast alcohol dehydrogenase

ring in turn: if the two views are different (as they are) an enzyme can, and probably will, concern itself with only one of the atoms or only one of the sides. By using the hydrogen isotopes deuterium or tritium as marking labels it was indeed possible to show that the hydrogen that is transferred occupies an unique stereochemical position in both substrate and coenzyme. A hydrogen not occupying one of these positions is not transferred at all. The stereochemistry has been worked out since then in a number of laboratories, including mine, and it is as shown (Scheme 3).

Further, when hydrogen was transferred from an unlabelled coenzyme to acetaldehyde in which the aldehydic hydrogen had been replaced by deuterium, the alcohol formed showed measurable (though small) rotation of polarized light, which made possible the correlation of its stereochemistry with this physical property. Finally, when the deuteriated ethanol was submitted to a purely chemical procedure: hydrolysis of its toluene-4-sulphonyl ester, a new specimen of deuteriated ethanol was obtained which, unlike its precursor, transferred deuterium and not hydrogen to the coenzyme in the presence of yeast alcohol dehydrogenase (Scheme 4).

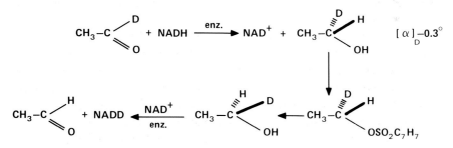

Scheme 4. Stereochemical inversion of 1-deuterioethanol

Now although this last experiment was not carried out for the purpose, it can be regarded as proving that the hydrolysis of a typical sulphonic ester of a primary alcohol proceeds with *inversion of configuration* at carbon; which was something never demonstrated before, although it had been shown to be true of secondary alcohols in which the asymmetry owed nothing to isotopic substitution. Knowledge of the stereochemistry of a chemical reaction is one of the most useful guides in elucidating the correct mechanism and in excluding alternatives.

I had been following this work with much interest, and perceived some of its potential importance for studying enzymic mechanisms; but I was engaged at the time with my biochemical colleague George Popják and our collaborators, on a problem of biosynthesis: by chemical degradation of cholesterol synthesized in rat liver preparations from acetic acid, we were showing the pattern of incorporation of the precursor into the ring structure of the sterol. Later, and especially when mevalonic acid emerged as the parent substance of steroids and terpenoids, we were able to plan experiments of greater subtlety, using mevalonic acid specifically labelled with carbon isotopes to decide details of the molecular rearrangement that takes place when the steroid ring

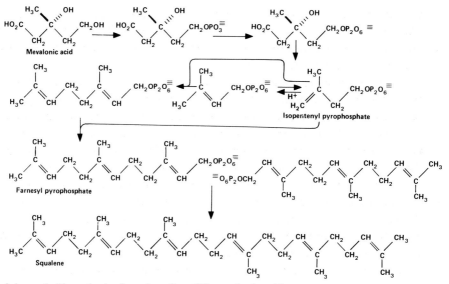

Scheme 5. Biosynthesis of squalene from 3*R*-mevalonic acid

structure is formed. At the same time, Konrad Bloch and Feodor Lynen were identifying the intermediate stages leading from mevalonic acid to the sterols in yeast, and Popják was demonstrating that the same intermediates were formed in rat liver. The sequence from mevalonic acid to squalene, the precursor of all steroids and triterpenoids, was mapped out as shown (Scheme 5).

In 1960 we were checking on the formation of squalene from two molecules of farnesyl pyrophosphate. This reaction looks like a symmetrical coupling of two identical halves: in fact, we found that the process is attended by the exchange of one, and only one, hydrogen atom from one of the carbon atoms that become joined together in squalene. This non-symmetrical synthesis of a symmetrical molecule roused further my curiosity about the mechanism of the whole process, from mevalonic acid to squalene.

Mevalonic acid has asymmetry of the ordinary kind, with C*abcd* substitution at the central carbon atom, but this type of asymmetry is soon lost in the biochemical sequence. But mevalonic acid has three C*aabc* centres (Scheme 6) and all of these undergo changes in bonding on the way to squalene. Each of these six hydrogen atoms in these three groups is stereochemically distinct in the Ogston sense, and so it was possible in principle to follow the fate of a hydrogen atom from any one of these six positions, until it

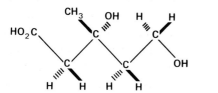

Scheme 6. Mevalonic acid and its six methylene hydrogens

was either lost in the aqueous medium of reaction or came to a specific, and stereospecific, destination in a molecule of squalene. This fate is uniquely determined by, and therefore throws light on, the stereochemistry of the enzymic reactions in the biosynthetic sequence.

Thus it became necessary to place a distinguishing mark on each of these six hydrogen atoms in turn, and this could not be done except by replacing normal hydrogen by one of its isotopes: the stable deuterium or the radioactive tritium. This problem was solved to a large extent by drawing on the vast store of organic chemical knowledge, especially that part which concerns the stereochemistry of reactions. In this way, the individual labelling of four out of the six hydrogen atoms was achieved by non-enzymic processes, and an organic reaction of known stereochemical preference was employed to define the stereochemistry of an enzyme—mevaldate reductase—which was used to generate a label on the fifth hydrogen: the sixth was also labelled, eventually, with the help of enzymes of known stereochemistry.

These labelled mevalonic acids were introduced into enzyme preparations made from rat or pig liver. According to the nature of the preparation and to the presence or absence of co-factors or inhibitors, it was possible to execute the whole sequence from mevalonic acid to squalene or to stop at various intermediate stages; phosphomevalonic acid, isopentenyl pyrophosphate, or farnesyl pyrophosphate (see Scheme 5). Especially in experiments where the label was deuterium, unusually large amounts of product—typically, 50 milligrams—had to be accumulated from these enzymic incubations, and Popják and I became familiar with the dialogue "How much can you make?" "How little do you need?"

After the enzymic transformations, the products were examined to find out what had happened to the labelling isotope. When it was a question of whether the isotope had been lost from, or retained in, the product the procedure was relatively simple; one examined the product in a mass spectrometer for the presence or absence of deuterium, or one measured the radioactivity assignable to tritium. When the absolute stereochemistry at a labelled position was needed, it was necessary to use deuterium as the label and to degrade the product chemically, by reactions that either left the labelled centre undisturbed or altered it in a well-defined manner, to a substance suitable for examination by polarimetry or mass spectrometry. It was fortunate that this work coincided with the development of polarimeters sensitive enough to measure, in favourable cases, optical activity due solely to substitution of hydrogen by deuterium in specimens of a few milligrams. Our first measurements of this sort were, indeed, made on a prototype machine at the National Physical Laboratory. Since contamination by optically active material of the usual type could have been damaging, we had to develop a technique for recrystallization in capillary tubes. This permitted the recrystallization of succinic acid, for example, in milligram quantities from about two parts of water. Once the optical rotation was known it could be compared with the rotation of a sample into which a known absolute configuration had been built by enzymic and chemical synthesis.

Thus the work required in unusual measure the harmonious blending of stereospecific synthesis, isotopic labelling, enzymology, chemical degradation on the centigram scale, and sensitive physical methods of analysis, into a single experimental sequence. In the end, we succeeded in demonstrating stereospecificity for all but one of the enzymic steps then known for squalene biosynthesis; the fate of individual hydrogen atoms was as shown in Scheme 7. This was about as far as we could get by introducing asymmetry into CH_2bc groups by isotopic substitution of one of the hydrogens, but the availability of these specifically labelled mevalonic acids was to prove, in our hands and in others', of considerable use in mapping the pathways of terpenoid biosynthesis in general.

One step in terpenoid biosynthesis is the (reversible) isomerization of isopentenyl pyrophosphate into dimethylallyl pyrophosphate (see Scheme 5). The addition of a proton from the aqueous medium to the terminal methylene group, if it is stereospecific, is to one side only of the terminal methylene group. Thanks to the work already done, labelled mevalonic acids were available which were known to give isopentenyl pyrophosphate having a geometrically defined deuterium or tritium label at either of the two hydrogens of the methylene group; but if this group was converted in a normal incubation to a methyl group, free rotation about its C-C bond would give indistinguishable products whatever the initial geometry of the label and whatever the direction of addition of the proton. The only way in which it seemed possible to retain the individuality of the three hydrogen atoms concerned was to use all three isotopes of hydrogen—protium, deuterium and tritium—in proper sequence. Then, if the isopentenyl pyrophosphate was stereospecifically la-

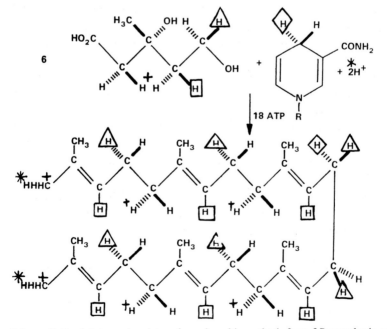

Scheme 7. Partial stereochemistry of squalene biosynthesis from 3*R*-mevalonic acid

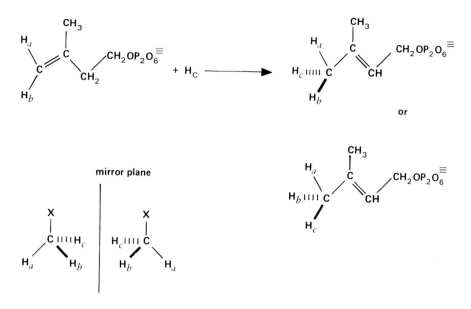

Scheme 8. Generation of a chiral methyl group

belled with two hydrogen isotopes and the third isotope was supplied in the water of incubation, a stereospecific addition of hydrogen from the water would give a *chiral methyl group,* the chirality of which would be determined by, and diagnostic of, the direction of proton addition (Scheme 8).

Chiral methyl groups were unknown at the time, and it was not obvious how their absolute configuration could be determined: optical rotation was an unlikely candidate for measurement since a substance having tritium in atomic proportion (instead of the usual small labelling concentration) would have a specific radioactivity of some 30,000 Curies per mole, and the rotatory power would probably be so small as to require large specimens having this order of radioactivity.

The solution of the problem grew from a suggestion made by Hermann Eggerer in 1967. He had been studying the enzyme malate synthase, which makes malic acid, an asymmetric substance having sinistral or S chirality according to the convention of Cahn, Ingold and Prelog, from glyoxylic acid and acetyl-coenzyme A; and he was led to favour a mechanism for the reaction which predicted a particular stereochemical relation between the hydrogen atom that is lost from the methyl group and the glycollic acid residue that replaces it (Scheme 9). If this mechanism was correct and if the reaction was attended by a normal "isotope effect", then protium should be displaced from a chiral methyl group more often than deuterium. Thus, the molecules of malic acid that contained tritium should comprise a larger proportion of the species containing deuterium and tritium than of the species containing tritium and protium; and in these two species the stereochemical location of the tritium must be different. An analytical method for determining this location was already available: the enzyme fumarase was known to catalyse the stereospecific *anti* elimination (Scheme 10) of water from S malic

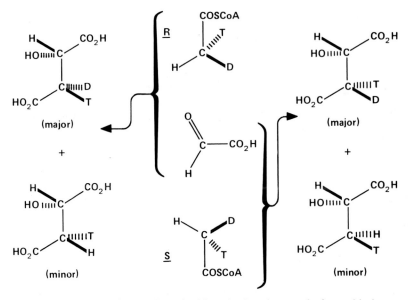

Scheme 9. Predicted distribution of tritium in *S* malate made from chiral acetates and glyoxylate on malate synthase, assuming retention of configuration at the chiral methyl group

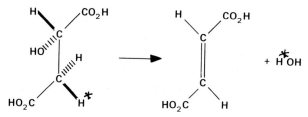

Scheme 10. Stereochemistry of loss of carbon-bound hydrogen from *S* malate on the enzyme fumarase

acid. So that if one carried out the sequence chiral acetate → acetyl-coenzyme A → malate → fumarate and measured the percentage loss of tritium in the last stage, this percentage should be different for the two chiral forms of acetate: the percentage retention for the one should equal the percentage loss for the other. And if the stereochemical mechanism for malate synthase was assumed correct, one could infer the chirality of the acetate from these measurements of radioactivity alone.

The final plan (Scheme 11) did not depend on this assumption (which was just as well, for it turned out to be wrong). Instead, we were able at Milstead to synthesize, purely by chemical methods, potassium acetates the chirality and absolute stereochemistry of which were defined rigorously by the method of synthesis. When these were put through the sequence acetate → acetyl-coenzyme A → malate → fumarate in Eggerer's laboratory, the malate derived from *S* acetate lost over three-quarters of its tritium on incubation with fumarase. In complementary contrast, malate derived from *R* acetate retained more than three-quarters of its tritium. Thus, without making any assumptions about the mechanisms of the enzymes used, we had a convenient meth-

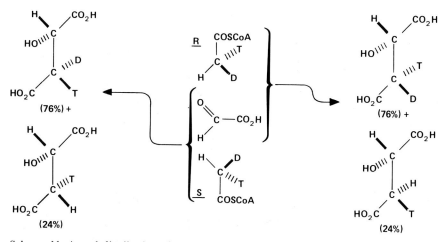

Scheme 11. Actual distribution of tritium in *S* malate made from synthetic chiral acetates and glyoxylate on malate synthase

od for determining whether a given specimen of chiral acetate was *R* or *S*. At ETH. Zürich, Duilio Arigoni and Janos Retey independently produced a very similar solution of the same problem. With this analytical method it became possible to solve not only the problem of the addition of hydrogen to isopentenyl pyrophosphate but also to deduce the stereochemistry of a large and still growing number of enzymic reactions in which a methyl group is either generated or transformed. When at last we knew the stereochemical origin of all fifty hydrogens in squalene biosynthesized from mevalonic acid, I had a three-dimensional model made to illustrate this. The last scheme summarizes the information conveyed (Scheme 12).

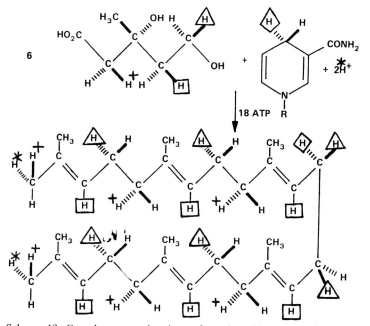

Scheme 12. Complete stereochemistry of squalene biosynthesis from 3*R*-mevalonic acid

Our adventure with the chiral methyl group reinforced the conviction that stereospecificity is something not just incidental, but essential, to enzymic catalysis. Life does depend on accurate replication of molecules and its complexity often requires that an enzyme shall accept one molecular species or type and transform it to equally specific products. But the hidden specificity that we have helped to reveal goes much further than this: an enzyme must, it seems, catalyse strictly stereospecific reactions even when this specificity is not required by the structural relation of substrate to product. Indeed, many examples are now available in which an enzyme can accept more than one molecular species as substrate but still transforms each of them with absolute, though hidden, control of the stereochemistry of reaction.

By combining chemical, biochemical and physical techniques it has thus become possible to investigate the nature of enzymic catalysis in a novel manner, complementary to the other approaches which have developed over the same period. The work required concentrated effort, and I owe much to the skill and dedication of many collaborators; but I call your attention to three. George Popják sustained the biochemical side of these investigations with exceptional insight, ability and resource until 1967. I was fortunate enough to be associated with him as colleague and partner for over twenty years and, after him, to enjoy collaboration with another great biochemist, Hermann Eggerer, whose co-operative spirit made light the difficulties of concerting experiments in laboratories a thousand kilometers apart. Thirdly, my wife Rita Cornforth, with patience and great experimental skill, executed much of the chemical synthesis on which the success of the work was founded. To her, in this as in other ways. I owe more than I can well express.

To my teacher and friend, Robert Robinson, whose death early this year sadly forestalled his presence on this occasion, I remain especially grateful, and could hope for nothing better than to retain, as he did to the end of a long and creative life, fresh curiosity and wonder at the chemistry of Nature.

Vladimir Prelog

VLADIMIR PRELOG

I was born on July 23rd, 1906 in Sarajevo in the province of Bosnia, which then belonged to the Austrian-Hungarian Monarchy and later, in 1918, became part of Yugoslavia. In the western world my birthplace has a somewhat sinister reputation that was characterized by an older tax-inspector in the Midwest of America as "the place where all that mess started". Actually, as an 8 years old boy I stood near to the spot where Archiduke Franz Ferdinand and his wife were assassinated. At the beginning of the first World War, in 1915, we moved to Zagreb, the capital of Croatia, where I attended the gymnasium. The period 1924 to 1929 was spent studying Chemistry at the Czech Institute of Technology in Prague, Czechoslovakia. The supervisor of my thesis was Professor Emil Votoček, one of the prominent founders of chemical research in Czechoslovakia. My mentor, however, was Rudolf Lukeš, then lecturer and later successor of Votoček to the chair of organic chemistry. To Lukeš I owe the greatest part of my early scientific education, and he remained my close friend until his premature death in 1960. In addition to these two "real" teachers I admired Robert Robinson, Christopher Ingold and Leopold Ruzicka, all of whom I considered as my "imaginary" teachers. In later years I was fortunate to become well acquainted with all three of these great chemists.

The close of my studies with a degree of a Dr. Ing. in 1929 coincided with the great economic crisis, and I was not able to find an academic position. I was therefore very grateful for a position in the newly created laboratory of G.J. Dříza in Prague where rare chemicals were produced on small scale. I had there also a modest opportunity to do some research, but I badly wanted to work in an academic environment. This is why I was so eager to accept the position of a lecturer at the University of Zagreb in 1935. I did not know that I had to fulfil there all the duties of a full professor and to live on a salary of an underpaid assistent, but it would probably not have affected my decision if I had known. With the help of a couple of enthusiastic young coworkers and of a developing small pharmaceutical factory, I had just managed to solve at least the most urgent problems for myself and my laboratory when the second World War broke out. After the German occupation of Zagreb in 1941 it became clear that I was likely to get into serious trouble if I remained there. At this critical point I received an invitation of Richard Kuhn to give some lectures in Germany, and shortly afterwards Leopold Ruzicka, whom I had asked for help, invited me to visit him on the way. With these two invitations, it was possible for me to escape with my wife to Switzerland. Through Ruzicka I soon obtained generous support of CIBA Ltd. and started work in the Organic Chemistry Laboratory at the Federal Institute of Technology (ETH) in Zürich. The cooperation with Ruzicka lasted many years and enabled me to make my slow progress up the academic hierarchical ladder. Starting as assistent, I became "Privat-Dozent", "Titularprofessor", associate (ausserordentlicher) professor and in 1952 full professor ad personam. Finally, in 1957, I succeeded Ruzicka as head of the Laboratory, a height that I never dreamt of when I was a student in Prague. In becoming director of the Laboratory I reached, according to Peter's principle, the level of my incompetence and I tried hard for several years to step down. Surrounded and supported by a group of very able young colleagues, I finally succeeded in introducing a rotating chairmanship from which I was exempted. So far this has worked very satisfactorily and it may have helped some of my colleagues to resist tempting offers from other Universities.

My main interests were natural compounds, from adamantane and alkaloids to rifamycins and boromycin. During the work on natural compounds stereochemical problems emerged from all sides. As E.L. Eliel pointed out, stereochemistry is not so much a branch of chemistry but rather a way of looking at chemistry. It was, and still is, great fun trying to find new points of view for it.

I travel a lot. Recently I counted that I have given lectures in more than 150 places, often several times. This in spite of the fact that I do not speak any language properly. I suspect that many people come to my lectures because they enjoy my strange accent and skill in managing without actually cheating.

I married my wife Kamila in Prague in 1933. A son Jan was born to us in Zürich in 1949.

For many years, when still a Yugoslav citizen, I was already a Swiss patriot and in 1959 I obtained Swiss citizenship. However, I consider myself a world citizen and I am very grateful to my adopted country that it allows me to be one.

The way from Sarajevo to Stockholm is a long one and I am fully aware that I have been very lucky to arrive there. The journey could not have been made without the generous help of friends, colleagues, coworkers and also of innumerable earlier chemists "on whose shoulders we stand".

CHIRALITY IN CHEMISTRY

Nobel Lecture, December 12, 1975
by VLADIMIR PRELOG
ETH, Laboratory of Organic Chemistry, Zürich, Switzerland

An object is chiral if it cannot be brought into congruence with its mirror image by translation and rotation. Such objects are devoid of symmetry elements which include reflexion: mirror planes, inversion centers or improper rotational axes.

The useful terms chiral and chirality were coined by W.H. Thompson (Lord Kelvin) in 1884 and are derived from cheir, the Greek word for a hand, indeed one of the most familiar chiral objects. The simplest chiral object of the three-dimensional perceptual space is, however, the chiral three-dimensional simplex, the irregular tetrahedron. As early as 1827 the famous German mathematician August Ferdinand Möbius (of the Möbius-strip) pointed out that the volume of a tetrahedron, expressed as a determinant involving the Cartesian coordinates of its labelled vertices, and of its mirror image have different signs, which are not dependent on the position of the tetrahedra but change by reflection.

Many objects of our three-dimensional perceptual world are not only chiral but appear in Nature in two versions, related at least ideally, as a chiral object and its mirror image. Such objects are called enantiomorphous or simply enantiomorphs. There are enantiomorphous quartz crystals (**Fig. 1**), pine cones, snail shells, screws, shoes etc.

Fig. 1

The genius who first suggested (on the basis of optical activity) that molecules can be chiral was around 1850 Louis Pasteur. He also showed by his famous experiments with tartaric acids that there is a connection between enantiomorphism of crystals and of molecules.

The Swiss painter Hans Erni has drawn for me the paraphernalia necessary for dealing with chirality (fig. 2): human intelligence, a left and a right hand and two enantiomorphous tetrahedra.

To grasp the essence of chirality, it is instructive to withdraw for a moment from the familiar three-dimensional world into a two-dimensional one, into a plane, and enquire what chirality means there. In doing this, we are following in the footsteps of E. A. Abbot, who published his well-known science fiction book "Flatland" about 70 years ago. The simplest chiral figure in "Flatland" is an irregular triangle, a scalene. A scalene can be located in a plane in two different ways so that it displays one or other of its two opposite faces. Two equal scalenes "oriented" differently in a plane cannot be brought into congruence by translation or rotation in two-dimensional space but only by reflection across a straight line, the mirror of "Flatland". They are two-dimensionally enantiomorphous. This holds for any triangle where the vertices are distinctly identified.

Let us now consider an intelligent chiral "Flatlander" who can distinguish right and left and who carries on his front side a device which allows him to receive signals from the identifiable vertices ABC of the two triangles, which are for him not transparent (fig. 3). He will perceive the signals of the first (colorless) triangle in the sequence ACB, CBA, BAC and from the second (black) one in the sequence ABC, BCA, CAB. Thus he will be able to distinguish the two enantiomorphs. However, if one takes them into three-dimensional space they will become indistinguishable. Their nonequivalence gets lost in three-dimensional space.

A planar geometrical figure with more than three vertices can be decomposed into a set of triangles and it can be reconstructed from a set of triangles. Two two-dimensionally chiral triangles can be combined together in a

Fig. 2

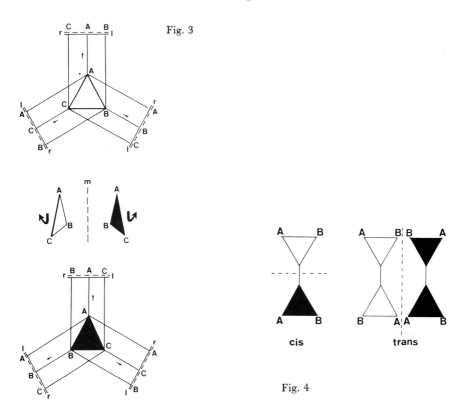

Fig. 3

cis   trans

Fig. 4

plane in two different ways (fig. 4). If they display the same face the combination is chiral. If their faces are different the combination can be made—depending on the symmetry of the combination operation—composite achiral. The two combinations, the chiral and the achiral one, cannot be made congruent, neither by translation and/or rotation nor by reflection; neither in two- nor in three-dimensional space. We call them diastereomorphous or diastereomorphs. Diastereomorphism is not lost in higher dimensions. Thus: chirality is a geometrical property. Enantiomorphism is due to the "orientability" of an object in an "orientable" space. Diastereomorphism is the result of the "mutual orientation" of at least two chiral objects.

These conclusions are valid not only for two-dimensional space but also for spaces of higher dimensions, e.g. our three-dimensional perceptual space, apart from the mathematically trivial limitation that we are not actually able to leave our three-dimensional world—at least the great majority of us.

Familiar planar objects in the two-dimensional information space are capital block letters (fig. 5). Some of them such as A, B, C are two-dimensionally achiral, the others e.g. F, G, J ... are chiral; they cannot be brought in a

ACHIRAL   ABCDEHIKMOTUVWXY

CHIRAL   FGJLNPQRSZ ƨƧЯOꟼИꞀႱႧꟻ

Fig. 5

plane into congruence with their mirror images. In the following discussion these three types of capital block letters will be used to represent all kinds of two- and three-dimensionally achiral and chiral objects and the enantiomorphs of the latter. In the text the somewhat inconvenient mirror image letters will be replaced by barred ones: $\bar{F}$, $\bar{G}$, $\bar{J}$ ... e.g. the shorthand representation for a scalene of for a chiral tetrahedron will be the letter F and for its enantiomorph $\bar{F}$. The chiral combination of two triangles or tetrahedra will be represented by F—F or $\bar{F}$—$\bar{F}$, the achiral one by F—$\bar{F}$.

But let us now switch over to the second part of the title of my lecture—to the chemistry. Chemistry takes a unique position among the natural sciences for it deals not only with material from natural sources but creates the major part of its objects by synthesis. In this respect as stated many years ago by Marcelin Berthelot, chemistry resembles the arts, the potential of its creativity is terrifying.

Although organic chemistry overlaps with inorganic chemistry and biochemistry it concentrates on compounds of the element carbon. So far, about 2 million of organic compounds are registered with innumerable reactions and interconversions, but the number of compounds obtainable by existing methods is astronomic.

Aldous Huxley writes in an essay: "Science is the reduction of the bewildering diversity of unique events to manageable uniformity within one of a number of symbol systems, and technology is the art of using these symbol systems so as to control and organize unique events. Scientific observation is always viewing of things through the refracting medium of a symbol system and technological praxis is always handling of things in ways that some symbol system has dictated. Education in science and technology is essentially education on the symbolic level". If we agree with Huxley, one of the most important aims of organic chemistry is to develop an efficient symbol or model system. Because biochemistry and biology use the same symbol system when working at the molecular level, every progress in this direction is also a progress of these sciences.

In spite of the great number of known and possible facts, chemistry has succeeded in developing in less than 10^{10} s (i.e. 200 years), a system which allows it to keep the "bewildering diversity of events" under control. Compared with the total evolution time of 10^{17} s (3 billion years) this is a remarkably short time, almost a miracle. If the system sometimes does not work perfectly the occasional flaws add to the appeal of organic chemistry for experimentalists and theoreticians in challenging them to improve it.

How does this symbol system work? Organic chemists are mainly interested in pure compounds, i.e. substances which consist of only one molecular species. In polymer chemistry, where this is sometimes not possible, we have to be content to work with compounds built from the same building blocks in a uniform manner. The first important information the organic chemist searches for in a compound is the composition or molecular formula i.e. the kind and number of atoms in the molecule. The second step is to determine the constitution, i.e. which atoms are bound to which and by what types of bond. The result is expressed by a planar graph (or the corresponding connectivity matrix), the constitutional formula introduced into chemistry by Couper around 1858. In constitutional formulae, the atoms are represented by letters and the bonds by lines. They describe the topology of the molecule. Compounds which have the same molecular formulae but different constitution are called isomers. In the late sixties of the last century it was clear that compounds exist which have the same constitution but different properties. One of my predecessors in Zürich, Johannes Wislicenus, expressed the implications of this in a prophetic sentence: "Die Thatsachen zwingen dazu, die verschiedenen Molecüle von gleicher Structurformel durch verschiedene Lagerung der Atome im Raume zu erklären". The prophecy was fulfilled a few years later when, almost simultaneously, a young Dutchman, Jacobus Hendricus van't Hoff (22), and a young Frenchman, Joseph Achilles Le Bel (27) came out with some simple but novel ideas about the "position of atoms in space". These ideas comprised concepts such as asymmetric atom, free rotation etc. Van't Hoff also introduced regular tetrahedra as atomic models from which molecular models could be constructed. This contributed substantially to the rapid propagation of these ideas about chemistry in space, called stereochemistry by Victor Meyer, another of my predecessors in Zürich. The different compounds having the same constitution were called by him stereoisomers, and he distinguished enantiomorphous stereoisomers, enantiomers, and diastereomorphous ones which he named diastereoisomers.

Let us illustrate this by using as example an antibiotic isolated in our laboratories and named boromycin. This is a compound of medium complexity and has the molecular formula $C_{45}H_{74}O_{15}BN$ (fig. 6). The van't Hoff—Le Bel model system allows an average student of chemistry to calculate that the constitutional formula of boromycin corresponds to 262,144 ($=2^{18}$) stereoisomers. This is a rather large number, compared with the 2 million organic compounds which have hitherto been isolated or synthesized by thousands of hard-working chemists during almost two centuries. If a chemist were to set off to synthesize boromycin, he would not get very far from a knowledge of its constitutional formula alone. To approach his goal, he has to know what is the invariant part of its spatial architecture. Moreover, he has to know processes, stereospecific reactions, which produce specifically the desired stereoisomers and not randomly all possible ones.

One problem in dealing with the multiplicity of stereoisomers is that of communication—how to transfer the information about their molecular architecture in space. This can be done, of course, by three-dimensional models (or their projections) constructed on the basis of coordinates obtainable by diffraction methods e.g. by X-ray crystal structure analysis. Such models that describe the complete molecular topography are invaluable for any detailed discussion of the molecules. However, they often include many structural details that are unnecessary for our purpose, i.e. specification of the particular stereoisomer. Indeed, some of these details may be dependent on the state (solid, liquid, vapour, solution) in which the molecule was observed. The very abundance of this information often makes it difficult to recognize, register,

$C_{45}H_{74}O_{15}BN$

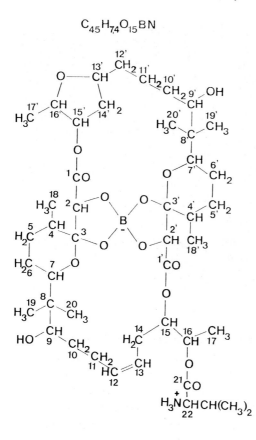

$2^{18} = 262144$ Fig. 6

and memorize that invariant aspect of the topography, the so-called primary structure, which is essential for specification and synthesis of the compound.

In 1954 I joined R. S. Cahn and Sir Christopher Ingold in their efforts to build up a system for specifying a particular stereoisomer by simple and unambiguous descriptors which could be easily assigned and deciphered. This system, which now carries our names, makes it possible to convey the essential information with the aid of a few conventions, letter symbols or numbers. In the rather complex model of boromycin (fig. 7) which contains 136 atoms corresponding to 408 coordinates the primary structure is specified by 18 descriptors. They are:

2R	3R	4R	7S	9S	15S	16R	22R	2'R	3'R	4'R
0	0	0	1	1	1	0	0	0	0	0

7'S	9'R	13'R	15'S	16'R	B R;	12,13 seq cis
1	0	0	1	0	0	(0)

Fig. 7

The letter symbols used in our system always occur in pairs (R, S; M, P; cis, trans) and hence they can be replaced by the numbers 0 and 1. If these numbers are ordered by using the conventional constitutional sequence of the atoms, we obtain a binary number, which can then be expressed in decimal form e.g. for boromycin by $(0)00100100000111000 \rightarrow 18488$. From this number the invariant part of the molecular architecture can easily be retrieved.

In the course of building up and improving our system, many problems emerged with regard to the basis of stereoisomerism and the fundamental concepts of stereochemistry. It was soon evident that by specifying most of the stereoisomers, especially those which were called optical isomers, one specifies their total or partial three-dimensional invariant chiralities. Somewhat later it was recognized that cis-trans isomerism (sometimes misleadingly called "geometrical" isomerism) is a two-dimensional diastereomorphism. For years, the important rôle of two-dimensional chirality had been hidden behind a variety of concepts and words, such as pseudoasymmetry, stereoheterotopy, prochirality, propseudoasymmetry, retention and inversion of configuration, etc. All these partially mysterious concepts can be illuminated by regarding them as manifestations of two-dimensional chirality.

The question "What about one-dimensional chirality, the chirality of Lineland?" can be easily answered. The enantiomorphism of Lineland already is lost in Flatland, and the diastereomorphs of Lineland must have different constitutions in one-dimensional space and are therefore not stereoisomers by definition.

Summarizing and extrapolating, one can claim that the duality inherent in the invisible, intangible two- and three-dimensional chiralities of stable molecules or of their parts is the geometrical basis of all stereoisomerism. Such an uniform point of view towards stereochemistry is not only gratifying for theoretical reasons but has also a heuristic value.

Ever since van't Hoff introduced the regular tetrahedra as a model of the carbon atom, chemists have been solving their daily stereochemical problems by inspection of molecular models. The exhaustive exploration of the possibilities of such models (which are essentially geometrical figures) allowed them to answer practically all questions with regard to the number and symmetry of stereoisomers encountered in their work. A good example is Emil Fischer's classical elucidation of the enigmatic diversity of sugars and their derivatives by applying van't Hoff-Le Bel ideas. This is nicely illustrated by the following paragraph from Fischer's autobiography: "I remember especially a stereochemical problem. During the winter 1890—91 I was busy with the elucidation of the configuration of sugars but I was not successful. Next spring in Bordighera (where Fischer was accompanied by Adolf von Baeyer) I had an idea that might solve the problem by establishing the relation of pentoses to trihydroxyglutaric acids. However, I was not able to find out how many of these acids are possible; so I asked Baeyer. He attacked such problems with great zeal and immediately constructed carbon atom models from bread crumbs and toothpicks. After many trials he gave up because the problem was seemingly too hard for him. Only later in Würzburg by long and careful inspection of good models did I succeed in finding the final solution".

Because of the indubitable success of "playing" with models, stereochemistry developed mainly as a pragmatic science. Several attempts to give it a more theoretical background, by F. M. Jaeger, G. Polya, J. K. Senior, E. Ruch, to mention only a few pioneers, had little influence on the experimentalists in the field.

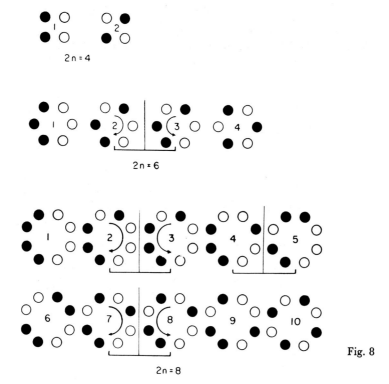

Fig. 8

If one tries to develop a universal system for specification of stereoisomers, as we did, it is somewhat embarrassing to find that one does not actually know what types of steroisomers are possible. During the century which had elapsed since the foundation of stereochemistry several types of stereoisomers were discovered, always as a kind of surprise. To mention only a few: the atropisomerism of polyphenyls and of ansa-compounds, due to the so-called secondary structure, i.e. hindered rotation around single bonds, "geometrical enantiomorphic" isomerism, etc. How many novel types still remained to be discovered? This question is especially relevant when one considers more complex classes of molecules that have not been so thoroughly investigated.

Several years ago Hans Gerlach and I discovered one such novel type, cyclo-stereoisomerism. Head-to-tail combination of equal numbers of enantiomeric building blocks such as (ABF) and (ABF̄) (represented in the following figs. by black and white dots) can lead to cyclic molecules which are either achiral or chiral, depending on the symmetry of the building pattern. Such patterns for the total number of building blocks n ≠ 4, 6 and 8 are shown on fig. 8. For one pattern with n = 6, two enantiomers are possible with different "sense" of the ring (Nos. 2 and 3). We call these cycloenantiomers. There are two pairs of cycloenantiomers with n = 8 (Nos. 2, 3 and 7, 8). With n = 10 (fig. 9) there are already 6 pairs of cycloenantiomers, but in addition to pat-

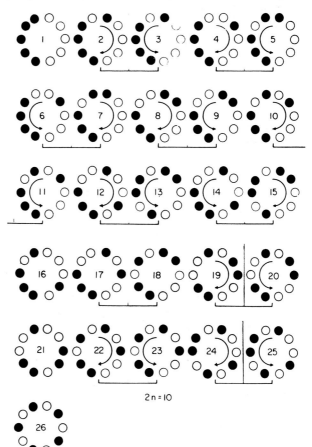

2n = 10

Fig. 9

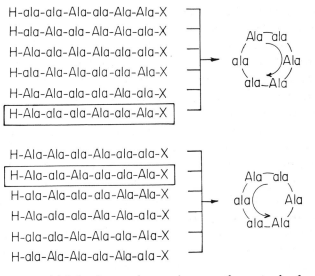

H-ala-ala-Ala-ala-Ala-Ala-X
H-ala-Ala-ala-Ala-Ala-ala-X
H-Ala-ala-Ala-Ala-ala-ala-X
H-ala-Ala-Ala-ala-ala-Ala-X
H-Ala-Ala-ala-ala-Ala-ala-X
H-Ala-ala-ala-Ala-ala-Ala-X

H-Ala-Ala-ala-Ala-ala-ala-X
H-Ala-ala-Ala-ala-ala-Ala-X
H-ala-Ala-ala-ala-Ala-Ala-X
H-Ala-ala-ala-Ala-Ala-ala-X
H-ala-ala-Ala-Ala-ala-Ala-X
H-ala-Ala-Ala-ala-Ala-ala-X

Fig. 10

terns which lead to cycloenantiomers others can be found that give diastereomers on changing the "sense" of the ring (Nos. 4, 6; 5, 7; 12, 14 and 13, 15); these are called cyclodiastereomers. Both types of cyclostereoisomers can be realized in the cyclopolypeptide series. For example, by cyclization of the corresponding penta-alanyl-alanines, two enantiomeric cyclo-hexa-alanyls can be obtained, as shown on fig. 10. With increasing number of building blocks, the number of possible stereoisomers increases considerably: with 15 pairs of enantiometric alanines 5,170,604 stereoisometric cyclo-trikosa-alanyls are possible with the same constitutional formula.

With this in mind, we thought that it might be useful to build up a catalogue of models, based on chirality, which would enable us not only to classify the known stereoisomers but also explore the extent of our present knowledge of stereoisomerism in certain areas.

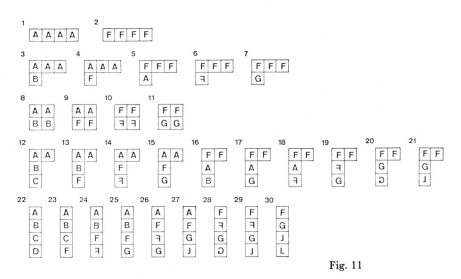

Fig. 11

By showing how to construct one rather trivial page of such a catalogue I may manage to illustrate the general principles. First, all possible different combinations of achiral and chiral objects, including the enantiomorphs of the latter, are selected with the help of partition diagrams, as shown on fig. 11 for number four. In partition diagrams equal objects are in horizontal rows, unequal ones in vertical columns. If the objects in question are parts of molecules we call them ligands. By ocuppying vertices of a polyhedron, in our case a regular tetrahedron, with all combinations of ligands, models are obtained which, according to their symmetry, can be divided into two classes: achiral and chiral. An additional classification into two subclasses is possible by introducing the criterion of permutability. Some of the models do not change if two ligands are permuted, the others are transformed by such a permutation either into their enantiomorphs or diastereomorphs.

Among the achiral models obtained by this procedure, Nos. 5 and 6 shown on fig. 12 are noteworthy because they are models of socalled prochiral and propseudoasymmetric atoms; Nos. 7 and 8 are models of pseudoasymmetric atoms.

achiral

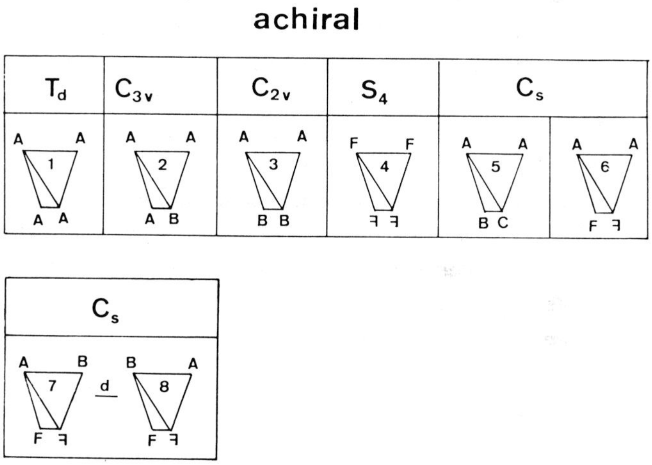

Fig. 12

The enantiomorphs of chiral models on fig. 13 are shown only if they arise by exchange of ligands, as in Nos. 24, 25 and 32, 33. Nos. 24 and 25 are models of the classical "asymmetric atom", the most familiar member of the subclass of "atoms" which are not invariant to permutation (Nos. 24—39).

If one considers that stereochemists have "played" with tetrahedra for more than a century, it is hardly surprising that this catalogue page contains only models of familiar stereoisomers. However, some generalizations are possible. Tetrahedral asymmetric atoms are also called centers of asymmetry or chirality, but such centers are not necessarily occupied by an asymmetric atom (Fig. 14). They can be occupied by atoms with rotational symmetry or the asymmetric atom can be replaced by a rigid atomic skeleton with tetrahedral symmetry such as the adamantane skeleton. The centre of the achiral skeleton of adamantane is a center of chirality which is not occupied by an atom.

chiral

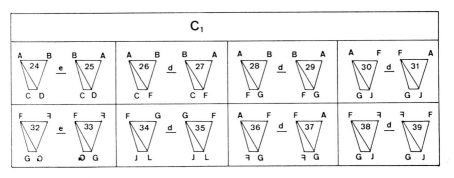

Fig. 13

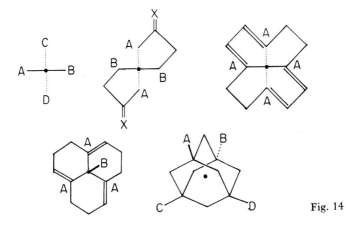

Fig. 14

Van't Hoff had already noticed that there are chiral molecules without centers of chirality and had postulated that allenes with the constitutional formula (AB)C = C = C(AB) are chiral. Models for such cases can be constructed by using as the basic geometrical figure tetrahedra of lower symmetry than that of a regular tetrahedron. Eight point-group symmetries (shown on fig. 15) are possible for a tetrahedron. Three of them (D_2, C_2 and C_1) are intrinsically chiral i.e. their chirality does not depend on how ligands occupy their vertices. The regular tetrahedron (T_d) itself and four others are achiral (D_{2d}, C_{3v}, C_{2v} amd C_s). By occupying the vertices of such tetrahedra with all combinations of four ligands new pages of the catalogue are obtained. These new pages contain some types of stereoisomerism which had escaped the notice of pragmatic stereochemists.

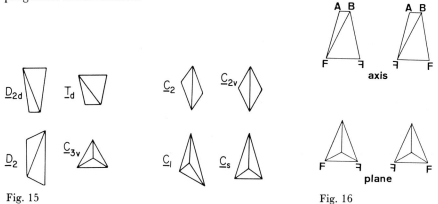

Fig. 15   Fig. 16

I should like to mention only generalized pseudoasymmetric cases with pseudoasymmetric axes and planes, models of which are shown in fig. 16. Examples of stereoisomeric molecules represented by these models have been prepared by Günter Helmchen in our laboratory (Fig. 17 and 18). It is noteworthy that many bilateral organisms including men are examples of planar pseudoasymmetry.

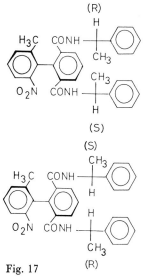

Fig. 17

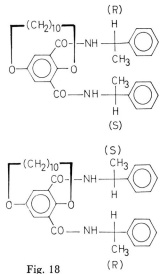

Fig. 18

I have limited the discussion to three-dimensional basic figures with 4 ligands because they are typical for organic stereochemistry. The same procedures can be applied to produce catalogues based on figures with five or more vertices but the multiplicity of models so obtained is larger and therefore more difficult to deal with in a lecture.

The time at my disposition also does not permit me to deal with the manifold biochemical and biological aspects of molecular chirality. Two of these must be mentioned, however, briefly. The first is the fact that although most compounds involved in fundamental life processes, such as sugars and amino acids, are chiral and although the energy of both enantiomers and the probability of their formation in an achiral environment are equal, only one enantiomer occurs in Nature; the enantiomers involved in life processes are the same in men, animals, plants and microorganisms, independent on their place and time on Earth. Many hypotheses have been conceived about this subject, which can be regarded as one of the first problems of molecular theology. One possible explanation is that the creation of living matter was an extremely improbable event, which occured only once.

The second aspect I would like to touch, the maintenance of enantiomeric purity, is less puzzling but nevertheless still challenging to chemists. Nature is the great master of stereospecificity thanks to the *ad hoc* tools, the special catalysts called enzymes, that she has developed. The stereospecificity of enzymic reactions can be imitated by chemists only in rare cases. The mystery of enzymic activity and specificity will not be elucidated without a knowledge of the intricate stereochemical details of enzymic reactions. The protagonist in this field is John Warcup Cornforth.

Chemistry 1976

WILLIAM N. LIPSCOMB

for his studies on the structure of boranes illuminating problems of chemical bonding

THE NOBEL PRIZE FOR CHEMISTRY

Speech by Professor GUNNAR HÄGG of the Royal Academy of Sciences
Translation from the Swedish text

Your Majesties, Your Royal Highnesses, Ladies and Gentlemen,
This year's Nobel Prize for Chemistry has been awarded to Professor William
Lipscomb for his studies on the structure of boranes illuminating problems of
chemical bonding.

A couple of days after the announcement of the chemistry Prize, a Swedish
newspaper carried a cartoon by a wellknown Swedish cartoonist showing an
elderly couple in front of their TV. The legend ran: "Can you remember
ever having seen a borane, Gustav?" This question is quite proper, indeed.
Gustav and his wife certainly had never seen a borane. Boranes do not exist
in nature and can hardly be found in other places than chemical laboratories.

The name borane is the collective term for compounds of hydrogen with
boron, the latter element forming part of among other things boric acid and
borax. A great number of boranes and related compounds are known, and it
is this whole group of substances which has been studied by Lipscomb. In the
eighties of the last century it was understood that such compounds exist in the
gas mixture that is formed when alloys between boron and certain metals are
decomposed by acids. But it was only from about 1912 that the German
chemist Alfred Stock succeeded in producing some pure boranes.

The structures and bonding conditions of boranes remained, however,
unknown until about 1950, and it is not without reason that they have been
considered problematical. The experimental study of the boranes has been
very difficult. They are in most cases unstable and chemically aggressive and
must, therefore, as a rule be investigated at very low temperatures. But it was
still more serious that their structure and bonding conditions were essentially
different from what was known for other compounds. Stock had found borane
molecules that, for example, consisted of in one case two boron and six
hydrogen atoms and in another case ten boron and fourteen hydrogen atoms.
But when the object was to determine how these atoms are bound to each
other, i.e. the appearance of the molecule, and also the nature of the bonds
which keep the atoms together within the molecule, one was left in the dark
for many years. One might suppose that these bonds were similar to those
between the atoms in the hydrogen compounds of carbon, for instance in the
hydrocarbons in liquefied petroleum gas. In these, the bond between two
neighbouring atoms usually involves two electrons, an electron pair. However,
boron does not have as many bonding electrons as carbon and, therefore, all
the bonds cannot be of this type. A new type of bond, where two electrons
co-operate in binding *three* atoms together, and which thus can master this
electron deficiency, was proposed in 1949 but it was not until the researches

of Lipscomb from 1954 and onwards that the problems of borane chemistry could begin to be solved satisfactorily.

Lipscomb has attacked these problems through skilful calculations of the possible combinations within the molecules of conceivable bond types and he has together with his collaborators determined the geometrical appearance of the molecules, above all using X-ray methods. But he has proceeded much farther than that in illuminating the binding conditions in detail through advanced theoretical computations. Thus it became possible to predict the stability of the molecules and their reactions under varying conditions. This has contributed to a marked development of preparative borane chemistry. These studies by Lipscomb have not only been applied to the proper, electrically neutral, borane molecules but also to charged molecules, i.e. ions, as well as other molecules related to the boranes.

It is rare that a single investigator builds up, almost from the beginning, the knowledge of a large subject field. William Lipscomb has achieved this. Through his theories and his experimental studies he has completely governed the vigorous growth which has characterized borane chemistry during the last two decades and which has given rise to a systematics of great importance for future development.

Professor Lipscomb,

You have attacked in an exemplary way the very difficult problems within an earlier practically unknown field of chemistry. You have worked on a broad front using both experimental and theoretical methods and the success of your efforts is shown by the fact that your results and your views have governed the recent development of borane chemistry.

In recognition of your services to science the Royal Swedish Academy of Sciences decided to award you this year's Nobel Prize for Chemistry. To me has been granted the privilege of conveying to you the most hearty congratulations of the Academy and of requesting you to receive your prize from the hands of his Majesty the King.

William N. Lipscomb

WILLIAM N. LIPSCOMB

Although born in Cleveland, Ohio, USA, on December 9, 1919, I moved to Kentucky in 1920, and lived in Lexington through my university years. After my bachelors degree at the University of Kentucky, I entered graduate school at the California Institute of Technology in 1941, at first in physics. Under the influence of Linus Pauling, I returned to chemistry in early 1942. From then until the end of 1945 I was involved in research and development related to the war. After completion of the Ph.D., I joined the faculty of the University of Minnesota in 1946, and moved to Harvard University in 1959. Harvard's recognitions include the Abbott and James Lawrence Professorship in 1971, and the George Ledlie Prize in 1971.

The early research in borane chemistry is best summarized in my book "Boron Hydrides" (W. A. Benjamin, Inc., 1963), although most of this and later work is in several scientific journals. Since about 1960, my research interests have also been concerned with the relationship between three-dimensional structures of enzymes and how they catalyze reactions or how they are regulated by allosteric transformations.

Besides memberships in various scientific societies, I have received the Bausch and Lomb honorary science award in 1937; and, from the American Chemical Society, the Award for Distinguished Service in the Advancement of Inorganic Chemistry, and the Peter Debye Award in Physical Chemistry. Local sections of this Society have given the Harrison Howe Award and Remsen Award. The University of Kentucky presented to me the Sullivan Medallion in 1941, the Distinguished Alumni Centennial Award in 1965, and an honorary Doctor of Science degree in 1963. A Doctor Honoris Causa was awarded by the University of Munich in 1976. I am a member of the National Academy of Sciences U.S.A. and of the American Academy of Arts and Sciences, and a foreign member of the Royal Netherlands Academy of Sciences and Letters.

My other activities include tennis and classical chamber music as a performing clarinetist.

THE BORANES AND THEIR RELATIVES

Nobel Lecture, December 11, 1976
by
WILLIAM N. LIPSCOMB
Harvard University, Cambridge, Massachusetts, USA

This year, 1976, the Nobel Prize in Chemistry has been awarded for research in pure inorganic chemistry, in particular the boranes. May I say that I am most pleased and profoundly grateful. My own orientation to this field has been, as it has in all of my studies, the relationships of the chemical behavior of molecules to their three-dimensional geometrical and electronic structures. The early work on the molecular structures of boranes by X-ray diffraction led to a reasonable basis for a theory of chemical bonding different from that which is typical in carbon chemistry, and yielded an understanding of the pleasing polyhedral-like nature of these compounds. Assimilated by the preparative chemists, the principles helped to establish a large body of a hitherto unknown chemistry, which made a reality of the expectation that boron, next to carbon in the periodic table, should indeed have a complex chemistry.

In these nearly thirty years both the theoretical and experimental methods have been applied by us and others to areas of inorganic, physical, organic and biochemistry. For examples, these areas include low temperature X-ray diffraction techniques, and the theoretical studies of multicentered chemical bonds including both delocalized and localized molecular orbitals. An early example is extended Hückel theory, originally developed for studies of the boranes, and even now one of the most widely applicable approximate methods for theoretical studies of bonding in complex molecules. More soundly based theories are presently in use by my research students for studying how enzymes catalyze reactions, details of which are based on the three-dimensional structures by X-ray diffraction methods. Besides illuminating particular problems, these developments may contribute toward the redefinition of areas of chemistry, and thereby broaden the chemist's view. Our research in the boranes and their related molecular species crosses areas of inorganic, experimental physical, theoretical and organic chemistry, and includes applications in biochemistry. More simply stated, the area is the study of the relationships of molecular structure to function.

BORANES, AND EARLY STRUCTURE STUDIES

By now, large numbers of chemical compounds related to polyborane chemistry exist: boron hydrides, carboranes, metalloboranes, metallocarboranes, mixed compounds with organic moieties, and others. These discoveries of preparative chemists are relatively recent. Long ago, Alfred Stock established borane chemistry. He developed the experimental techniques which were required for the preparation of the volatile and potentially explosive compounds, B_2H_6, B_4H_{10}, B_5H_9, B_5H_{11} and B_6H_{10}, and the relatively stable white crystalline $B_{10}H_{14}$. This work, beautifully summarized in his Baker Lectures,[1] was celebrated at the Third International Meeting on Boron Chemistry this past July in Munich, 100 years after his birth. Sidgwick[2] wrote, "All statements about the hydrides of boron earlier than 1912, when Stock began to work on them, are untrue."

Aside from the classification as either an N+4 or an N+6 series, no simple basis for these unusual formulas was foreseen before their structures were established. A B_6 octahedron (Fig. 1) was known in certain crystalline borides, and the B_{12} iscosahedron (Fig. 2) was found in boron carbide, but no one realized that a systematic description of the boron arrangements in these hydrides might be based on fragments of these polyhedra. Most electron diffraction work before 1940 supported more open structures, which Pauling[3] described in terms of resonating one-electron bonds. In 1940—1941 Stitt[4] produced infrared and thermodynamic evidence for the bridge structure of B_2H_6 (Fig. 3). More general realization of the correctness of this structure followed the theoretical studies,[5-8] and especially the infrared work of Price.[9] The three-center bridge BHB bond was clearly formulated by Longuet-Higgins.[8]

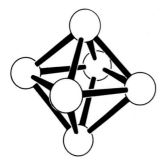

Fig. 1. The B_6 octahedron, which occurs in certain metal borides, in which each boron atom is bonded externally to a boron atom of another octahedron. In $B_6H_6^{-2}$ or $C_2B_1H_6$, a terminal hydrogen atom is bonded externally to each B or C atom.

Fig. 2. The B_{12} icosahedron, which occurs in boron carbide $B_{12}C_3$, in elementary boron, and in $B_{12}H_{12}^{-2}$. The three isomers of $C_2B_{10}H_{12}$ also have this icosahedral arrangement in which there is one externally bonded hydrogen on each B or C atom.

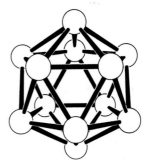

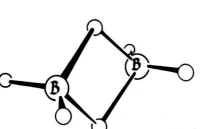

Fig. 3. The geometrical structure of B_2H_6.

The first of the higher hydrides to be structurally characterized was $B_{10}H_{14}$ (Fig. 4). Kasper, Lucht and Harker[10] showed that the boron arrangement was a fragment of a B_{12} icosahedron in which there are four bridge BHB bonds in the open face. Next was B_5H_9,[11, 12] (Fig. 5), a fragment of the B_6 octahedron, and then B_4H_{10}[13, 14] (Fig. 6), a B_4 unit from these polyhedra. Both of these structures were established by X-ray diffraction in our laboratory, and by electron diffraction at the California Institute of Technology. Our X-ray diffraction results on B_5H_{11}[15] (Fig. 7) and tetrahedral B_4Cl_4[16] then set the stage for the beginning of the theory of bonding. The B_6H_{10} structure,[17, 18] (Fig. 8) completing the compounds found by Stock, was to be one of our later X-ray diffraction studies, which were to include many other boranes and related molecules.

It was actually in 1946 that I decided to enter this area of inorganic chemistry. At that time, no reliable methods were generally available for accumulation of large amounts of X-ray diffraction data at low temperatures. Our methods[19] paralleled those in Fankuchen's laboratory[20] at the then Polytechnic Institute of Brooklyn, but were quite independent. Because of the special difficulties of working with the volatile boranes, I chose among other topics, a series of X-ray diffraction studies of single crystals grown at low temperatures from low melting liquids for which putative residual entropy problems

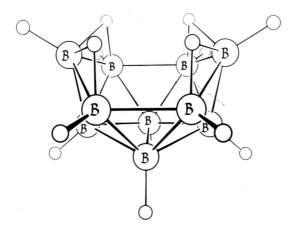

Fig. 4. $B_{10}H_{14}$.

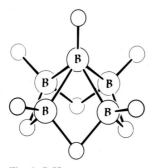

Fig. 5. B_5H_9.

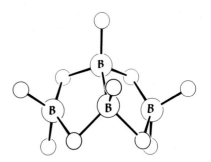

Fig. 6. B_4H_{10}.

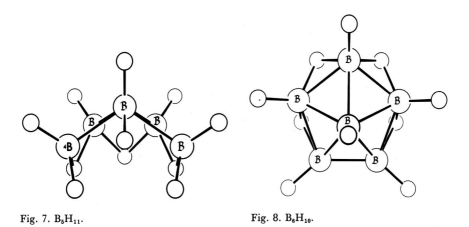

Fig. 7. B_5H_{11}.   Fig. 8. B_6H_{10}.

existed: N_2O_2, CH_3NH_2, CH_3OH, N_2H_4, $COCl_2$ and H_2O_2. The subsequent low temperature studies of single crystals of these volatile and unstable boranes were not without hazards. Vacuum line techniques were learned as we needed them. Fortunately, no serious injuries were incurred as a result of several explosions resulting from cracks in these vacuum systems. I was relieved, on one occasion, when I had taken Russell Grimes to a hospital in Cambridge after one of these explosions to hear the doctor tell me, "Louis Fieser sends me much more interesting cases than you do." I still have in my office the air gun which I, or my young son, used on a number of occasions to destroy a cracked vacuum system from a safe distance. We also had chemical surprises, for example, when we found a presumed B_8 hydride to be B_9H_{15}.[21] Our only chemical analysis of this compound was the count of the numbers of boron and hydrogen atoms in the electron density map, which was calculated in those days from rough visual estimates of intensities of diffraction maxima on films.

THREE-CENTER BONDS AMONG BORON ATOMS

At the fortunate time in 1953 of W. H. Eberhardt's sabbatical, he, Crawford and I examined[22] the open boron hydrides B_2H_6, B_4H_{10}, B_5H_9, B_5H_{11} and $B_{10}H_{14}$ from the viewpoint of three-center bonds; and we studied B_5H_9, the unknown polyhedral molecule B_4H_4, the then hypothetical ions $B_6H_6^{-2}$ (Fig. 1) and $B_{12}H_{12}^{-2}$ (Fig. 2) from the viewpoint of molecular orbitals. Longuet-Higgins[23] also, independently, formed an early molecular orbital description almost like ours. One of the simple consequences of these studies was that electron deficient molecules, defined as having more valence orbitals than electrons, are not really electron deficient. I mean by this non-sequitur that the three-center two-electron bonds make possible a simple description of these molecules and ions as filled orbital species. Filled molecular orbitals were later extended to closed polyhedral compounds B_nH_n for all values of n from 5 to 24, all of which have a formal charge of -2, even though hypothetical B_4H_4 is neutral. In fact, all of the experimentally known ions, for $6 \leqslant n \leqslant 12$, do have -2 charge.

In addition to $B_6H_6^{-2}$ (Fig. 1) and $B_{12}H_{12}^{-2}$ (Fig. 2), those polyhedra for $5 \leqslant n \leqslant 10$ are shown in Fig. 9. The isoelectronic series $C_2B_{n-2}H_n$ is known for $5 \leqslant n \leqslant 12$.

Equations for the atom, orbital and electron balance were formulated in our paper of 1954, allowing prediction of many new chemical species. One simple form of these rules is exemplified for a neutral hydride formula B_pH_{p+q} in which each boron atom has at least one terminal hydrogen atom. Define the number of BHB bridges as s, the number of three-center BBB bonds as t, the number of two-center BB bonds as y and the number of extra terminal hydrogens on each BH unit as x. Then

$$s + x = q$$

$$s + t = p$$

$$p = t + y + q/2$$

The first equation is the hydrogen balance. The second comes from the fact that each of the p boron atoms supplies four orbitals and only three electrons, and the extra orbital is utilized in one of the two types of three-center bond. Finally, if each BH unit is recognized as contributing a pair of electrons, these p pairs are used in BBB or BB bonds, or in half of the pairs required for the extra hydrogens, s and x. These rules, and the accompanying valence structures, are especially helpful in describing those polyboron compounds which are open, but they are also useful for closed polyhedral molecules and ions.

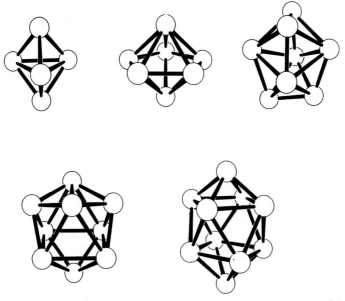

Fig. 9. Structures for the boron and carbon arrangements in some of the polyhedral molecules $C_2B_{n-2}H_n$, which are known for $5 \leqslant n \leqslant 12$. The isoelectronic anions are known for $6 \leqslant n \leqslant 12$.

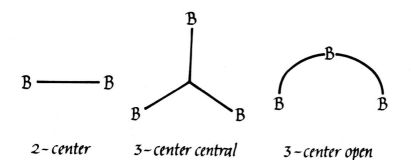

2-center   3-center central   3-center open

Fig. 10. The three types of bonds within the boron framework of a borane structure, or an equivalent carborane structure, are a BB bond, and the two types of three-center bonds. As shown below, the open three-center bond is known only for BCB bonds, not BBB bonds.

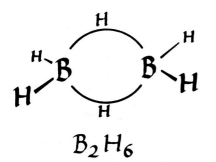

B_2H_6

Fig. 11. Bonds in B_2H_6, according to three-center bond theory.

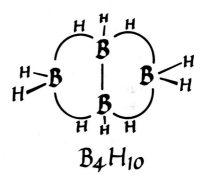

B_4H_{10}

Fig. 12. Bonds in B_4H_{10}.

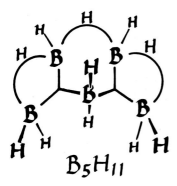

B_5H_{11}

Fig. 13. Bonds in B_5H_{11}.

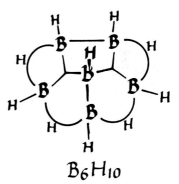

B_6H_{10}

Fig. 14. Bonds in B_6H_{10}.

There are two kinds of three-center bonds among three boron atoms (Fig. 10). The central three-center bond involves positive overlap among hybrid orbitals from each of three boron atoms, while the open three-center bond involves, on the central atom, a π orbital which overlaps in a bonding manner with an orbital from each of the adjacent boron atoms. The less compact B_2H_6, B_4H_{10}, B_5H_{11} and B_6H_{10} structures (Figs. 11—14) are well described with the use of these three-center bonds, omitting open three-center BBB bonds for the moment. However, B_5H_9 (Fig. 15) requires a resonance hybrid of four valence structures, related by the four-fold symmetry; and $B_{10}H_{14}$ (Fig. 16) requires a resonance hybrid of 24 valence structures.

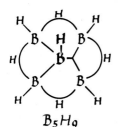

B_5H_9

One of 4 resonance forms

Fig. 15. One of four resonance structures for B_5H_9. The other three are obtained by reorientation of the framework bonds by 90° about the molecular four-fold axis.

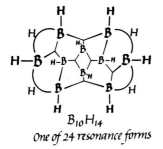

$B_{10}H_{14}$

One of 24 resonance forms

Fig. 16. One of the 24 resonance structures for $B_{10}H_{14}$, omitting open three-center bonds.

Three-center bond theory was further developed in the following 22 years.[24] I realized that the valence rules implied the existence of a large body of boron chemistry, and then ventured predictions, some of which were actually verified experimentally. Dickerson[25] and I formalized my intuitive approach into a theory of connectivity of various bonding patterns within the three-center description. Also, geometrical constraints were introduced in order to avoid overcrowding of hydrogen atoms and to preserve known bonding angles about boron atoms among the boranes.[26] More recently, Epstein[27] and I reformulated the topology using central three-center BBB bonds, to the exclusion of open three-center BBB bonds. The recent work on localized molecular orbitals which has led to the restriction of open three-center bonds, so far, to BCB bonds is exemplified in a later part of this manuscript.

MOLECULAR ORBITAL STUDIES OF BORANES

Molecular orbitals are more appropriate for describing the valence structures of the polyhedral molecules and ions, and the more compact polyhedral fragments. Two simple examples will suffice. In B_5H_9 there are three pairs of electrons in the boron framework, which has four-fold symmetry. The bonding is beautifully described without resonance, by a simple σ and π set of molecular orbitals[22, 24, 28] (Fig. 17). These orbitals are similar to the bonding orbitals, for example, between planar four-fold cyclobutadiene (C_4H_4) and a CH^+ placed along the four-fold axis to give the carbonium ion $C_5H_5^+$.

A similar σ, π situation occurs[28] if a BH unit is removed from icosahedral $B_{12}H_{12}^{-2}$ leaving $B_{11}H_{11}^{-2}$ having five atomic orbitals containing four electrons (Fig. 18). This set of five orbitals gives σ, π molecular orbitals which can bond to a similar set of σ, π orbitals from another atom or group of atoms supplying two more electrons. I suggested, conceptually, adding H_3^+ to predict $B_{11}H_{14}^-$, for example. Although recognizing the similarity of this set of orbitals to those in C_5H_5, which was known to form ferrocene $Fe(C_5H_5)_2$, I did not then go quite so far as to suggest bonding of this $B_{11}H_{11}^{-2}$ fragment to a transition metal. Later, Hawthorne did so, using these ideas as a starting point, and thereby created the large family of metalloboranes and metallocarboranes[29] (Fig. 19).

In the early 1960's, when large scale computing facilities became available to us, Roald Hoffmann and Lawrence Lohr independently programmed the

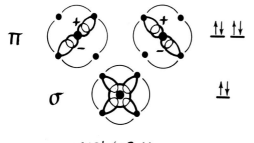

MO's in B_5H_9

Fig. 17. Symmetry molecular orbitals in the boron framework of B_5H_9. The σ molecular orbital is a five-centered one, and each of the π components is actually an open three-center bond.

Fig. 18. Five atomic orbitals, containing four electrons remain when a neutral BH unit is removed from the apex of $B_{12}H_{12}^{-2}$, or $C_2B_{10}H_{12}$. These five atomic orbitals form a σ(bonding), π (bonding) and δ (antibonding) set of five molecular orbitals, like those in the π electron system of $C_5H_5^-$.

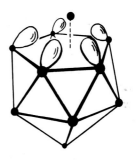

Five AO's to apex in $B_{12}H_{12}^{-2}$

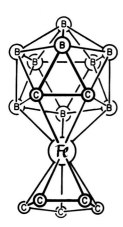

Fig. 19. The structure of $(B_9C_2H_{11})Fe(C_5H_5)$, a metallocarborane derivative.

extended Hückel method in my research group. Its first application was to boron chemistry,[30] where, particularly in the carboranes (compounds of boron, carbon and hydrogen), the charge distributions predicted sites of electrophilic and nucleophilic substitution. One of the rather simple rules which emerged was that nucleophilic attack occurs in a polyhedral carborane at a boron atom closest to carbon, while electrophilic attack was preferred at a boron furthest removed from carbon. Experimental studies of boranes and carboranes in which one or more hydrogens had been replaced by a halogen or by an amine group confirmed these predictions. The extended Hückel method became for a time the most widely used program for the study of molecular theory and reactions in complex organic and inorganic molecules. For example, the Woodward-Hoffmann rules, and related orbital concepts, were tested by their inventors[31] with the use of this theory.

I thought then that progress in structure determination, for new polyborane species and for substituted boranes and carboranes, would be greatly accelerated if the [11]B nuclear magnetic resonance spectra, rather than X-ray diffraction, could be used. One approach was empirical,[32] while the other was purely theoretical. This latter approach required the development of a theory for fairly reliable calculation of chemical shifts of [11]B from first principles of quantum mechanics.[33] This theory, the coupled Hartree-Fock method for a molecule in a magnetic or electric field, yielded molecular constants good to a few to several percent[34] for many diatomic molecules. A striking result was the prediction of the paramagnetic nature of diatomic BH.[35] However, the application of this method to the complex boranes still lies in the future, even after 13 years of effort. While we understand some of the contributions to chemical shift, diamagnetic and large temperature-independent paramagnetic effects, the use of this method for structure determination of complex polyboranes is still somewhat limited.

These programs yielded accurate self-consistent field molecular orbitals, which were explored in other areas of chemistry, as well as in the boranes and carboranes. One example is the first accurate calculation of the barrier to internal rotation in ethane.[36] At this point, my research students and I set out to explore the gap between extended Hückel theory and self-consistent field theory. We achieved by stages[37, 38] a molecular orbital theory[39] still being extended, which was applicable to large polyboranes and other molecules, and which had essentially the accuracy of self-consistent field methods. Molecular orbital methods which do not go beyond symmetry orbitals tend to make each molecule a separate case. Hence we began only a few years ago to explore the connections between these symmetry orbitals and the three-center bonds of the previous section.

LOCALIZED MOLECULAR ORBITALS

Ordinarily, molecular orbitals are classified according to their symmetry types. However, it is possible to make linear combinations of molecular orbitals of different symmetries in such a way that the total electron density of the molecule remains invariant. The most popular methods are those of Edmiston and Ruedenberg[40] who maximize

$$\sum_i \iint \Phi_i(1)\,\Phi_i(1)\frac{1}{r_{12}}\,\Phi_i(2)\,\Phi_i(2)\,dV_1 dV_2$$

and of Boys[41] who minimizes

$$\sum_i \iint \Phi_i(1)\,\Phi_i(1)r^2{}_{12}\,\Phi_i(2)\,\Phi_i(2)\,dV_1 dV_2$$

These procedures, respectively, maximize the repulsions of those electron pairs within each molecular orbital, and minimize the orbital self-extension of each electron pair. Thus, in slightly different manner, the symmetry orbitals are converted into linear combinations which are a good approximation to the localized electron-pair bond. These objective procedures, without adjustable parameters, have been compared in some detail.[42] They provide support of three-center bond descriptions within a theoretical chemical framework. Also, they test the level at which two-center and three-center bonds require some further delocalization. These studies have also provided preferred descriptions of valence structures, often eliminating or reducing the need for resonance hybrids.

Open three-center and central three-center BBB bonds are almost equivalent descriptions. Including both types there are 111 resonance structures for $B_{10}H_{14}$, of which 24 are based only on central three-center bonds. In B_5H_{11} (Fig. 20) two slightly unsymmetrical central three-center BBB bonds are

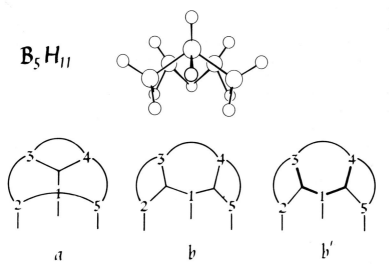

Fig. 20. Three valence structures for B_5H_{11} include (a) the disfavored open three-center bond, (b) the symmetrical central three-center bonds, and (b') the localized slightly un-symmetrical three-center bonds. For clarity, one terminal B-H bond has been omitted at each boron site in (a), (b) and (b').

found[43] by both localization procedures (Fig. 20b'), whereas the open three-center description is not favored. In a series of studies of localized molecular orbitals in all of the boron hydrides, carboranes, and ions of known geometry, we have never found an open three-center BBB bond. However, open three-center BCB bonds do exist. For example, the open three-center BCB bond[44] occurs twice in $1,2-C_2B_4H_6$ (Fig. 21), when the Edmiston-Ruedenberg procedure is used. A comparison of these results with those of the Boys procedure is given below. In the simplest molecular orbital description of an open three-center bond, the electron pair is distributed as e/2, e, e/2 among these three atoms. It is probable that the extra nuclear charge of carbon stabilizes this distribution, to give open three-center bonds rather than nearly equivalent central three-center bonds, when carbon is the middle atom.

Another new general result is that almost every single bond within a triangulated borane or carborane framework shows some donation to the nearest adjacent atoms (Fig. 22). This tendency for multi-centered bonding usually involves 10 per cent or less of the electron pair. In B_4H_{10}, about 0.2 e is donated[43] from the single BB bond to each of atoms B_1 and B_4, which themselves are relatively electron deficient because of the open three-center bonds of the hydrogen bridges (Fig. 23). This donation then causes these

Fig. 21. Localized molecular orbitals are shown in the framework of $1,2-C_2B_4H_6$. The open three-center bonds go through carbon. An external hydrogen has been omitted at each B or C atom.

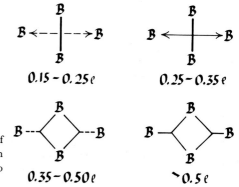

Fig. 22. Notation is shown for amounts of single bond donation to adjacent boron atoms. Unsymmetrical donation may also occur.

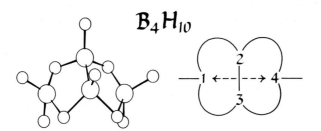

Fig. 23. Single bond donation in B_4H_{10} is about 0.19e from the B_2B_3 bond to each of the outer doubly bridged BH_2 groups of the molecule.

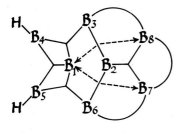

$$B_8 H_{13}^-$$

Fig. 24. Single bond donation is shown in $B_8H_{13}^-$.

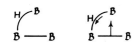

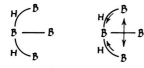

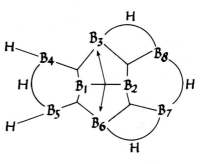

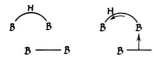

$$B_8 H_{14}$$

Fig. 25. Donation occurs from the B_1B_2 single bond in B_8H_{14}.

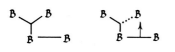

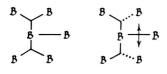

Fig. 26. Some common patterns of single bond donation cause hydrogen atom displacement and electron withdrawal from the atom to which donation occurs.

hydrogen bridges to become unsymmetrical, displaced toward B_2 and B_3, in accord with the results obtained from our X-ray diffraction study. More generally, this donation causes electron withdrawal along other bonds in the molecule, whether or not hydrogen bridges are available for accommodating this electron displacement. Two additional examples are illustrated in $B_8H_{13}^-$ and in the predicted structure of B_8H_{14}, respectively[45] (Fig. 24 and 25). I have abstracted from these examples some typical modes of single bond donation, and accompanying electron withdrawal or hydrogen displacement, in Fig. 26. These and similar valence diagrams may be useful in understanding intra-molecular distortions and reaction mechanisms.

A new type of bonding, which is conceptually transferable among these molecules, is described as fractional three-center bonds.[46] Its simplest interpretation is the replacement of a pair of resonance structures by a single valence structure (Fig. 27). The fractional use of orbitals, indicated by dotted lines, increases the apparent number of bonds to a given boron above the usual four, in an element in the first row of the periodic table. However, the Pauli exclusion principle is not violated because less than a full atomic orbital is required at a fractional bond. Indeed, the localized molecular orbitals are themselves derived from wavefunctions for which the exclusion principle is rigorously introduced. The use of fractional bonds to reduce the number of resonance hybrids, often to a single preferred structure, is not limited to boranes and their relatives. Our rather extensive studies of bonding among atoms other than boron have indicated that these simplified localized molecular orbitals may be an informative and useful alternative to the more conventional valence bond and molecular orbital descriptions in other parts of the periodic table. A very simple example, carboxylate anion,[47] is shown in Fig. 28.

Fractional bonds to atom B_2 in 4,5—$C_2B_4H_8$ are preferred[48] over a resonance hybrid of a single and a central three-center bond, and also are preferred over the open three-center bond (Fig. 29). Also, the bonding of the two carbon atoms in this carborane by both a single bond and a central three-center bond has not been found in boranes, and, in particular, does not occur in B_6H_{10} which

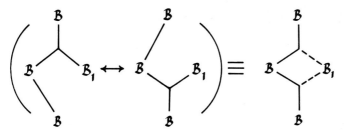

Fig. 27. Equivalence is shown of the two fractional three-center bonds to a resonance hybrid of a single bond and a three-center bond.

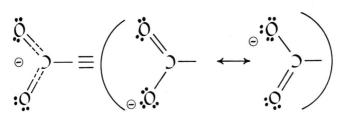

Fig. 28. The four localized molecular orbitals in carboxylate anion have a charge distribution which shows displacement toward oxygen. On the left, the equivalent resonance hybrid is given.

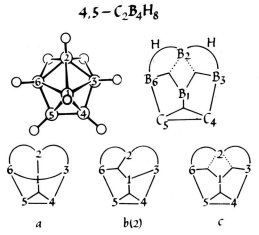

$4,5-C_2B_4H_8$

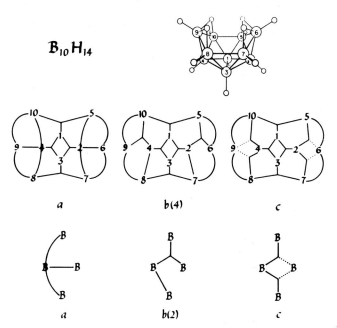

Fig. 29. Localized orbitals in $4,5\text{-}C_2B_4H_8$ show fractional donation to atom B_2 as dotted lines.

$B_{10}H_{14}$

Fig. 30. The pattern of fractional bonds, dotted toward B_6 and B_9, in $B_{10}H_{11}$ can be expected when there is resonance of the type shown in the central diagram, or when one is tempted to draw open three-center BBB bonds as shown on the left. The fractional bond description is preferred.

is isoelectronic with $4,5\text{---}C_2B_4H_8$. Fractional bonds give a particularly simple valence description[45] of bonding in $B_{10}H_{14}$ (Fig. 30), where the single valence structure (Fig. 30c) replaces a resonance hybrid of 24 central three-center bond structures. A very similar simplification[45] occurs in $B_{10}H_{14}^{-2}$, where the two pairs of fractional bonds are dotted toward atoms B_2 and B_4, rather than toward B_6 and B_9.

Similar pairs of fractional bonds are found by Boys' localization procedure in the polyhedral carboranes $1,2-C_2B_4H_6$ (Fig. 31) and $1,7-C_2B_{10}H_{12}$ (Fig. 32). Perhaps this procedure tends to exaggerate the separation of charge centroids of bonds when they lie at or near a single center. For this reason, we tend to prefer[42] the almost equivalent open three-center bonds, as found by the Edmiston-Ruedenberg procedure in $1,2-C_2B_4H_6$. When it becomes economically feasible to test this alternative in $1,7-C_2B_{10}H_{12}$, I would guess that localized open three-center bonds will be found, centered at carbon atoms, in this molecule, like those in $1,2-C_2B_4H_6$.

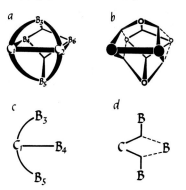

Fig. 31. In $1,2-C_2B_4H_6$ the Edmiston-Ruedenberg localization yields open three-center bonds at carbon, while the Boys localization gives fractional three-center bonds. The latter procedure gives greater emphasis to separation of orbital centroids when they are on the same atom.

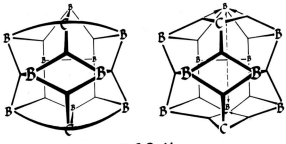

1,7 $C_2B_{10}H_{12}$

Fig. 32. A situation parallel to that of Fig. 31 may occur in $1,7-C_2B_{10}H_{12}$, where the Boys localization (right side) shows fractional bonds. The open three-center bond description may be found by the Edmiston-Ruedenberg procedure, if and when the calculations can be made.

Fractional three-center bonds are not always unique, particularly in aromatic hydrocarbons, or in those boranes which have a valence pattern similar to that in the aromatics. An example of this orientational ambiguity occurs in the boron framework bonds in B_5H_9 (Fig. 33). Actually, the two valence structures represent extremes over a $45°$ range of orientation angle, about the four-fold axis. There is a continuum of valence structures between these extremes. Moreover, this ambiguity continues throughout the whole $360°$, in accord with the four-fold symmetry which the electron density must have in this molecule. All of these valence structures are equally preferred, and each has a total electron density which is consistent with the four-fold molecular symmetry.

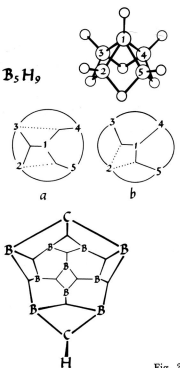

B_5H_9

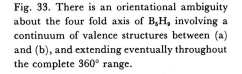

Fig. 33. There is an orientational ambiguity about the four fold axis of B_5H_9 involving a continuum of valence structures between (a) and (b), and extending eventually throughout the complete 360° range.

a b

$C_2B_{10}H_{13}^-$

Fig. 34. There is little fractional bonding in the localized orbitals of the nearly polyhedral $C_2B_{10}H_{13}^-$ ion.

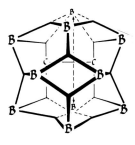

$1,2\ C_2B_{10}H_{12}$

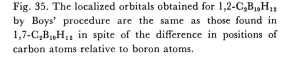

Fig. 35. The localized orbitals obtained for $1,2\text{-}C_2B_{10}H_{12}$ by Boys' procedure are the same as those found in $1,7\text{-}C_2B_{10}H_{12}$ in spite of the difference in positions of carbon atoms relative to boron atoms.

In more complex molecules one can find various degrees of simplicity in the bonding patterns of localized molecular orbitals. In $C_2B_{10}H_{13}^-$, which has a somewhat open near-icosahedral C_2B_{10} arrangement, the bonding is especially simple,[49] not requiring resonance or appreciable fractional bonding (Fig. 34). In $1,2\text{—}C_2B_{10}H_{12}$, another simple idea occurs: the bonding, as found by Boys' procedure, is to a good approximation just like that in $1,7\text{—}C_2B_{10}H_{12}$ in spite of the very different positions of the two carbon atoms relative to each other and to their boron neighbors (Fig. 35). Thus there is some tendency for bonding invariance in closely related geometrical structures. In iso-$B_{18}H_{22}$, we find[50] both single bond donation, from B_4B_8 and $B_{14}H_{18}$, and fractional bonds, to B_6 and B_{16} (Fig. 36). Here, the single bond donations are greater toward B_9, which is relatively electron deficient, than toward B_3 and B_{13}. In $B_{20}H_{16}$ there is a remarkable amount of single-bond donation, particularly toward the electron deficient B_9—B_{16} atoms, inclusive (Fig. 37), and the valence structure of each half is clearly like that in $B_{10}H_{14}^{-2}$ (Fig. 38) rather than like that in $B_{10}H_{14}$ (Fig. 30).

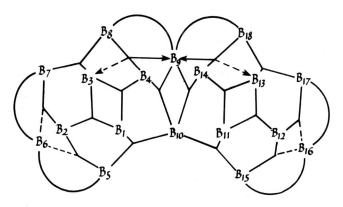

Fig. 36. Single bond donation and fractional bonds occur in iso-$B_{18}H_{22}$, a molecule having a two-fold axis only, and no terminal hydrogen atoms on atoms B_9 and B_{10}.

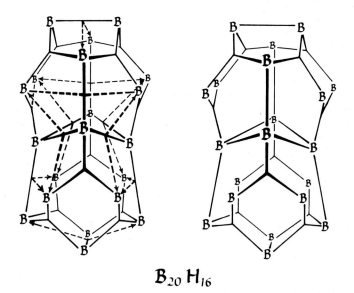

$$B_{20}H_{16}$$

Fig. 37. Donation and non-donation localized orbitals are shown for $B_{20}H_{16}$, in which the four borons nearest the equatorial belt have no terminal hydrogens. Boron atoms B_9B_{16}, inclusive, are particulary electron deficient, and receive substantial donation from single bonds. Each of these eight boron atoms has only two framework bonds in the figure at the right.

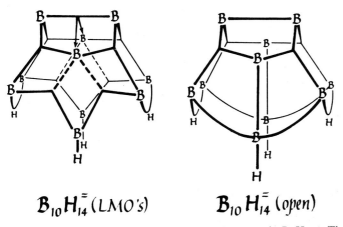

$$B_{10}H_{14}^{=} (LMO's) \qquad B_{10}H_{14}^{=} (open)$$

Fig. 38. A relationship similar to that in $B_{20}H_{16}$ occurs in $B_{10}H_{14}^{-2}$. The preferred localized orbital structure is shown on the left side, while the open three-center bond structure without donation is an oversimplified valence structure.

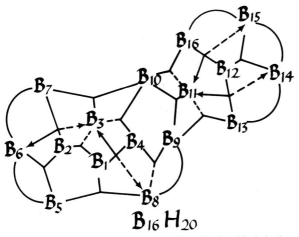

$$B_{16}H_{20}$$

Fig. 39. A rather complex pattern of localized orbitals is shown for $B_{16}H_{20}$. Bond donations can be withdrawn by the reader in order to discover the preferred valence bond structures of the simpler theory. This single valence structure replaces a resonance hybrid of 216 valence structures based upon central three-center bonds and single bonds. Atoms B_9 and B_{10} do not have terminal hydrogens.

The rather complex localized molecular orbitals in $B_{16}H_{20}$ (Fig. 39) are dominated by the three-center bond approximation; in addition they are modified by single bond donation accompanied by electron withdrawal from atoms to which donation occurs.[50] With recognition that atoms B_9 and B_{10} have no terminal hydrogens attached to them, I offer the reader a challenge to find the close relative valence structures which do not have fractional bonding or single bond donation in this molecule.

Finally, the bonding in reaction intermediates is a new area of study, primarily by purely theoretical methods. In BH_5 (Fig. 40) there is only a very weak donor bond from H_2 to the vacant orbital of BH_3.[51-55] This weakness is probably due to the absence of a pathway for back donation. On the other hand, in the dimerization of two BH_3 molecules to form B_2H_6, donation of electron density from one terminal H to B of the other BH_3 group is balanced by a symmetrically related donation toward the first BH_3 group[56] (Fig. 41). Our other recent progress in theoretical studies of bonding in reaction intermediates, such as B_3H_7 and B_4H_8, shows that the more stable transient species may have a vacant orbital[57] on one boron atom, because of the strain involved when that vacant orbital is filled by converting a terminal BH bond to a

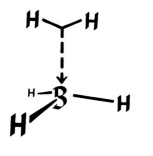

Fig. 40. Fractional donation of a single bond between two hydrogen atoms to the vacant orbital of a slightly pyramidal, nearly planar, BH_3 group. The resulting BH_5 molecule is a very short-lived reaction intermediate.

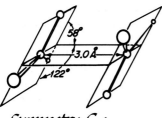

Symmetry C_{2h}

Fig. 41. Approximate geometry of the transition state when two BH_3 molecules form B_2H_6. This geometry is favored over that in an unsymmetrical approach of two BH_3 molecules.

bridge BHB bond. Actually less is known of the detailed reaction mechanisms for reactions of boranes and their related compounds than is known for organic reactions of comparable complexity. Surely, this is a fruitful area for future research.

I shall stop here, omitting descriptions of bonding in large polyhedral borane anions and other related compounds. Also, polyhedral rearrangements, hydrogen atom tautomerism, and particularly the use of bonding theory in bringing some degree of order to chemical transformations of the boranes have been omitted. Attention has thus been concentrated on those aspects of chemical bonding which have been especially illuminated by the molecular and crystal structures that we and others have studied over these many years.

My original intention in the late 1940's was to spend a few years understanding the boranes, and then to discover a systematic valence description of the vast numbers of "electron deficient" intermetallic compounds. I have made little progress toward this latter objective. Instead, the field of boron chemistry has grown enormously, and a systematic understanding of some of its complexities has now begun.

It remains to give credits where they really belong, to my research associates: graduate students, undergraduates, postdoctoral fellows and my other colleagues who have coauthored nearly all of these studies. For the figures of this manuscript, and of the lecture, I thank Jean Evans. I am most grateful to the Office of Naval Research who supported this research during the period from 1948 to 1976, a remarkably long time. I am most aware of the great influence of Linus Pauling on my whole scientific career. Finally, this manucript is dedicated to the memory of my sister, Helen Porter Lipscomb, composer, teacher and performer.

REFERENCES

1. Stock, A. 1933. *Hydrides of Boron and Silicon*. The George Fisher Baker Non-Resident Lectureship in Chemistry at Cornell University. Cornell University Press, Ithaca, New York.

2. Sidgwick, N. V. 1950. *The Chemical Elements and their Compounds*. The Clarendon Press, Oxford, p. 338.

3. Pauling, L. 1940. *The Nature of the Chemical Bond*. The George Fisher Baker Non-Resident Lectureship in Chemistry at Cornell University. Cornell University Press, Ithaca, New York.

4. Stitt, F. 1940. "The Gaseous Heat Capacity and Restricted Internal Rotation of Diborane." J. Chem. Phys. *8*, 981—986. Stitt, F. 1941. "Infra-Red and Raman Spectra of Polyatomic Molecules. XV. Diborane." J. Chem. Phys. *9*, 780—785.

5. Longuet-Higgins, H. C. and R. P. Bell. 1943. "The Structure of the Boron Hydrides." J. Chem. Soc., 250—255.

6. Pitzer, K. S. 1945. "Electron Deficient Molecules. I. The Principles of Hydroboron Structures." J. Am. Chem. Soc. *67*, 1126—1132.

7. Mulliken, R. S. 1947. "The Structure of Diborane and Related Molecules." Chem. Rev. *41*, 207—217.

8. Longuet-Higgins. 1949. "Substances Hydrogenées avec Défaut d'Electrons." J. Chim Phys. *46*, 268—275.

9. Price, W. C. 1947. "The Structure of Diborane." J. Chem. Phys. *15*, 614.
   Price, W. 1948. "The Absorption Spectrum of Diborane." J. Chem. Phys. *16*, 894—902.

10. Kasper, J. S., C. M. Lucht and D. Harker. 1950. "The Crystal Structure of Decaborane, $B_{10}H_{11}$." Acta Cryst. *3*, 436—455.

11. Dulmage, W. J. and W. N. Lipscomb. 1951. "The Molecular Structure of Pentaborane." J. Am. Chem. Soc. *73*, 3539. Dulmage, W. J. and W. N. Lipscomb. 1952. "The Crystal and Molecular Structure of Pentaborane." Acta Cryst. *5*, 260—264.

12. Hedberg, K., M. E. Jones and V. Schomaker. 1951. "On the Structure of Stable Pentaborane." J. Am. Chem. Soc. *73*, 3538—3539.
    Hedberg, K., M. E. Jones and V. Schomaker. 1952. "The Structure of Stable Pentaborane." Proc. Nat. Acad. Sci. *38*, 679—686.

13. Nordman, C. E. and W. N. Lipscomb. 1953. "The Molecular Structure of B_4H_{10}." J. Am. Chem. Soc. *75*, 4116—4117.
    Nordman, C. E. and W. N. Lipscomb. 1953. "The Crystal and Molecular Structure of Tetraborane." J. Chem. Phys. *21*, 1856—1864.

14. Jones, M. E., K. Hedberg and V. Schomaker. 1953. "On the Structure of Tetraborane." J. Am. Chem. Soc. *75*, 4116.

15. Lavine, L. and W. N. Lipscomb. 1954. "The Crystal and Molecular Structure of B_5H_{11}." J. Chem. Phys. *22*, 614—620.

16. Atoji, M. and W. N. Lipscomb. 1953. "The Molecular Structure of B_4Cl_4." J. Chem. Phys. *21*, 172.

Atoji, M. and W. N. Lipscomb. 1953. "The Crystal and Molecular Structure of B_4Cl_4." Acta Cryst. *6*, 547—550.

17. Eriks, K., W. N. Lipscomb and R. Schaeffer. 1954. "The Boron Arrangement in a B_6 Hydride." J. Chem. Phys. *22*, 754—755.

18. Hirshfeld, F. L., K. Eriks, R. E. Dickerson, E. L. Lippert, Jr. and W. N. Lipscomb. 1958. "Molecular and Crystal Structure of B_6H_{10}." J. Chem. Phys. *28*, 56—61.

19. Abrahams, S. C., R. L. Collin, W. N. Lipscomb and T. B. Reed. 1950. "Further Techniques in Single-Crystal X-ray Diffraction Studies at Low Temperatures." Rev. Sci. Instr. *21*, 396—397.

20. Kaufman, H. S. and I. Fankuchen. 1949. "A Low Temperature Single Crystal X-ray Diffraction Technique." Rev. Sci. Instr. *20*, 733—734.

21. Dickerson, R. E., P. J. Wheatley, P. A. Howell and W. N. Lipscomb. 1957. "Crystal and Molecular Structure of B_9H_{15}." J. Chem. Phys. *27*, 200—209.

22. Eberhardt, W. H., B. Crawford, Jr. and W. N. Lipscomb. 1954. "The Valence Structure of the Boron Hydrides." J. Chem. Phys. *22*, 989.

23. Longuet-Higgins, H. C. and M. de V. Roberts. 1954. "The Electronic Structure of the Borides MB_6." Proc. Roy. Soc. (London) *A224*, 336—347.
    Longuet-Higgins, H. C. and M. de V. Roberts. 1955. "The Electronic Structure of an Icosahedron of Boron Atoms." Proc. Roy. Soc. (London) *A230*, 110—119.

24. Lipscomb, W. N., *Boron Hydrides*. (W. A. Benjamin, Inc. 1963).

25. Dickerson, R. E. and W. N. Lipscomb. 1957. "Semitopological Approach to Boron-Hydride Structures." J. Chem. Phys. *27*, 212—217.

26. Lipscomb, W. N. 1964. "Geometrical Theory of Boron Hydrides." Inorg. Chem. *3*, 1683—1685.

27. Epstein, I. R. and W. N. Lipscomb. 1971. "Boron Hydride Valence Structures: A Topological Approach." Inorg. Chem. *10*, 1921—1928.

28. Moore, E. B., Jr., L. L. Lohr, Jr. and W. N. Lipscomb. 1961. "Molecular Orbitals in Some Boron Compounds." J. Chem. Phys. *35*, 1329—1334.

29. Hawthorne, M. F. 1968. "The Chemistry of the Polyhedral Species Derived from Transition Metals and Carboranes." Accounts of Chem. Res. *1*, 281—288.

30. Hoffmann, R. and W. N. Lipscomb. 1962. "Boron Hydrides: LCAO-MO and Resonance Studies." J. Chem. Phys. *37*, 2872—2883.
    Hoffmann, R, and W. N. Lipscomb. 1962. "Theory of Polyhedral Molecules. I. Physical Factorizations of the Secular Equation." J. Chem. Phys. *36*, 2179—2189.
    Hoffmann, R. and W. N. Lipscomb. 1962. "Theory of Polyhedral Molecules. III. Population Analyses and Reactivities for the Carboranes." J. Chem. Phys. *36*, 3489—3493.

31. Woodward, R. B. and R. Hoffmann. 1970. *The Conservation of Orbital Symmetry*. Verlag Chemie GmbH, Academic Press.

32. Eaton, G. R. and W. N. Lipscomb. 1969. *NMR Studies of Boron Hydrides and Related Compounds*. W. A. Benjamin, Inc.

33. Stevens, R. M., R. M. Pitzer and W. N. Lipscomb. 1963. "Perturbed Hartree-Fock Calculations. I. Magnetic Susceptibility and Shielding in the LiH Molecule." J. Chem. Phys. *38*, 550—560.

34. Lipscomb, W. N. 1972. "Molecular Properties." MTP International Review of Science, Theoretical Chemistry. Physical Chemistry Series One. Volume 1. Editors: A. D. Buckingham and W. Byers Brown. Butterworths, London, England. pp. 167—196.

35. Hegstrom, R. A. and W. N. Lipscomb. 1968. "Paramagnetism in Closed-Shell Molecules." Rev. Mod. Phys. *40*, 354—358.

36. Pitzer, R. M. and W. N. Lipscomb. 1963. "Calculation of the Barrier to Internal Rotation in Ethane." J. Chem. Phys. *39*, 1995—2004.

37. Newton, M. D., F. P. Boer and W. N. Lipscomb. 1966. "Molecular Orbital Theory for Large Molecules. Approximation of the SCF LCAO Hamiltonian Matrix." J. Am. Chem. Soc. *88*, 2353—2360.

38. Boer, F. P., M. D. Newton and W. N. Lipscomb. 1966. "Molecular Orbitals for Boron Hydrides Parameterized from SCF Model Calculations." J. Am. Chem. Soc. *88*, 2361—2366.

39. Halgren, T. A. and W. N. Lipscomb. 1972. "Approximations to Self-Consistent Field Molecular Wavefunctions." Proc. Nat. Acad. Sci. USA *69*, 652—656.
    Halgren, T. A. and W. N. Lipscomb. 1973. "Self-Consistent Field Wavefunctions for Complex Molecules. The Approximations of Partial Retention of Diatomic Differential Overlap. J. Chem. Phys. *58*, 1569—1591.

40. Edmiston, C. and K. Ruedenberg. 1963. "Localized Atomic and Molecular Orbitals." Rev. Mod. Phys. *35*, 457—465.

41. Boys, S. F. 1966. *Quantum Theory of Atoms, Molecules and the Solid State.* (P. O. Löwdin, Ed.). Academic Press, New York. pp. 253—262.

42. Kleier, D. A., T. A. Halgren, J. H. Hall, Jr. and W. N. Lipscomb. 1974. "Localized Molecular Orbitals for Polyatomic Molecules. I. A Comparison of the Edmiston-Ruedenberg and Boys Localization Methods." J. Chem. Phys. *61*, 3905—3919.

43. Switkes, E., W. N. Lipscomb and M. D. Newton. 1970. "Localized Bonds in Self-Consistent-Field Wave Functions for Polyatomic Molecules. II. Boron Hydrides." J. Am. Chem. Soc. *92*, 3847—3853.

44. Epstein, I. R., D. S. Marynick and W. N. Lipscomb. 1973. "Localized Orbitals for 1,2- and 1,6-Dicarbahexaborane(6). The Open Three-Center Bond, and Implications for Carborane Topology." J. Am. Chem. Soc. *95*, 1760—1766.

45. Hall, J. H., Jr., D. A. Dixon, D. A. Kleier, T. A. Halgren, L. D. Brown and W. N. Lipscomb. 1975. "Localized Molecular Orbitals for Polyatomic Molecules. II. Structural Relationships and Charge Distributions for Open Boron Hydrides and Ions." J. Am. Chem. Soc. *97*, 4202—4212.

46. Marynick, D. S. and W. N. Lipscomb. 1972. "Self-Consistent Field Wave Function and Localized Orbitals for 2,4-Dicarbaheptaborane(7). The Fractional Three-Center Bond." J. Am. Chem. Soc. *94*, 8692—8699.

47. Dixon, D. A. and W. N. Lipscomb. 1976. "Electronic Structure and Bonding of the Amino Acids Containing First Row Atoms." J. Biol. Chem. *251*, 5992—6000.

48. Marynick, D. S. and W. N. Lipscomb. 1972. "A Self-Consistent Field and Localized Orbital Study of 4,5-Dicarbahexaborane(8)." J. Am. Chem. Soc. *94*, 8699—8706.

49. Tolpin, E. I. and W. N. Lipscomb. 1973. "Crystal and Molecular Structure of Tetramethylammonium C,C'-Diphenylundecahydrodicarbanido-dodecaborate(1-)." Inorg. Chem. *12*, 2257—2262.

50. Dixon, D. A., D. A. Kleier, T. A. Halgren and W. N. Lipscomb. 1976. "Localized Molecular Orbitals for Polyatomic Molecules. IV. Large Boron Hydrides." J. Am. Chem. Soc. *98*, 2086—2096.

51. Kreevoy, M. M. and J. E. C. Hutchins. 1972. "H_2BH_3 as an Intermediate in Tetrahydridoborate Hydrolysis." J. Am. Chem. Soc. *94*, 6371—6376.

52. Pepperberg, I. M., T. A. Halgren and W. N. Lipscomb. 1976. "A Molecular Orbital Study of the Role of BH_5 in the Hydrolysis of BH_4." J. Am. Chem. Soc. *98*, 3442—3451.

53. Hoheisel, C. and W. Kutzelnigg. 1975. "Ab Initio Calculation Including Electron Correlation of the Structure and Binding Energy of BH_5 and $B_2H_7^-$." J. Am. Chem. Soc. *97*, 6970—6975.

54. Collins, J. B., P. v. R. Schleyer, J. S. Binkley, J. A. Pople and L. Radom. 1976. J. Am. Chem. Soc., *98*, 3436—3441 (1976).

55. Hariharan, P. C., W. A. Latham and J. A. Pople. 1972. "Molecular Orbital Theory of Simple Carbonium Ions." Chem. Phys. Lett. *14*, 385—388.

56. Dixon, D. A., I. M. Pepperberg and W. N. Lipscomb. 1974. "Localized Molecular Orbitals and Chemical Reactions. II. A Study of Three-Center Bond Formation in the Borane-Diborane Reaction." J. Am. Chem. Soc. *96*, 1325—1333.

57. Dupont, J. A. and R. Schaeffer. 1960. "Interconversion of Boranes. I. A. Kinetic Study of the Conversion of Tetraborane-10 to Pentaborane-11. "J. Inorg. Nucl. Chem. *15*, 310—315.

Chemistry 1977

ILYA PRIGOGINE

for his contributions to non-equilibrium thermodynamics, particularly the theory of dissipative structures

THE NOBEL PRIZE FOR CHEMISTRY

Speech by Professor STIG CLAESSON of the Royal Academy of Sciences
Translation from the Swedish text

Your Majesties, Your Royal Highnesses, Ladies and Gentlemen,

The discoveries for which Ilya Prigogine has been awarded this year's Nobel Prize for Chemistry come within the field of thermodynamics, which represents one of the most sophisticated branches of scientific theory and is of enourmous practical relevance.

The history of thermodynamics dates back to the early years of the nineteenth century. With the acceptance of the atomic theory—due above all to the works of John Dalton—the view began to be commonly accepted that what we term heat is merely the movement of the smallest components of matter. Then the invention of the steam engine created an increasingly urgent need for exact mathematical study of the interaction between heat and mechanical work.

A number of brilliant scientists, whose names have survived not only in the annals of science but also as terms for important units, contributed to the rapid development of thermodynamics during the nineteenth century. Apart from Dalton himself, we have Watt, Joule and Kelvin, who have given their names to the units of atomic weight, power, energy and absolute temperature calculated from absolute zero. Important work was also done by Helmholtz, Clausius and Gibbs, who adopted a statistical approach to the movement of atoms and molecules and created that synthesis of thermodynamics and statistics which we call statistical thermodynamics. Their names have been given to several important natural laws.

This process of development attained something of a conclusion in the early years of the present century, and thermodynamics began to be regarded as a branch of science whose evolution was essentially complete. However, it was subject to certain limitations. For the most part it could only deal with reversible processes, that is, processes occurring via states of equilibrium. Even an irreversible system as simple as the thermocouple, with its simultaneous conduction of heat and electricity, could not be satisfactorily treated until Onsager developed the reciprocity relations which earned him the 1968 Nobel Prize for Chemistry. The reciprocity relations were a great step forward in the development of a thermodynamics of irreversible processes, but they presupposed a linear approximation, which can only be employed relatively close to equilibrium.

Prigogine's great contribution lies in his successful development of a satisfactory theory of non-linear thermodynamics in states which are far removed from equilibrium. In doing so he has discovered phenomena and structures of completely new and completely unexpected types, with the result that this generalized, nonlinear and irreversible thermodynamics has already been given surprising applications in a wide variety of fields.

Prigogine has been particularly captivated by the problem of explaining how ordered structures—biological systems, for example—can develop from disorder. Even if Onsager's relations are utilized, the classical principles of equilibrium in thermodynamics still show that *linear* systems close to equilibrium always develop into states of *disorder* which are stable to perturbations and cannot explain the occurrence of *ordered structures*.

Prigogine and his assistants chose instead to study systems which follow non-linear kinetic laws and which, moreover, are in contact with their surroundings so that energy exchange can take place—*open systems,* in other words. If these systems are driven far from equilibrium, a completely different situation results. New systems can then be formed which display order in both time and space and which are stable to perturbations. Prigogine has called these systems *dissipative systems,* because they are formed and maintained by the dissipative processes which take place because of the exchange of energy between the system and its environment and because they disappear if that exchange ceases. They may be said to live in symbiosis with their environment.

The method which Prigogine has used to study the stability of the dissipative structures to perturbations is of very great general interest. It makes it possible to study the most varied problems, such as city traffic problems, the stability of insect communities, the development of ordered biological structures and the growth of cancer cells to mention but a few examples.

Three of the persons who have assisted Prigogine for many years, above all P. Glansdorff but also R. Lefever and G. Nicolis, deserve special mention in this context, because of the important and original contributions they have made towards the development of science.

Thus Prigogine's researches into irreversible thermodynamics have fundamentally transformed and revitalized the science, given it a new relevance and created theories to bridge the gaps between chemical, biological and social scientific fields of inquiry. His works are also distinguished by an elegance and a lucidity which have earned him the epithet "the poet of thermodynamics".

Professor Prigogine, I have tried briefly to describe your great contribution to non-linear irreversible thermodynamics, and it is now my privilege and pleasure to convey to you the heartiest congratulations of the Swedish Royal Academy of Sciences and to ask you ro receive your Nobel prize from the hands of His Majesty the King.

Ilya Prigogine

ILYA PRIGOGINE

Translation from the French text

In his memorable series "*Etudes sur le temps humain*", Georges Poulet devoted one volume to the "*Mesure de l'instant*" (1). There he proposed a classification of authors according to the importance they give to the past, present and future. I believe that in such a typology my position would be an extreme one, as I live mostly in the future. And thus it is not too easy a task to write this autobiographical account, to which I would like to give a personal tone. But the present explains the past.

In my Nobel Lecture, I speak much about fluctuations; maybe this is not unrelated to the fact that during my life I felt the efficacy of striking coincidences whose cumulative effects are to be seen in my scientific work.

I was born in Moscow, on the 25th of January, 1917 — a few months before the revolution. My family had a difficult relationship with the new regime, and so we left Russia as early as 1921. For some years (until 1929), we lived as migrants in Germany, before we stayed for good in Belgium. It was at Brussels that I attended secondary school and university. I acquired Belgian nationality in 1949.

My father, Roman Prigogine, who died in 1974, was a chemical engineer from the Moscow Polytechnic. My brother Alexander, who was born four years before me, followed, as I did myself, the curriculum of chemistry at the Université Libre de Bruxelles. I remember how much I hesitated before choosing this direction; as I left the classical (Greco-Latin) section of Ixelles Athenaeum, my interest was more focused on history and archaeology, not to mention music, especially piano. According to my mother, I was able to read musical scores before I read printed words. And, today, my favourite pastime is still piano playing, although my free time for practice is becoming more and more restricted.

Since my adolescence, I have read many philosophical texts, and I still remember the spell "*L'évolution créatrice*" cast on me. More specifically, I felt that some essential message was embedded, still to be made explicit, in Bergson's remark:

"The more deeply we study the nature of time, the better we understand that duration means invention, creation of forms, continuous elaboration of the absolutely new."

Fortunate coincidences made the choice for my studies at the university. Indeed, they led me to an almost opposite direction, towards chemistry and physics. And so, in 1941, I was conferred my first doctoral degree. Very soon, two of my teachers were to exert an enduring influence on the orientation of my future work.

I would first mention Théophile De Donder (1873–1957) (2). What an amiable character he was! Born the son of an elementary school teacher, he began his career in the same way, and was (in 1896) conferred the degree of

Doctor of Physical Science, without having ever followed any teaching at the university.

It was only in 1918 — he was then 45 years old — that De Donder could devote his time to superior teaching, after he was for some years appointed as a secondary school teacher. He was then promoted to professor at the Department of Applied Science, and began without delay the writing of a course on theoretical thermodynamics for engineers.

Allow me to give you some more details, as it is with this very circumstance that we have to associate the birth of the Brussels thermodynamics school.

In order to understand fully the originality of De Donder's approach, I have to recall that since the fundamental work by Clausius, the second principle of thermodynamics has been formulated as an inequality: "uncompensated heat" is positive — or, in more recent terms, entropy production is positive. This inequality refers, of course, to phenomena that are *irreversible*, as are any natural processes. In those times, these latter were poorly understood. They appeared to engineers and physico-chemists as "parasitic" phenomena, which could only hinder something: here the productivity of a process, there the regular growth of a crystal, without presenting any intrinsic interest. So, the usual approach was to limit the study of thermodynamics to the understanding of equilibrium laws, for which entropy production is zero.

This could only make thermodynamics a "thermostatics". In this context, the great merit of De Donder was that he extracted the entropy production out of this "sfumato" when related it in a precise way to the pace of a chemical reaction, through the use of a new function that he was to call "affinity" (3).

It is difficult today to give an account of the hostility that such an approach was to meet. For example, I remember that towards the end of 1946, at the Brussels IUPAP meeting (4), after a presentation of the thermodynamics of irreversible processes, a specialist of great repute said to me, in substance: "I am surprised that you give more attention to irreversible phenomena, which are essentially transitory, than to the final result of their evolution, equilibrium."

Fortunately, some eminent scientists derogated this negative attitude. I received much support from people such as Edmond Bauer, the successor to Jean Perrin at Paris, and Hendrik Kramers in Leyden.

De Donder, of course, had precursors, especially in the French thermodynamics school of Pierre Duhem. But in the study of chemical thermodynamics, De Donder went further, and he gave a new formulation of the second principle, based on such concepts as affinity and degree of evolution of a reaction, considered as a chemical variable.

Given my interest in the concept of time, it was only natural that my attention was focused on the second principle, as I felt from the start that it would introduce a new, unexpected element into the description of physical world evolution. No doubt it was the same impression illustrious physicists such as Boltzmann (5) and Planck (6) would have felt before me. A huge part of my scientific career would then be devoted to the elucidation of macroscopic as well as microscopic aspects of the second principle, in order to extend its validity to new situations, and to the other fundamental approaches of theoretical physics, such as classical and quantum dynamics.

Before we consider these points in greater detail, I would like to stress the influence on my scientific development that was exerted by the second of my teachers, Jean Timmermans (1882–1971). He was more an experimentalist, specially interested in the applications of classical thermodynamics to liquid solutions, and in general to complex systems, in accordance with the approach of the great Dutch thermodynamics school of van der Waals and Roozeboom (7).

In this way, I was confronted with the precise application of thermodynamical methods, and I could understand their usefulness. In the following years, I devoted much time to the theoretical approach of such problems, which called for the use of thermodynamical methods; I mean the solutions theory, the theory of corresponding states and of isotopic effects in the condensed phase. A collective research with V. Mathot, A. Bellemans and N. Trappeniers has led to the prediction of new effects such as the isotopic demixtion of helium $He^3 + He^4$, which matched in a perfect way the results of later research. This part of my work is summed up in a book written in collaboration with V. Mathot and A. Bellemans, *The Molecular Theory of Solutions* (8).

My work in this field of physical chemistry was always for me a specific pleasure, because the direct link with experimentation allows one to test the intuition of the theoretician. The successes we met provided the confidence which later was much needed in my confrontation with more abstract, complex problems.

Finally, among all those perspectives opened by thermodynamcis, the one which was to keep my interest was the study of irreversible phenomena, which made so manifest the "arrow of time". From the very start, I always attributed to these processes a constructive role, in opposition to the standard approach, which only saw in these phenomena degradation and loss of useful work. Was it the influence of Bergson's "*L'évolution créatrice*" or the presence in Brussels of a performing school of theoretical biology (9)? The fact is that it appeared to me that living things provided us with striking examples of systems which were highly organized and where irreversible phenomena played an essential role.

Such intellectual connections, although rather vague at the beginning, contributed to the elaboration, in 1945, of the theorem of minimum entropy production, applicable to non-equilibrium stationary states (10). This theorem gives a clear explanation of the analogy which related the stability of equilibrium thermodynamical states and the stability of biological systems, such as that expressed in the concept of "homeostasy" proposed by Claude Bernard. This is why, in collaboration with J. M. Wiame (11), I applied this theorem to the discussion of some important problems in theoretical biology, namely to the energetics of embryological evolution. As we better know today, in this domain the theorem can at best give an explanation of some "late" phenomena, but it is remarkable that it continues to interest numerous experimentalists (12).

From the very beginning, I knew that the minimum entropy production was valid only for the linear branch of irreversible phenomena, the one to which the famous reciprocity relations of Onsager are applicable (13). And, thus, the question was: What about the stationary states far from equilibrium, for which Onsager relations are not valid, but which are still in the scope of macroscopic description? Linear relations are very good approximations for the study of transport phenomena (thermical conductivity, thermodiffusion, etc.), but are generally not valid for the conditions of chemical kinetics. Indeed, chemical equilibrium is ensured through the compensation of two antagonistic processes, while in chemical kinetics — far from equilibrium, out of the linear branch — one is usually confronted with the opposite situation, where one of the processes is negligible.

Notwithstanding this local character, the linear thermodynamics of irreversible processes had already led to numerous applications, as shown by people such as J. Meixner (14), S. R. de Groot and P. Mazur (15), and, in the area of biology, A. Katchalsky (16). It was for me a supplementary incentive when I had to meet more general situations. Those problems had confronted us for more than twenty years, between 1947 and 1967, until we finally reached the notion of "dissipative structure" (17).

Not that the question was intrinsically difficult to handle; just that we did not know how to orientate ourselves. It is perhaps a characteristic of my scientific work that problems mature in a slow way, and then present a sudden evolution, in such a way that an exchange of ideas with my colleagues and collaborators becomes necessary. During this phase of my work, the original and enthusiastic mind of my colleague Paul Glansdorff played a major role.

Our collaboration was to give birth to a general evolution criterion which is of use far from equilibrium in the non-linear branch, out of the validity domain of the minimum entropy production theorem. Stability criteria that resulted were to lead to the discovery of critical states, with branch shifting and possible appearance of new structures. This quite unexpected manifestation of "disorder–order" processes, far from equilibrium, but conforming to the second law of thermodynamics, was to change in depth its traditional interpretation. In addition to classical equilibrium structures, we now face dissipative coherent structures, for sufficient far-from-equilibrium conditions. A complete presentation of this subject can be found in my 1971 book co-authored with Glansdorff (18).

In a first, tentative step, we thought mostly of hydrodynamical applications, using our results as tools for numerical computation. Here the help of R. Schechter from the University of Texas at Austin was highly valuable (19). Those questions remain wide open, but our centre of interest has shifted towards chemical dissipative systems, which are more easy to study than convective processes.

All the same, once we formulated the concept of dissipative structure, a new path was open to research and, from this time, our work showed striking acceleration. This was due to the presence of a happy meeting of circumstances; mostly to the presence in our team of a new generation of clever young scientists. I cannot mention here all those people, but I wish to stress the important role played by two of them, R. Lefever and G. Nicolis. It was with them that we were in a position to build up a new kinetical model, which would prove at the same time to be quite simple and very instructive — the "Brusselator", as J. Tyson would call it later — and which would manifest the amazing variety of structures generated through diffusion–reaction processes (20).

This is the place to pay tribute to the pioneering work of the late A. Turing (21), who, since 1952, had made interesting comments about structure formation as related to chemical instabilities in the field of biological morphogenesis. I had met Turing in Manchester about three years before, at a time when M. G. Evans, who was to die too soon, had built a group of young scientists, some of whom would achieve fame. It was only quite a while later that I recalled the comments by Turing on those questions of stability, as, perhaps too concerned about linear thermodynamics, I was then not receptive enough.

Let us go back to the circumstances that favoured the rapid development of the study of dissipative structures. The attention of scientists was attracted to coherent non-equilibrium structures after the discovery of experimental oscillating chemical reactions such as the Belusov–Zhabotinsky reaction (22); the explanation of its mechanism by Noyes and his co-workers (23); the study of oscillating reactions in biochemistry (for example the glycolytic cycle, studied by B. Chance (24) and B. Hess) (25); and eventually the important research led by M. Eigen (26). Therefore, since 1967, we have been confronted with a huge number of papers on this topic, in sharp contrast with the total absence of interest which prevailed during previous times.

But the introduction of the concept of dissipative structure was also to have other unexpected consequences. It was evident from start that the structures were evolving out of fluctuations. They appeared in fact as giant fluctuations, stabilized through matter and energy exchanges with the outer world. Since the formulation of the minimum entropy production theorem, the study of non-equilibrium fluctuation had attracted all my attention (27). It was thus only natural that I resumed this work in order to propose an extension of the case of far-from-equilibrium chemical reactions.

This subject I proposed to G. Nicolis and A. Babloyantz. We expected to find for stationary states a Poisson distribution similar to the one predicted for equilibrium fluctuations by the celebrated Einstein relations. Nicolis and Babloyantz developed a detailed analysis of linear chemical reactions and were able to confirm this prediction (28). They added some qualitative remarks which suggested the validity of such results for any chemical reaction.

Considering again the computations for the example of a non-linear biomolecular reaction, I noticed that this extension was not valid. A further analysis, where G. Nicolis played a key role, showed that an unexpected phenomenon appeared while one considered the fluctuation problem in nonlinear systems far from equilibrium: the distribution law of fluctuations depends on their scale, and only "small fluctuations" follow the law proposed by Einstein (29). After a prudent reception, this result is now widely accepted, and the theory of non-equilibrium fluctuations is fully developing now, so as to allow us to expect important results in the following years. What is already clear today is that a domain such as chemical kinetics, which was considered conceptually closed, must be thoroughly rethought, and that a brand-new discipline, dealing with non-equilibrium phase transitions, is now appearing (30, 31, 32).

Progress in irreversible phenomena theory leads us also to reconsideration of their insertion into classical and quantum dynamics. Let us take a new look at the statistical mechanics of some years ago. From the very beginning of my research, I had had occasion to use conventional methods of statistical mechanics for equilibrium situations. Such methods are very useful for the study of thermodynamical properties of polymer solutions or isotopes. Here we deal mostly with simple computational problems, as the conceptual tools of equilibrium statistical mechanics have been well established since the work of Gibbs and Einstein. My interest in non-equilibrium would by necessity lead me to the problem of the foundations of statistical mechanics, and especially to the microscopic interpretation of irreversibility (33).

Since the time of my first graduation in science, I was an enthusiastic reader of Boltzmann, whose dynamical vision of physical becoming was for me a model of intuition and penetration. Nonetheless, I could not but notice some unsatisfying aspects. It was clear that Boltzmann introduced hypotheses foreign to dynamics; under such assumptions, to talk about a dynamical justification of thermodynamics seemed to me an excessive conclusion, to say the least. In my opinion, the identification of entropy with molecular disorder could contain only one part of the truth if, as I persisted in thinking, irreversible processes were endowed with this constructive role I never cease to attribute to them. For another part, the applications of Boltzmann's methods were restricted to diluted gases, while I was most interested in condensed systems.

At the end of the forties, great interest was aroused in the generalization of kinetic theory to dense media. After the pioneering work by Yvon (34), publications of Kirkwodd (35), Born and Green (36), and of Bogoliubov (37) attracted a lot of attention to this problem, which was to lead to the birth of non-equilibrium statistical mechanics. As I could not remain alien to this movement, I proposed to G. Klein, a disciple of Fürth who came to work with me, to try the application of Born and Green's method to a concrete, simple example, in which the equilibrium approach did not lead to an exact solution. This was our first tentative step in non-equilibrium statistical mechanics (38). It was eventually a failure, with the conclusion that Born and Green's formalism did not lead to a satisfying extension of Boltzmann's method to dense systems.

But this failure was not a total one, as it led me, during a later work, to a first question: Was it possible to develop an "exact" dynamical theory of irreversible phenomena? Everybody knows that according to the classical point of view, irreversibility results from supplementary approximations to fundamental laws of elementary phenomena, which are strictly reversible. These supplementary approximations allowed Boltzmann to shift from a dynamical, reversible description to a probabilistic one, in order to establish his celebrated H theorem.

We still encountered this negative attitude of "passivity" imputed to irreversible phenomena, an attitude that I could not share. If — as I was prepared to think — irreversible phenomena actually play an active, constructive role, their study could not be reduced to a description in terms of supplementary approximations. Moreover, my opinion was that in a good theory a viscosity coefficient would present as much physical meaning as a specific heat, and the mean life duration of a particle as much as its mass.

I felt confirmed in this attitude by the remarkable publications of Chandrasekhar and von Neumann, which were also issued during the forties (39). That was why, still with the help of G. Klein, I decided to take a fresh look at an example already studied by Schrödinger (40), related to the description of a system of harmonic oscillators. We were surprised to see that, for all such a simple model allowed us to conclude, this class of systems tend to equilibrium. But how to generalize this result to non-linear dynamical systems?

Here the truly historic performance of Léon van Hove opened for us the way (1955) (41). I remember, with a pleasure that is always new, the time—which was too short— during which van Hove worked with our group. Some of his works had a lasting effect on the whole development of statistical physics; I mean not only his study of the deduction of a "master equation" for anharmonic systems, but also his fundamental contribution on phase transitions, which was to lead to the branch of statistical mechanics that deals with so-called "exact" results (42).

This first study by van Hove was restricted to weakly coupled anharmonic systems. But, anyway, the path was open, and with some of my colleagues and collaborators, mainly R. Balescu, R. Brout, F. Hénin and P. Résibois, we achieved a formulation of non-equilibrium statistical mechanics from a purely dynamical point of view, without any probabilistic assumption. The method we used is summed up in my 1962 book (43). It leads to a "dynamics of correlations", as the relation between interaction and correlation constitutes the essential component of the description. Since then, these methods have led to numerous applications. Without giving more detail, here, I will restrict myself to mentioning two recent books, one by R. Balescu (44), the other by P. Résibois and M. De Leener (45).

This concluded the first step of my research in non-equilibrium statistical mechanics. The second is characterized by a very strong analogy with the approach of irreversible phenomena which led us from linear thermodynamics to non-linear thermodynamics. In this tentative step also, I was prompted by a feeling of dissatisfaction, as the relation with thermodynamics was not established by our work in statistical mechanics, nor by any other method. The H theorem of Boltzmann was still as isolated as ever, and the question of the nature of dynamics systems to which thermodynamics applies was still without answer.

The problem was by far more wide and more complex than the rather technical considerations that we had reached. It touched the very nature of dynamical systems, and the limits of Hamiltonian description. I would never have dared approach such a subject if I had not been stimulated by discussions with some highly competent friends such as the late Léon Rosenfeld from Copenhagen, or G. Wentzel from Chicago. Rosenfeld did more than give me advice; he was directly involved in the progressive elaboration of the concepts we had to explore if we were to build a new interpretation of irreversibility. More than any other stage of my scientific career, this one was the result of a collective effort. I could not possibly have succeeded had it not been for the help of my colleagues M. de Haan, Cl. George, A. Grecos, F. Henin, F. Mayné, W. Schieve and M. Theodosopulu. If irreversibility does not result from supplementary approximations, it can only be formulated in a theory of transformations which expresses in "explicit" terms what the usual formulation of dynamics does "hide". In this perspective, the kinetic equation of Boltzmann corresponds to a formulation of dynamics in a new representation (46, 47, 48, 49).

In conclusion: dynamics and thermodynamics become two complementary descriptions of nature, bound by a new theory of non-unitary transformation. I came so to my present concerns; and, thus, it is time to end this intellectual autobiography. As we started from specific problems, such as the thermodynamic signification of non-equilibrium stationary states, or of transport phenomena in dense systems, we have been faced, almost against our will, with problems of great generality and complexity, which call for reconsideration of the relation of physico-chemical structures to biological ones, while they express the limits of Hamiltonian description in physics.

Indeed, all these problems have a common element: time. Maybe the orientation of my work came from the conflict which arose from my humanist vocation as an adolescent and from the scientific orientation I chose for my university training. Almost by instinct, I turned myself later towards problems of increasing complexity, perhaps in the belief that I could find there a junction in physical science on one hand, and in biology and human science on the other.

In addition, the research conducted with my friend R. Herman on the theory of car traffic (50) gave me confirmation of the supposition that even human behaviour, with all its complexity, would eventually be susceptible of a mathematical formulation. In this way the dichotomy of the "two cultures" could and should be removed. There would correspond to the breakthrough of biologists and anthropologists towards the molecular description or the "elementary structures", if we are to use the formulation by Lévi-Strauss, a complementary move by the physico-chemist towards complexity. Time and complexity are concepts that present intrinsic mutual relations.

During his inaugural lecture, De Donder spoke in these terms (51): "Mathematical physics represents the purest image that the view of nature may generate in the human mind; this image presents all the character of the product of art; it begets some unity, it is true and has the quality of sublimity; this image is to physical nature what music is to the thousand noises of which the air is full . . ."

Filtrate music out of noise; the unity of the spiritual history of humanity, as was stressed by M. Eliade, is a recent discovery that has still to be assimilated (52). The search for what is meaningful and true by opposition to noise is a tentative step that appears to be intrinsically related to the coming into consciousness of man facing a nature of which he is a part and which it leaves.

I have many times advocated the necessary dialogue in scientific activity, and thus the vital importance of my colleagues and collaborators in the journey that I have tried to describe. I would also stress the continuing support that I received from institutions which have made this work a feasible one, especially the Université Libre de Bruxelles and the University of Texas at Austin. For all of the development of these ideas, the International Institute of Physics and Chemistry founded by E. Solvay (Brussels, Belgium) and the Welch Foundation (Houston, Texas) have provided me with continued support.

The work of a theoretician is related in a direct way to his whole life. It takes, I believe, some amount of internal peace to find a path among all successive bifurcations. This peace I owe to my wife, Marina. I know the frailty of the present, but today, considering the future, I feel myself to be a happy man.

REFERENCES

1. G. Poulet, *Etudes sur le temps humain*, Tone 4, Edition 10/18, Paris, 1949.
2. See the note on De Donder in the Florilège (pendant le XIXe siècle et le début du XXe), Acad. Roy. Belg., Bull. Cl. Sc., page 169, 1968.
3. Th. De Donder (Rédaction nouvelle par P. Van Rysselberghe), Paris, Gauthier-Villars, 1936.
   See also:
   I. Prigogine and R. Defay: Thermodynamique Chimique conformément aux méthodes de Gibbs et De Donder (2 Tomes), Liège, Desoer, 1944–1946.
   Or the translation in English:
   Chemical Thermodynamics, translated by D. H. Everett, Langmans 1954, 1962.
4. See Colloque de Thermodynamique, Union Intern. de Physique pure et appliquée (I. U. P. A. P.), 1948.
5. Boltzmann, L., Wien, Ber. *66*, 275,1872.
6. Planck, M., Vorlesungen über Thermodynamik, Walter de Gruyter, Berlin, Leipzig, 1930.
7. Timmermans, J., Les Solutions Concentrées, Masson et Cie, Paris, 1936.
   Let us also quote his thesis on experimental research on demixtion in liquid mixtures.
8. Prigogine, I., The Molecular Theory of Solutions, avec A. Bellemans et V. Mathot; North-Holland Publ. Company, Amsterdam, 1957.
   See also: Prigogine and Defay, Ref. 3.
9. Let us quote some remarkable works of this School:
   Brachet, A., La Vie créatrice des formes, Alcan, Paris, 1927.
   Dalcq, A., L'Oeuf et son dynamisme organisateur, Alban Michel, Paris, 1941.
   Brachet, J., Embryologie Chimique, Desoer, Liège et Masson, Paris, 1946.
   I was also much interested in the beautiful book by Marcel Florkin: L'Evolution biochimique, Desoer, Liège, 1944.
10. Prigogine, I., Acad. Roy. Belg. Bull. Cl. Sc. *31*, 600, 1945.
    —Etude thermodynamique des phénomènes irréversibles. Thèse d'agrégation présentée en 1945 à l'Université Libre de Bruxelles.
    Desoer, Liège, 1947.
    —Introduction à la Thermodynamique des processus irréversibles, traduit de l'anglais par J. Chanu, Dunod, Paris, 1968.
11. Prigogine, I. and Wiame, J. M., Experientia, *2*, 451, 1946.
12. Nicolis, G. and Prigogine, I., Self-Organization in Non-Equilibrium Systems (Chaps. III and IV), J. Wiley and Sons, New York, 1977.
13. Onsager, L., Phys. Rev., *37*, 405, 1931.
14. Meixner, J., Ann. Physik, (5), *35*, 701,1939; *36*, 103, 1939; *39*, 333,1941; *40*, 165,1941; Zeitsch Phys. Chim. B *53*, 235, 1943.
15. de Groot, S. R. and Mazur, P., Non-Equilibrium Thermodynamics, North-Holland, Amsterdam, 1962.
16. Katchalsky, A. and Curran, P. F., Non-Equilibrium Thermodynamics in Biophysics, Harvard Univ. Press, Cambridge, Mass., 1964.
17. Prigogine, I., Structure, Dissipation and Life.
    Theoretical Physics and Biology, Versailles, 1967.
    North-Holland Publ. Co, Amsterdam, 1969.
    It is in this communication that the term "structure dissipative" is used for the first time.
18. Glansdorff, P. and Prigogine, I., Structure, Stabilité et Fluctuations, Masson, Paris, 1971.
    —Thermodynamic Theory of Structure Stability and Fluctuations, Wiley and Sons, London, 1971.
    —Traduction en langue russe: Mir, Moscou, 1973.
    —Traduction en langue japonaise: Misuzu Shobo, 1977.
    This book presents in detail the original work by the two authors, which led to the concept of dissipative structure. For a brief historical account, see also:
    Acad. Roy. Belg., Bull. des Cl. Sc., LIX, *80*, 1973.

19. Schechter, R. S., The Variational Method in Engineering, McGraw-Hill, New York, 1967.

20. Tyson, J., Journ. of Chem. Physics, *58*, 3919, 1973.

21. Turing, A., Phil. Trans. Roy. Soc. London, Ser B, *237*, 37, 1952.

22. Belusov, B. P., Sb. Ref. Radiat. Med. Moscow, 1958.
    Zhabotinsky, A. P., Biofizika, *9*, 306, 1964.
    Acad. Sc. U.R.S.S. Moscow (Nauka), 1967.

23. Noyes, R. M. et al., Ann. Rev. Phys. Chem. *25*, 95, 1974.

24. Chance, B., Schoener, B. and Elsaesser, S., Proc. Nat. Acad. Sci. U.S.A. *52*, 337–341, 1964.

25. Hess, B., Ann. Rev. Biochem, *40*, 237, 1971.

26. Eigen, M., Naturwissenschaften, *58*, 465, 1971.

27. Prigogine, I. and Mayer, G., Acad. Roy. Belg., Bull. Cl. Sc., *41*, 22, 1955.

28. Nicolis, G. and Babloyantz, A., Journ. Chem. Phys., *51*, 6, 2632, 1969.

29. Nicolis, G. and Prigogine, I., Proc. Nat. Acad. Sci. U.S.A. *68*, 2102, 1971.

30. Prigogine, I., Proc. 3rd Symp. Temperature, Washington D.C., 1954.
    Prigogine, I. and Nicolis, G., Proc. 3rd. Intern. Conference: From Theoretical Physics to Biology, Versailles, France, 1971.

31. Nicolis, G. and Turner, J. W., Proc. of the Conference on Bifurcation Theory, New York, 1977. To appear.

32. Prigogine, I. and Nicolis, G., Non-Equilibrium Phase Transitions and Chemical Reactions, Scientific American. To appear.

33. Prigogine, I., Non-Equilibrium Statistical Mechanics, Interscience Publ., New York, London, 1962–1966. (For a brief history and original references.)

34. Yvon, J., Les Corrélations et l'Entropie en Mécanique Statistique Classique. Dunod, Paris, 1965.

35. Kirkwood, J. G., Journ. Chem. Physics, *14*, 180, 1946.

36. Born, M. and Green, H.S., Proc. Roy. Soc. London, A *188*, 10, 1946 and A *190*, 45, 1947.

37. Bogoliubov, N. N., Jour. Phys. U.S.S.R., *10*, 257, 265, 1946.

38. Klein, G. and Prigogine, I., Physica XIX 74–88; 88–100; 1053–1071, 1953.

39. Chandrasekhar, S., Stochastic Problems in Physics and Astronomy; Rev. of Mod. Physics, 15, no 1, 1943.

40. Schrödinger, E., Ann. der Physik, *44*, 916, 1914.

41. Van Hove, L., Physica, *21*, 512 (1955).

42. Van Hove, L., Physica, *16*, 137 (1950).

43. Prigogine, I., cf. Ref. 33.

44. Balescu, R., Equilibrium and Non-Equilibrium Statistical Mechanics, Wiley, Interscience, 1975.

45. Résibois, P. and De Leener, M., Classical Kinetic Theory of Fluids, Wiley, Interscience, New York, 1977.

46. Prigogine, I., George, C., Henin, F. and Rosenfeld, L., Chemica Scripta, *4*, 5–32, 1973.

47. Prigogine, I., George, C. and Henin, F., Physica *45*, 418–434, 1969.

48. Prigogine, I. and Grecos, A. P., The Dynamical Theory of Irreversible Processes, Proc. Intern. Conf. on Frontiers of Theor. Phys., New Delhi, 1976.
    Kinetic Theory and Ergodic Properties in Quantum Mechanics, Abhandlungen der Akad. der Wiss., der D.D.R. Nr 7 N Berlin, Jahrgang, 1977.

49. Grecos, A. P. and Prigogine, I., Thirteenth IUPAP Conference on Statistical Physics, Haifa, August 1977.

50. Prigogine, I. and Herman, R., Kinetic Theory of Vehicular Traffic, Elsevier, 1971.

51. For the reference, see note 2.

52. Mircéa Eliade, Histoire des croyances et fies idées religieuses Vol. I., p. 10, Payot, Paris, 1976.

TIME, STRUCTURE AND FLUCTUATIONS

Nobel Lecture, 8 December, 1977
by
ILYA PRIGOGINE
Université Libre de Bruxelles, Brussels, Belgium
and the University of Texas at Austin, Austin, Texas, USA

1. INTRODUCTION

The problem of time in physics and chemistry is closely related to the formulation of the second law of thermodynamics. Therefore another possible title of this lecture could have been: "the macroscopic and microscopic aspects of the second law of thermodynamics".

It is a remarkable fact that the second law of thermodynamics has played in the history of science a fundamental role far beyond its original scope. Suffice it to mention Boltzmann's work on kinetic theory, Planck's discovery of quantum theory or Einstein's theory of spontaneous emission, which were all based on the second law of thermodynamics.

It is the main thesis of this lecture that we are only at the beginning of a new development of theoretical chemistry and physics in which thermodynamic concepts will play an even more basic role. Because of the complexity of the subject we shall limit ourselves here mainly to conceptual problems. The conceptual problems have both macroscopic and microscopic aspects. For example, from the macroscopic point of view classical thermodynamics has largely clarified the concept of equilibrium structures such as crystals.

Thermodynamic equilibrium may be characterized by the minimum of the Helmholtz free energy defined usually by

$$F = E - TS \qquad (1.1)$$

Are most types of "organisations" around us of this nature? It is enough to ask such a question to see that the answer is negative. Obviously in a town, in a living system, we have a quite different type of functional order. To obtain a thermodynamic theory for this type of structure we have to show that that non-equilibrium may be a *source of order*. Irreversible processes may lead to a new type of dynamic states of matter which I have called "dissipative structures". Sections 2—4 are devoted to the thermodynamic theory of such structures.

These structures are today of special interest in chemistry and biology. They manifest a coherent, supermolecular character which leads to new, quite spectacular manifestations; for example in biochemical cycles involving oscillatory enzymes.

How do such coherent structures appear as the result of reactive collisions. This question is briefly discussed in Section 5. We emphasize that conventional chemical kinetics corresponds to a "mean field" theory very similar to the Van der Waals theory of the equation of state or Weiss' theory of ferromagnetism. Exactly as in these cases the mean field theory breaks down near the instability where the new dissipative structures originate. Here (as in equilibrium theory) fluctuations play an essential role.

In the last two sections we turn to the microscopic aspects. We briefly review the recent work done by our group in this direction. This work leads to a microscopic definition of irreversible processes. However this is only possible through a transformation theory which allows one to introduce new non-unitary equations of motion that explicitly display irreversibility and approach to thermodynamic equilibrium.

The inclusion of thermodynamic elements leads to a reformulation of (classical or quantum) dynamics. This is a most surprising feature. Since the beginning of this century we were prepared to find new theoretical structures in the microworld of elementary particles or in the macroworld of cosmological dimensions. We see now that even for phenomena on our own level the incorporation of thermodynamic elements leads to new theoretical structures. This is the price we have to pay for a formulation of theoretical methods in which time appears with its full meaning associated with irreversibility or even with "history", and not merely as a geometrical parameter associated with motion.

2. ENTROPY PRODUCTION

At the very core of the second law of thermodynamics we find the basic distinction between "reversible" and "irreversible processes" (1). This leads ultimately to the introduction of entropy S and the formulation of the second law of thermodynamics. The classical formulation due to Clausius refers to isolated systems exchanging neither energy nor matter with the outside world. The second law then merely ascertains the existence of a function, the entropy S, which increases monotonically until it reaches its maximum at the state of thermodynamic equilibrium,

$$\frac{dS}{dt} \geq 0. \tag{2.1}$$

It is easy to extend this formulation to systems which exchange energy and matter with the outside world. (see fig. 2.1).

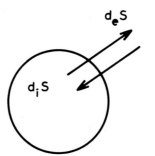

Fig. 2.1. The exchange of entropy between the outside and the inside.

We have then to distinguish in the entropy change dS two terms: the first, d_eS, is the transfer of entropy across the boundaries of the system, and the second d_iS, is the entropy produced within the system. The second law assumes then that the entropy production inside the system is positive (or zero)

$$d_iS \geqslant 0. \tag{2.2}$$

The basic distinction here is between "reversible processes" and "irreversible processes". Only irreversible processes contribute to entropy production. Obviously, the second law expresses the fact that irreversible processes lead to one-sidedness of time. The positive time direction is associated with the increase of entropy. Let us emphasize the strong and very specific way in which the one-sidedness of time appears in the second law. According to its formulation it leads to the existence of a function having quite specific properties as expressed by the fact that for an isolated system it can only increase in time. Such functions play an important role in modern theory of stability as initiated by the classic work of Lyapounov. For this reason they are called Lyapounov functions (or functionals).

The entropy S is a Lyapounov function for isolated systems. As shown in all textbooks thermodynamic potentials such as the Helmholtz or Gibbs free energy are also Lyapounov functions for other "boundary conditions" (such as imposed values of temperature and volume).

In all these cases the system evolves to an equilibrium state characterized by the existence of a thermodynamic potential. This equilibrium state is an "attractor" for non-equilibrium states. This is an essential aspect which was rightly emphasized by Planck (1).

However thermodynamic potentials exist only for exceptional situations. The inequality (2.2), which does not involve the total differential of a function, does not in general permit one to define a Lyapounov function. Before we come back to this question let us emphasize that one hundred fifty years after its formulation, the second law of thermodynamics still appears more as a program than a well defined theory in the usual sense, as nothing precise (except the sign) is said about the entropy production. Even the range of validity of this inequality is left unspecified. This is one of the main reasons why the applications of thermodynamics were essentially limited to equilibrium.

To extend thermodynamics to non-equilibrium processes we need an explicit expression for the entropy production. Progress has been achieved along this line by supposing that even outside equilibrium entropy depends only on the same variables as at equilibrium. This is the assumption of "local" equilibrium (2). Once this assumption is accepted we obtain for P, the entropy production per unit time,

$$P = \frac{d_iS}{dt} = \sum_\rho J_\rho X_\rho \geqslant 0 \tag{2.3}$$

where the J_ρ are the rates of the various irreversible processes involved (chemical reactions, heat flow, diffusion . . .) and the X_ρ the corresponding generalized

forces (affinities, gradients of temperature, of chemical potentials . . .). This
is the basic formula of macroscopic thermodynamics of irreversible processes.

Let us emphasize that we have used supplementary assumptions to derive
the explicit expression (2.3) for the entropy production. This formula can only
be established in some neighborhood of equilibrium (see Ref. 3). This neighborhood defines the region of "local" equilibrium, which we shall discuss from
the point of view of statistical mechanics in Section 7.

At thermodynamic equilibrium we have simultaneously for *all* irreversible
processes,

$$J_\rho = 0 \quad \text{and} \quad X_\rho = 0. \tag{2.4}$$

It is therefore quite natural to assume, at least near equilibrium, linear
homogeneous relations between the flows and the forces. Such a scheme
automatically includes empirical laws such as Fourier's law, which expresses
that the flow of heat is proportional to the gradient of temperature, of Fick's
law for diffusion, which states that the flow of diffusion is proportional to the
gradient of concentration. We obtain in this way the linear thermodynamics
of irreversible processes characterized by the relations (4),

$$J_\rho = \sum_{\rho'} L_{\rho\rho'} X_{\rho'}. \tag{2.5}$$

Linear thermodynamics of irreversible processes is dominated by two
important results. The first is expressed by the Onsager reciprocity relations (5),
which state that

$$L_{\rho\rho'} = L_{\rho'\rho}. \tag{2.6}$$

When the flow J_ρ, corresponding to the irreversible process ρ, is influenced
by the force $X_{\rho'}$ of the irreversible process ρ', then the flow $J_{\rho'}$ is also influenced
by the force X_ρ through the *same* coefficient.

The importance of the Onsager relations resides in their generality. They
have been submitted to many experimental tests. Their validity has for the
first time shown that nonequilibrium thermodynamics leads, as does equilibrium thermodynamics, to general results independent of any specific
molecular model. The discovery of the reciprocity relations corresponds really
to a turning point in the history of thermodynamics.

A second interesting theorem valid near equilibrium is the theory of minimum
entropy production (6). It states that for steady states sufficiently close to
equilibrium entropy production reaches its minimum. Time-dependent states
(corresponding to the same boundary conditions) have a higher entropy
production. It should be emphasized that the theorem of minimum entropy
production requires even more restrictive conditions than the linear relations
(2.5). It is valid in the frame of a "strictly" linear theory in which the deviations from equilibrium are so small that the phenomenological coefficients
$L_{\rho\rho'}$ may be treated as constants.

The theorem of minimum entropy production expresses a kind of "inertial"
property of nonequilibrium systems. When given boundary conditions prevent
the system from reaching thermodynamic equilibrium (that is, zero entropy

production) the system settles down to the state of "least dissipation".

It was clear since the formulation of this theorem that this property is strictly valid only in the neighborhood of equilibrium. For many years great efforts were made to generalize this theorem to situations further away from equilibrium. It came as a great surprise when it was finally shown that far from equilibrium the thermodynamic behavior could be quite different, in fact, even opposite to that indicated by the theorem of minimum entropy production.

It is remarkable that this new type of behavior appears already in typical situations studied in classical hydrodynamics. The example which was first analyzed from this point of view is the so-called "Bénard instability". Consider a horizontal layer of fluid between two infinite parallel planes in a constant gravitational field, and let us maintain the lower boundary at temperature T_1 and the higher boundary at temperature T_2 with $T_1 > T_2$. For a sufficiently large value of the "adverse" gradient $(T_1-T_2)/(T_1+T_2)$, the state of rest becomes unstable and convection starts. The entropy production is then increased as the convection provides a new mechanism of heat transport. Moreover the state of flow, which appears beyond the instability, is a state of organization as compared to the state of rest. Indeed a macroscopic number of molecules have to move in a coherent fashion over macroscopic times to realize the flow pattern.

We have here a good example of the fact that non-equilibrium may be a source of order. We shall see in Sections 3 and 4 that this situation is not limited to hydrodynamic situations but also occurs in chemical systems when well-defined conditions are imposed on the kinetic laws.

It is interesting to notice that Boltzmann's order principle as expressed by the canonical distribution would assign almost zero probability to the occurrence of Bénard convection. Whenever new coherent states occur far from equilibrium, the very concept of probability, as implied in the counting of number of complexions, breaks down. In the case of Bénard convection, we may imagine that there are always small convection currents appearing as fluctuations from the average state; but below a certain critical value of the temperature gradient, these fluctuations are damped and disappear. However, above some critical value certain fluctuations are amplified and give rise to a macroscopic current. A new supermolecular order appears which corresponds basically to a giant fluctuation stabilized by exchanges of energy with the outside world. This is the order characterized by the occurrence of "dissipative structures".

Before we discuss further the possibility of dissipative structures, let us briefly review some aspects of thermodynamic stability theory in relation to the theory of Lyapounov functions.

3. THERMODYNAMIC STABILITY THEORY

The states corresponding to thermodynamic equilibrium, or the steady states corresponding to a minimum of entropy production in linear non-equilibrium thermodynamics, are automatically stable. We have already introduced in Section 2 the concept of a Lyapounov function. According to the theorem of minimum entropy production the entropy production is precisely such a Lyapounov function in the strictly linear region around equilibrium. If the system is perturbed, the entropy production will increase, but the system reacts by coming back to the minimum value of the entropy production.

Similarly, closed equilibrium states are stable when corresponding to the maximum of entropy. If we perturb the system around its equilibrium value, we obtain

$$S = S_0 + \delta S + \tfrac{1}{2} \delta^2 S. \tag{3.1}$$

However, because the equilibrium state was a maximum, the first order term vanishes, and therefore the stability is given by the sign of the second order term $\delta^2 S$.

As Glansdorff and the author have shown $\delta^2 S$ is a Lyapounov function in the neighborhood of equilibrium *independently of the boundary conditions* (7).

Classical thermodynamics permits us to calculate explicitly this important expression. One obtains (8)

$$T \ \delta^2 S = - \left[\frac{C_v}{T} (\delta T)^2 + \frac{\rho}{\varkappa} (\delta v)^2_{N_\gamma} + \sum_{\gamma\gamma'} \mu_{\gamma\gamma'} \ \delta N_\gamma \ \delta N_{\gamma'} \right] < 0. \tag{3.2}$$

Here ρ is the density, $v = 1/\rho$ the specific volume (the index N_γ means that composition is maintained constant in the variation of v) $\varkappa$ the isothermal compressibility, N_γ the mole fraction of component γ and $\mu_{\gamma\gamma'}$ the derivative

$$\mu_{\gamma\gamma'} = \left(\frac{\partial \mu_\gamma}{\partial N_{\gamma'}} \right)_{pT}. \tag{3.3}$$

The basic stability conditions of classical thermodynamics first formulated by Gibbs are:

$$C_v > 0 \text{ (thermal stability)},$$
$$\varkappa > 0 \text{ (mechanical stability)},$$
$$\sum_{\gamma\gamma'} \mu_{\gamma\gamma'} \ \delta N_\gamma \ \delta N_{\gamma'} > 0 \text{ (stability with respect to diffusion)}.$$

These conditions imply that $\delta^2 S$ is a negative quadratic function. Moreover it can be shown by elementary calculations that the time derivative of $\delta^2 S$ is related to the entropy production P through (7) (see 2.3),

$$\frac{1}{2} \frac{\partial}{\partial t} \delta^2 S = \sum J_\rho X_\rho = P > 0. \tag{3.4}$$

It is precisely because of inequalities (3.2) and (3.4) that $\delta^2 S$ is a Lyapounov function. Its existence ensures the damping of all fluctuations. That is the reason why near equilibrium a macroscopic description for large systems is sufficient. Fluctuations can only play a subordinate role, appearing as corrections to the macroscopic laws which can be neglected for large systems.

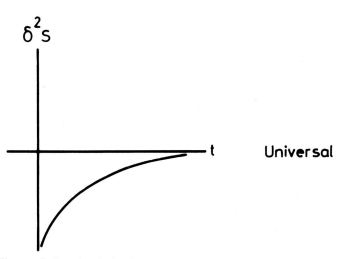

Fig. 3.1. Time evolution of second-order excess entropy (δ^2S)   around equilibrium.

We are now prepared to investigate the fundamental questions: Can we extrapolate this stability property further away from equilibrium? Does δ^2S play the role of a Lyapounov function when we consider larger deviations from equilibrium but still in the frame of macroscopic description? We again calculate the perturbation δ^2S but now around a nonequilibrium state. The inequality (3.2) still remains valid in the range of macroscopic description. However, the time derivative of δ^2S is no longer related to the total entropy production as in (3.4) but to the perturbation of this entropy production. In other words we now have (9),

$$\frac{1}{2}\frac{\partial}{\partial t}\delta^2S = \sum \delta J_\rho \, \delta X_\rho. \tag{3.5}$$

The right-hand side is what we called the "excess entropy production". Let us again emphasize that the δJ_ρ and δX_ρ are the deviations from the values J_ρ and X_ρ at the stationary state, the stability of which we are testing through a perturbation. Now contrary to what happens for equilibrium or near-equilibrium situations, the right-hand side of (3.5) corresponding to the excess entropy production has generally not a well-defined sign. If for all t larger than t_0, where t_0 is the starting time of the perturbation we have,

$$\sum_\rho \delta J_\rho \, \delta X_\rho \geqslant 0, \tag{3.6}$$

then δ^2S is indeed a Lyapounov function and stability is ensured (see fig. 3.2). Note that in the linear range the excess entropy production has the same sign as the entropy production itself and we recover the same result as with the theorem of minimum entropy production. However the situation changes in the far-from-equilibrium range. There the form of chemical kinetics plays an essential role.

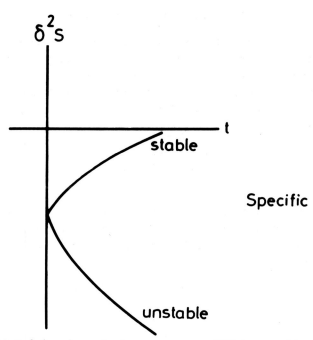

Fig. 3.2. Time evolution of second-order excess entropy ($\delta^2 S$) in case of (asymptotically) stable, marginally stable, and unstable situations.

In the next section we shall consider a few examples. For appropriate types of chemical kinetics the system may become unstable. This shows that there is an essential difference between the laws of equilibrium and the laws far away from equilibrium. The laws of equilibrium are universal. However, far from equilibrium the behavior may become very specific. This is of course a welcome circumstance, because it permits us to introduce a distinction in the behavior of physical systems which would be incomprehensible in an equilibrium world.

Note that all these considerations are very general. They may be extended to systems in which macroscopic motion may be generated or to problems involving surface tension or the effect of external field (10). For example in the case in which we include macroscopic motion we have to consider the expression (see Glansdorff and Prigogine (9)),

$$\delta^2 Z = \delta^2 S - \frac{1}{2} \int \frac{\rho u^2}{T} dV \leqslant 0, \qquad (3.7)$$

where u are the macroscopic convection velocities. We have integrated over the volume to take into account the space dependence of all u. We may again calculate the time derivative of $\delta^2 Z$ which takes now a more complicated form. As the result may be found elsewhere (9) we shall not reproduce it here. Let us only mention that spontaneous excitation of internal convection cannot be generated from a state at rest which is at thermodynamic equilibrium. This applies of course as a special case to the Bénard instability we have mentioned in Section 2.

Let us now return to the case of chemical reactions.

4. APPLICATION TO CHEMICAL REACTIONS

A general result is that to violate inequality (3.6) we need autocatalytic reactions. More precisely autocatalytic steps are necessary (but not sufficient) conditions for the breakdown of the stability of the thermodynamical branch. Let us consider a simple example. This is the so-called "Brusselator", which corresponds to the scheme of reactions (11),

$$
\begin{align}
\text{A} &\to \text{X} \tag{a} \\
2\text{X} + \text{Y} &\to 3\text{X} \tag{b} \\
\text{B} + \text{X} &\to \text{Y} + \text{D} \tag{c} \\
\text{X} &\to \text{E} \tag{d}
\end{align} \tag{4.1}
$$

The initial and final products are A, B, D, E, which are maintained constant while the concentrations of the two intermediate components, X and Y, may change in time. Putting the kinetic constants equal to one, we obtain the system of equations,

$$
\frac{d\text{X}}{dt} = \text{A} + \text{X}^2\text{Y} - \text{BX} - \text{X}
$$

$$
\frac{d\text{Y}}{dt} = \text{BX} - \text{X}^2\text{Y}, \tag{4.2}
$$

which admits the steady state

$$
\text{X}_0 = \text{A}, \qquad \text{Y}_0 = \frac{\text{B}}{\text{A}} \tag{4.3}
$$

Using the thermodynamic stability criterion or normal mode analysis we may show that solution (4.3) becomes unstable whenever

$$
\text{B} > \text{B}_c = 1 + \text{A}^2 \tag{4.4}
$$

Beyond this critical value of B we have a "limit cycle", that is, any initial point in the space X, Y tends to the same periodic trajectory. The important point is therefore that in contrast with oscillating chemical reactions of the Lotka-Volterra type the frequency of oscillation is a well defined function of the macroscopic variables such as concentrations, temperatures . . . The chemical reaction leads to coherent time behavior; it becomes a chemical clock. In the literature this is often called a Hopf bifurcation.

When diffusion is taken into account the variety of instabilities becomes quite amazing and for this reason the reaction scheme (4.1) has been studied by many authors over the past years. A special name has even been introduced—it is generally called the Brusselator. In the presence of diffusion, equations (4.2) now become

$$
\frac{\partial \text{X}}{\partial t} = \text{A} + \text{X}^2\text{Y} - \text{BX} - \text{X} + \text{D}_\text{X}\frac{\partial^2 \text{X}}{\partial r^2},
$$

$$
\frac{\partial \text{Y}}{\partial t} = \text{BX} - \text{X}^2\text{Y} + \text{D}_\text{Y}\frac{\partial^2 \text{Y}}{\partial r^2}. \tag{4.5}
$$

In addition to the limit cycle we have now the possibility of nonuniform steady states. We may call it the "Turing bifurcation" as Turing was the first to notice the possibility of such bifurcations in chemical kinetics in his classic paper on morphogenesis in 1952 (12). In the presence of diffusion the limit cycle may also become space dependent and lead to chemical waves.

Some order can be brought into the results by considering as the "basic solution" the one corresponding to the thermodynamic branch. Other solutions may then be obtained as successive bifurcations from this basic one, or as higher order bifurcations from a non-thermodynamic branch, taking place when the distance from equilibrium is increased.

A general feature of interest is that dissipative structures are very sensitive to global features which characterize the environment of chemical systems, such as their size and form, the boundary conditions imposed on their surface and so on. All these features influence in a decisive way the type of instabilities which lead to dissipative structures.

Far from equilibrium, there appears therefore an unexpected relation between chemical kinetics and the "space-time structure" of reacting systems. It is true that the interactions which determine the values of the relevant kinetic constants and transport coefficients result from short range interactions (valency forces, hydrogen bonds, Van der Waals forces). However, the solutions of the kinetic equations depend in addition on global characteristics. This dependence, which on the thermodynamic branch, near equilibrium, is rather trivial, becomes decisive in chemical systems working under far-from-equilibrium conditions. For example, the occurrence of dissipative structures generally requires that the system's size exceeds some critical value. The latter is a complex function of the parameters describing the reaction-diffusion processes. Therefore we may say that chemical instabilities involve long range order through which the system acts as a whole.

There are three aspects which are always linked in dissipative structures: the function as expressed by the chemical equations, the space-time structure, which results from the instabilities, and the fluctuations, which trigger the instabilities. The interplay between these three aspects

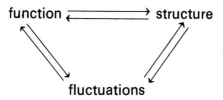

leads to most unexpected phenomena, including "order through fluctuations" which we shall analyze in the next sections.

Generally we have successive bifurcations when we increase the value of some characteristic parameter (like B in the Brusselator scheme).

solutions

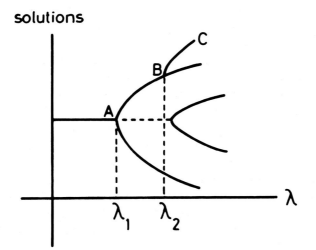

Fig. 4.1. Successive bifurcations.

On the Fig. 4.1. we have a single solution for the value λ_1, but multiple solutions for the value λ_2.

It is interesting that bifurcation introduces in a sense "history" into physics. Suppose that observation shows us that the system whose bifurcation diagram is represented by Fig. 4.1 is in the state C and came there through an increase of the value of λ. The interpretation of this state X implies the knowledge of the prior history of the system, which had to go through the bifurcation points A and B. In this way we introduce in physics and chemistry an "historical" element, which until now seemed to be reserved only for sciences dealing with biological, social, and cultural phenomena.

Every description of a system which has bifurcations will imply both deterministic and probabilistic elements. As we shall see in more detail in Section 5, the system obeys deterministic laws, such as the laws of chemical kinetics, between two bifurcations points, while in the neighborhood of the bifurcation points fluctuations play an essential role and determine the "branch" that the system will follow.

We shall not go here into the theory of bifurcations and its various aspects such as, for example, the theory of catastrophes due to René Thom (13). These questions are discussed in my recent monograph in collaboration with G. Nicolis (11). We shall also not enumerate the examples of coherent structures in chemistry and biology which are at present known. Again many examples may be found in Ref. 11.

5. THE LAW OF LARGE NUMBERS AND THE STATISTICAL DESCRIPTION OF CHEMICAL REACTIONS

Let us now turn to the statistical aspects of the formation of dissipative structures. Conventional chemical kinetics is based on the calculation of the average number of collisions and more specifically on the average number of reactive collisions. These collisions occur at random. However, how such a chaotic behaviour can ever give rise to coherent structures? Obviously a new feature has to come in. Briefly, this is the breakdown of the conditions of validity of the law of large numbers; as a result the distribution of reactive particles near instabilities is no more at random.

Let us first indicate what we mean by the law of large numbers. To do so we consider a typical probability description of great importance in many fields of science and technology, the Poisson distribution. This distribution involves a variable X which may take integer values $X = 0, 1, 2, 3, \ldots$ According to the Poisson distribution the probability of X is given by

$$\text{pr}(X) = e^{-<X>} \frac{<X>^X}{X!} \tag{5.1}$$

This law is found to be valid in a wide range of situations such as the distribution of telephone calls, waiting time in restaurants, fluctuations of particles in a medium of given concentration. In Eq. (5.1), $<X>$ corresponds to the average value of X. An important feature of the Poisson distribution is that $<X>$ is the only parameter which enters in the distribution. The probability distribution is entirely determined by its mean.

From (5.1), one obtains easily the so-called "variance" which gives the dispersion around the mean

$$<(\delta X)^2> = <(X - <X>)^2>. \tag{5.2}$$

The characteristic feature is that according to the Poisson distribution the dispersion is equal to the average itself,

$$<(\delta X)^2> = <X>. \tag{5.3}$$

Let us consider a situation in which X is an extensive quantity proportional to the number of particles N (in a given volume) or to the volume V. We then obtain for the *relative* fluctuations the famous square root law

$$\frac{\sqrt{<(\delta X)^2>}}{<X>} = \frac{1}{\sqrt{<X>}} \sim \frac{1}{\sqrt{N}} \text{ or } \frac{1}{\sqrt{V}}. \tag{5.4}$$

The order of magnitude of the relative fluctuation is inversely proportional to the square root of the average. Therefore, for extensive variables of order N we obtain relative deviations of order $N^{-1/2}$. This is the characteristic feature of the law of large numbers. As a result we may disregard fluctuations for large systems and use a macroscopic description.

For other distributions the mean square deviation is no more equal to the average as in (5.3). But whenever the law of large numbers applies, the order of magnitude of the mean square deviation is still the same, and we have

$$\frac{<(\delta X)^2>}{V} \sim \text{finite for } V \to \infty. \tag{5.5}$$

Let us now consider a stochastic model for chemical reactions. As has been done often in the past, it is natural to associate a Markov chain process of the "birth and death" type to a chemical reaction (14). This leads immediately to a Master Equation for the probability $P(X, t)$ of finding X molecules of species X at time t,

$$\frac{dP(X, t)}{dt} = \sum_r W(X - r \to X)\, P(X - r, t) - \sum_r W(X \to X + r)\, P(X, t). \quad (5.6)$$

On the right-hand side we have a competition between "gain" and loss terms. A characteristic difference with the classical Brownian motion problem is that the transition probabilities, $W(X - r \to X)$ or $W(X \to X + r)$, are non-linear in the occupation numbers. Chemical games are non-linear and this leads to important differences. For example it can be easily shown that the stationary distribution of X corresponding to the linear chemical reaction

$$A \rightleftharpoons X \leftrightharpoons F, \quad (5.7)$$

is given by a Poisson distribution (for given average values of A and F) (15). But it came as a great surprise when Nicolis and the author showed in 1971 (16) that the stationary distribution of X which appears as an intermediate in the chain,

$$\begin{aligned} A + M &\to X + M \\ 2X &\to E + D, \end{aligned} \quad (5.8)$$

is no more given by the Poisson distribution.

This is very important from the point of view of the macroscopic kinetic theory. Indeed, as has been shown by Malek-Mansour and Nicolis (17), the macroscopic chemical equations have generally to be corrected by terms associated with deviations from the Poissonian. This is the basic reason why today so much attention is devoted to the stochastic theory of chemical reactions.

For example the Schlögl reaction (18),

$$\begin{aligned} A + 2X &\leftrightharpoons 3X \\ X &\leftrightharpoons B, \end{aligned} \quad (5.9)$$

has been studied extensively by Nicolis and Turner (19) who have shown that this model leads to a "non equilibrium phase transition" quite similar to that described by the classical Van der Waals equation. Near the critical point as well as near the coexistence curve the law of large numbers as expressed by (5.5) breaks down, as $<(\delta X)^2>$ becomes proportional to a higher power of the volume. As in the case of equilibrium phase transitions, this breakdown can be expressed in terms of critical indices.

In the case of equilibrium phase transitions, fluctuations near the critical point have not only a large amplitude but they also extend over large distances. Lemarchand and Nicolis (20) have investigated the same problem for non-equilibrium phase transitions. To make the calculations possible, they considered a sequence of boxes. In each box the Brusselator type of reaction (4.1) is taking place. In addition, there is diffusion between one box and the other.

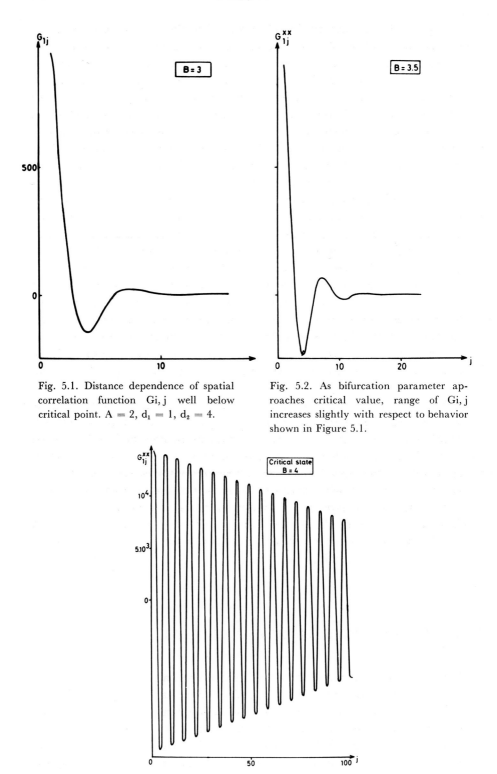

Fig. 5.1. Distance dependence of spatial correlation function Gi, j well below critical point. A = 2, d₁ = 1, d₂ = 4.

Fig. 5.2. As bifurcation parameter aproaches critical value, range of Gi, j increases slightly with respect to behavior shown in Figure 5.1.

Fig. 5.3. Critical behavior of spatial correlation function Gi, j for same values of parameters as in Fig. 5.1. Correlation function displays both linear damping with distance and spatial oscillations with wavelength equal to that of macroscopic concentration pattern.

Using the Markov method they then calculated the correlation between the occupation numbers of X in two different boxes. One would expect that chemical inelastic collisions together with diffusion would lead to a chaotic behavior. But that is not so. In Figures 5.1—5.3, the correlation functions for below and near the critical state are represented graphically. It is clearly seen that near the critical point we have long range chemical correlations. Again the system acts as a *whole* inspite of the short-range character of the chemical interactions. Chaos gives rise to order. Moreover numerical simulations indicate that it is only in the limit of number of particles, $N \rightarrow \infty$, that we tend to "long range" temporal order.

To understand at least qualitatively this result let us consider the analogy with phase transitions. When we cool down a paramagnetic substance, we come to the so-called Curie point below which the system behave slike a ferromagnet. Above the Curie point, all directions play the same role. Below, there is a privileged direction corresponding to the direction of magnetization.

Nothing in the macroscopic equation determines which direction the magnetization will take. In principle, all directions are equally likely. If the ferromagnet would contain a finite number of particles, this privileged direction would not be maintained in time. It would rotate. However, if we consider an infinite system, then no fluctuations whatsoever can shift the direction of the ferromagnet. The long-range order is established once and for all.

There is a striking similarity with the case of oscillating chemical reactions. When we increase the distance from equilibrium, the system begins to oscillate. It will move along the limit cycle. The phase on the limit cycle is determined by the initial fluctuation, and plays the same role as the direction of magnetization. If the system is finite, fluctuations will progressively take over and perturb the rotation. However, if the system is infinite, then we may obtain a long-range temporal order very similar to the long-range space order in the ferromagnetic system. We see therefore, that the appearance of a periodic reaction is a time-symmetry breaking process exactly as ferromagnetism is a space-symmetry breaking one.

6. THE DYNAMIC INTERPRETATION OF THE LYAPOUNOV FUNCTION

We shall now consider more closely the dynamic meaning of the entropy and more specifically of the Lyapounov function $\delta^2 S$ we have used previously.

Let us start with a very brief summary of Boltzmann's approach to this problem. Even today Boltzmann's work appears as a milestone. It is well known that an essential element in Boltzmann's derivation of the H-theorem was the replacement of the exact dynamic equations (as expressed by the Liouville equation to which we shall come back later) by this kinetic equation for the velocity distribution function f of the molecules,

$$\frac{\partial f}{\partial t} + v \frac{\partial f}{\partial x} = \int dw \, dv \; \sigma[f'f_1' - f f_1] \qquad (6.1)$$

where the notation is standard.

Once this equation is admitted it is easy to show that Boltzmann's H quantity,

$$H = \int dv \, f \log f, \qquad (6.2)$$

satisfies the inequality

$$\frac{dH}{dt} < 0, \qquad (6.3)$$

and plays therefore the role of a Lyapounov function.

The progress achieved through Boltzmann's approach is striking. Still many difficulties remain (21). First we have "practical difficulties", as for example the difficulty to extend Boltzmann's results to more general situations (for example dense gases). Kinetic theory has made striking progress in the last years; yet when one examines recent texts on kinetic theory or non-equilibrium statistical mechanics one does not find anything similar to Boltzmann's H-theorem which remains valid in more general cases. Therefore Boltzmann's result remains quite isolated, in contrast to the generality we attribute to the second law of thermodynamics.

In addition we have "theoretical difficulties". The most serious is probably Loschmidt's reversibility paradox. In brief, if we reverse the velocities of the molecules, we come back to the initial state. During this approach to the initial state Boltzmann's H-theorem (6.3) is violated. We have "anti-thermodynamic behavior". This conclusion can be verified, for example by computer simulations.

The physical reason for the violation of Boltzmann's H-theorem lies in the long-range correlations introduced by the velocity inversion. One would like to argue that such correlations are exceptional and may be disregarded. However, how should one find a criterion to distinguish between "abnormal" correlations and normal correlations especially when dense systems are considered?

The situation becomes even worse when we consider, instead of the velocity distribution, a Gibbs ensemble corresponding to phase density ρ. Its time evolution is given by the Liouville equation,

$$i \frac{\partial \rho}{\partial t} = L\rho, \qquad (6.4)$$

where

$$L\rho \Big< \begin{array}{l} \text{Poisson bracket } i\{H, \rho\} \text{ in classical dynamics} \\ \text{the commutator } [H, \rho] \text{ in quantum mechanics.} \end{array} \qquad (6.4')$$

If we consider positive convex functionals such as

$$\Omega = \int \rho^2 \, dp \, dq > 0, \qquad (6.5)$$

or in quantum mechanics,

$$\Omega = \mathrm{tr}\, \rho^+ \rho > 0, \tag{6.6}$$

it is easily shown that, as a consequence of Liouville's equation (6.4),

$$\frac{d\Omega}{dt} = 0. \tag{6.7}$$

Therefore, Ω as defined in (6.5) or (6.6) is not a Lyapounov function, and the laws of classical or quantum dynamics seem to prevent us from constructing at all a Lyapounov functional which would play the role of the entropy.

For this reason it has often been stated that irreversibility can only be introduced into dynamics through supplementary approximation such as coarse-graining added to the laws of dynamics (22).

I have always found it difficult to accept this conclusion especially because of the constructive role of irreversible processes. Can dissipative structures be the result of mistakes?

We obtain a hint about the direction in which the solution of this paradox may lie by inquiring why Boltzmann's kinetic equation permits one to derive an H-theorem while Liouville equation does not. Liouville's equation (6.4) is obviously Lt-invariant. If we reverse both the sign of L (this can be done in classical dynamics by velocity inversion) and the sign of t, the Liouville equation remains invariant. On the other hand, it can be easily shown (21) that the collision term in the Boltzmann equation breaks the Lt-symmetry as it is even in L. We may therefore rephrase our question by asking: How can we break the Lt-symmetry inherent in classical or quantum mechanics? Our point of view has been the following: The dynamical and thermodynamical descriptions are, in a certain sense, "equivalent" representations of the evolution of the system connected by a non-unitary transformation. Let us briefly indicate how we may proceed. The method which we follow has been developed in close collaboration with my colleagues in Brussels and Austin (23, 24, 25).

7. NON-UNITARY TRANSFORMATION THEORY

As the expression (6.6) has proved inadequate we start with a Lyapounov function of the form,

$$\Omega = \mathrm{tr}\, \rho^+ M\rho \geqslant 0, \tag{7.1}$$

(where M is a positive operator), with a non-increasing time derivative,

$$\frac{d\Omega}{dt} \leqslant 0 \tag{7.2}$$

This is certainly not always possible. In simple dynamical situations when the motion is periodic either in classical or quantum mechanics, no Lyapounov function may exist as the system returns after some time to its initial state. The existence of M is related to the type of spectrum of the Liouville operator. In the frame of classical ergodic theory this question has been recently studied by Misra (26). Here we shall pursue certain consequences of the possible existence of the operator M in (7.1) which may be considered as a "microscopic representation of entropy". As this quantity is positive, a general theorem permits us to represent it as a product of an operator, say Λ^{-1}, and its hermitian conjugate $(\Lambda^{-1})^+$ (this corresponds to taking the "square root" of a positive operator),

$$M = (\Lambda^{-1})^+ \Lambda^{-1}. \tag{7.3}$$

Inserting this in (7.1) we get,

$$\Omega = \text{tr } \tilde{\rho}^+ \tilde{\rho}, \tag{7.4}$$

with

$$\tilde{\rho} = \Lambda^{-1} \rho. \tag{7.5}$$

This is a most interesting result, because expression (7.4) is precisely of the type that we were looking for in the first place. But we see that this expression can only exist in a "new" representation related to the preceding by the transformation (7.5).

First let us write the new equations of motion. Taking into account (7.5), we obtain

$$i \frac{\partial \tilde{\rho}}{\partial t} = \Phi \tilde{\rho}, \tag{7.6}$$

with

$$\Phi = \Lambda^{-1} L\Lambda. \tag{7.7}$$

Now let us use the solution of the equations of motion (6.4). We may replace (7.1) and (7.2) by the more explicit inequalities

$$\Omega(t) = \text{tr } \rho^+(o)e^{iLt} M e^{-itL} \rho(o) \leqslant 0, \tag{7.8}$$

$$\frac{d\Omega}{dt} = - \text{ tr } \rho^+(o)^{iLt} i(ML - LM)e^{-iLt} \rho(o) \leqslant 0. \tag{7.9}$$

The microscopic "entropy operator" M may therefore not commute with L. The commutator represents precisely what could be called the "microscopic entropy production".

We are of course reminded of Heisenberg's uncertainty relations and Bohr's complementarity principle. It is most interesting to find here also a non-commutativity, but now between dynamics as expressed by the operator L and "thermodynamics" as expressed by M. We therefore have a new and most interesting type of complementarity between dynamics, which implies the knowledge of trajectories or wave functions, and thermodynamics, which implies entropy.

When the transformation to the new representation is performed, we obtain for the entropy production (7.9),

$$\frac{d\Omega}{dt} = -\text{tr } \tilde{\rho}^+(o) \, e^{i\Phi^+ t} \, i(\Phi - \Phi^+) \, e^{-i\Phi t} \, \tilde{\rho}(o) \leqslant 0. \qquad (7.10)$$

This implies that the difference between Φ and its hermitian adjoint Φ^+ does not vanish,

$$i(\Phi - \Phi^+) \geqslant 0 \qquad (7.11)$$

Therefore we reach the important conclusion that the new operator of motion which appears in the transformed Liouville equation (7.6) can no longer be hermitian as was the Liouville operator L. This shows that we have to leave the usual class of unitary (or anti-unitary) transformations and to proceed to an extension of the symmetry of quantum mechanical operators. Fortunately, it is easy to determine the class of transformations which we have to consider now. Average values can be calculated both in the old and the new representation. The result should be the same; in other words, we require that

$$<A> = \text{tr } A^+ \rho = \text{tr } \tilde{A}^+ \tilde{\rho}. \qquad (7.12)$$

Moreover, we are interested in transformations which will depend explicitly on the Liouville operator. This is indeed the very physical motivation of the theory. We have seen, that the Boltzmann type equations have a broken Lt-symmetry. We want to realize precisely this new symmetry through our transformation (23). This can only be done by considering L-dependent transformations $\Lambda(L)$. Using finally the fact that the density ρ and the observables have the same equations of motion, but with L replaced by $-L$, we obtain the basic condition,

$$\Lambda^{-1}(L) = \Lambda^+(-L), \qquad (7.13)$$

which replaces here the usual condition of unitarity imposed on quantum mechanical transformations.

It is not astonishing that we do find a non-unitary transformation law. Unitary transformations are very much like changes in coordinates, which do not affect the physics of the problem. Whatever the coordinate system, the physics of the system remains unaltered. But here we are dealing with quite a different problem. We want to go from one type of description, the dynamic one, to another, the "thermodynamic" one. This is precisely the reason why we need a more drastic type of change in representation as expressed by the new transformation law (7.13).

We have called this transformation a "star-unitary" transformation and introduced the notation,

$$\Lambda^* (L) = \Lambda^+(-L). \qquad (7.14)$$

We shall call Λ^* the "star-hermitian" operator associated to Λ (star always means the inversion $L \to -L$). Then (7.13) shows that for star-unitary transformations, the inverse of the transformation is equal to its star-hermitian conjugate.

Let us now consider (7.7). Using the fact that L as well as (7.13) and (7.14), are hermitian, we obtain

$$\Phi^* = \Phi^+(-L) = -\Phi(L) \tag{7.15}$$

or

$$(i\,\Phi)^* = i\,\Phi \tag{7.16}$$

The operator of motion is "star-hermitian". This is a most interesting result. To be a star-hermitian, an operator may be either even under L-inversion (that is, it does not change sign when L is replaced by −L) or anti-hermitian and odd (odd means that it changes sign when L is replaced by −L). A general star-hermitian operator can therefore be written as,

$$i\,\Phi = (i\,\overset{e}{\Phi}) + (i\,\overset{o}{\Phi}). \tag{7.17}$$

Here the superscripts e and o refer respectively to the even and the odd part of the new time evolution operator Φ. The condition of dissipativity (7.11), which expresses the existence of a Lyapounof function Ω, now becomes

$$i\,\overset{e}{\Phi} > 0. \tag{7.18}$$

It is the even part which gives the "entropy production".

Let us summarize what has been achieved. We obtain a new form of microscopic equation (as is the Liouville equation in classical or quantum mechanics), but which displays explicitly a part which may be associated to a Lyapounov function. In other words, the equation

$$i\,\frac{\partial\tilde{\rho}}{\partial t} = (\overset{o}{\Phi} + \overset{e}{\Phi})\tilde{\rho} \tag{7.19}$$

contains a "reversible" part $\overset{o}{\Phi}$ and an "irreversible" part $\overset{e}{\Phi}$. The symmetry of this new equation is exactly the one of Boltzmann's phenomenological kinetic equation, as the flow term is odd and the collision term even in L-inversion.

The macroscopic thermodynamic distinction between reversible and irreversible processes has in this way been transposed into the microscopic description. We have obtained what could be considered as the "missing link" between microscopic reversible dynamics and macroscopic irreversible thermodynamics. The scheme is as follows:

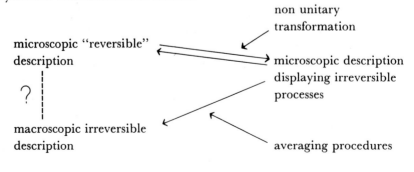

The effective construction of the Lyapounov function Ω, (7.1), through the transformation Λ involves a careful study of the singularities of the resolvent corresponding to the Liouville operator (24).

For small deviations from thermodynamical equilibrium it can be shown, as has been done recently together with Grecos and Theodosopulu (27), that the Lyapounov functional Ω, (7.1), reduces precisely to the macroscopic quantity $\delta^2 S$, (3.2), when in addition only the time evolution of conserved quantities is retained. We therefore have now established in full generality the link between non-equilibrium thermodynamics and statistical mechanics at least in the linear region. This is the extension of the result which was obtained long time ago in the frame of Boltzmann's theory, valid for dilute gases (28).

8. CONCLUDING REMARKS

Now a few concluding remarks.

The inclusion of thermodynamic irreversibility through a non-unitary transformation theory leads to a deep alteration of the structure of dynamics. We are led from groups to semigroups, from trajectories to processes. This evolution is in line with some of the main changes in our description of the physical world during this century.

One of the most important aspects of Einstein's theory of relativity is that we cannot discuss the problems of space and time independently of the problem of the velocity of light which limits the speed of propagation of signals. Similarly the elimination of "unobservables" has played an important role in the basic approach to quantum theory initiated by Heisenberg.

The analogy between relativity and thermodynamics has been often emphasized by Einstein and Bohr. We cannot propagate signals with arbitrary speed, we cannot construct a *perpetuum mobile* forbidden by the second law.

From the microscopic point of view this last interdiction means that quantities which are well defined from the point of view of mechanics cannot be observables if the system satisfies the second law of thermodynamics.

For example the trajectory of the system as a whole cannot be an observable. If it would, we could at every moment distinguish two trajectories and the concept of thermal equilibrium would lose its meaning. Dynamics and thermodynamics limit each other.

It is interesting that there are other reasons which at the present time seem to indicate that the relation between dynamic interaction and irreversibility may play a deeper role than was conceived till now.

In the classical theory of integrable systems, which has been so important in the formulation of quantum mechanics, all interactions can be eliminated by an appropriate canonical transformation. Is this really the correct prototype of dynamic systems to consider, especially when situations involving elementary particles and their interactions are considered? Do we not have first to go to a non-canonical representation which permits us to disentangle reversible and irreversible processes on the microscopic level and then only to eliminate the reversible part to obtain well defined but still interacting units?

These questions will probably be clarified in the coming years.

But already now the development of the theory permits us to distinguish various levels of time: time as associated with classical or quantum dynamics, time associated with irreversibility through a Lyapounov function and time associated with "history" through bifurcations. I believe that this diversification of the concept of time permits a better integration of theoretical physics and chemistry with disciplines dealing with other aspects of nature.

ACKNOWLEDGEMENTS

This lecture gives a survey of results which have been obtained in close collaboration with my colleagues in Brussels and Austin. It is impossible to thank them all individually. I want however to express my gratitude to Professor G. Nicolis and Professor J. Mehra for their help in the preparation of the final version of this lecture.

ABSTRACT

We have dealt with the fundamental conceptual problems that arise from the macroscopic and microscopic aspects of the second law of thermodynamics. It is shown that non-equilibrium may become a source of order and that irreversible processes may lead to a new type of dynamic states of matter called "dissipative structures". The thermodynamic theory of such structures is outlined. A microscopic definition of irreversible processes is given and a transformation theory is developed that allows one to introduce non-unitary equations of motion that explicitly display irreversibility and approach to thermodynamic equilibrium. The work of the Brussels group in these fields is briefly reviewed. We believe that we are only at the beginning of a new development of theoretical chemistry and physics in which thermodynamics concepts will play an ever increasing role.

REFERENCES

1. Planck, M., *Vorlesungen über Thermodynamik*, Leipzig, 1930 (English translation, Dover).
2. Prigogine, I., *Etude thermodynamique des phénomènes irréversibles*, Thèse, Bruxelles, 1945; published by Desoer, Liege, 1947.
3. This assumption is discussed in P. Glansdorff and Prigogine, I., *Thermodynamics of Structure, Stability and Fluctuations*, Wiley-Interscience, New York, 1971, Chapter II, p. 14.
4. The standard reference for the linear theory of irreversible processes is the monograph by de Groot, S. R. & Mazur, P., *Non-Equilibrium Thermodynamics*, North-Holland Publishing Co., Amsterdam, 1969.
5. Onsager, L., Phys. Rev. *37*, 405 (1931).
6. Prigogine, I., Bull. Acad. Roy. Belg. Cl. Sci. *31*, 600 (1945).
7. Glansdorff, P. and Prigogine, I., *Thermodynamics of Structure, Stability and Fluctuations*, Wiley-Interscience, New York, 1971, Chapter V.
8. Loc. cit., Ref. 7, p. 25.
9. Loc. cit., Ref. 7, Chapter VII.
10. Defay, R., Prigogine, I., and Sanfeld, A., Jr. of Colloid and Interface Science, *58*, 498 (1977).
11. The original references to, as well as many of the properties of, this reaction scheme can be found in Nicolis, G. and Prigogine, I., *Self-Organization in Nonequilibrium Systems*, Wiley-Interscience, New York, 1977. See especially Chapter VII.
12. Turing, A. M., Phil. Trans. Roy. Soc. Lond., *B237*, 37 (1952).
13. Thom, R., *Stabilité Structurelle et Morphogénèse*, Benjamin, New York, 1972.
14. A standard reference is Barucha-Reid, A. T., *Elements of the Theory of Markov Processes and Their Applications*, Mc Graw-Hill, New York, 1960.
15. Nicolis, G. and Babloyantz, A., J. Chem. Phys. *51*, 2632 (1969).
16. Nicolis, G. and Prigogine, I., Proc. Natl. Acad. Sci. (U.S.A.), *68*, 2102 (1971).
17. Malek-Mansour, M. and Nicolis, G., J. Stat. Phys. *13*, 197 (1975).
18. Schlögl, F., Z. Physik, *248*, 446 (1971); Z. Physik, *253*, 147 (1972).
19. Nicolis, G. and Turner, J. W., Physica *A* (1977).
20. Lemarchand, H. and Nicolis, G., Physica *82A*, 521 (1976).
21. Prigogine, I., The Statistical Interpretation of Entropy, in *The Boltzmann Equation*, (Eds.) Cohen, E. G. D. and Thirring, W., Springer-Verlag, 1973, pp. 401—449.
22. An elegant presentation of this point of view is contained in Uhlenbeck, G. E., Problems of Statistical Physics, in *The Physicist's Conception of Nature*, (Ed.) Mehra, J., D. Reidel Publishing Co., Dordrecht, 1973, pp. 501—513.
23. Prigogine, I., George, C., Henin, F., and Rosenfeld, L., Chemica Scripta, *4*, 5 (1973).
24. Grecos, A., Guo, T. and Guo, W., Physica *80A*, 421 (1975).
25. Prigogine, I., Mayné, F., George, C., and de Haan, M., Proc. Natl. Acad. Sci. (U.S.A.), *74*, 4152 (1977).
26. Misra, B., Proc. Natl. Acad. Sci. (U.S.A), *75*, 1629 (1978).
27. Theodosopulu, M., Grecos, A., and Prigogine, I., Proc. Natl. Acad. Sci. (U.S.A.), *75*, 1632 (1978); also Grecos, A. and Theodosopulu, M., Physica, to appear.
28. Prigogine, I., Physica *14*, 172 (1949); Physica *15*, 272 (1949).

Chemistry 1978

PETER MITCHELL

for his contribution to the understanding of biological energy transfer through the formulation of the chemiosmotic theory

THE NOBEL PRIZE FOR CHEMISTRY

Speech by Professor LARS ERNSTER of the Royal Academy of Sciences
Translation from the Swedish text

Your Majesties, Your Royal Highnesses, Ladies and Gentlemen,

The discoveries for which Peter Mitchell has been awarded this year's Nobel Prize for Chemistry relate to a field of biochemistry often referred to in recent years as bioenergetics, which is the study of those chemical processes responsible for supplying energy to living cells.

All living organisms need energy to survive. Muscular work, thought processes, growth, and reproduction are all examples of biological activities that require energy. We know today that every living cell is capable, by means of suitable catalysts, of deriving energy from its environment, converting this energy into a biologically useful chemical form, and utilizing it for various energy-requiring processes.

Green plants and other photosynthetic organisms derive energy directly from sunlight – the ultimate source of energy for all life on Earth – and utilize this energy to convert carbon dioxide and water into organic compounds. Other organisms, including all animals and many bacteria, are dependent for their existence on organic compounds which they take up as nutrients from their environment. Through a process called cell respiration these compounds are oxidized by atmospheric oxygen to carbon dioxide and water with a concomitant release of energy.

Both respiration and photosynthesis involve a series of oxidation-reduction (or electron-transport) reactions in which energy is liberated and utilized for the synthesis of adenosine triphosphate (ATP) from adenosine diphosphate (ADP) and inorganic phosphate. These processes are usually called oxidative and photosynthetic phosphorylation. Both processes are typically associated with cellular membranes. In higher cells, they take place in special, membrane-enclosed organelles, called mitochondria and chloroplasts, while, in bacteria, both processes are associated with the cell membrane.

ATP serves as a universal energy currency for living cells. This compound is split by a variety of specific enzymes and the energy released is used for various energy-requiring processes. The regeneration of ATP by way of oxidative and photosynthetic phosphorylation thus plays a fundamental role in the energy supply of living cells.

The above concepts had been broadly outlined by about the middle of the 1950's, but the exact mechanisms by which electron transport is coupled to ATP synthesis in oxidative and photosynthetic phosphorylation remained unknown. Many hypotheses were formulated, most of which postulated the ocurrence of 'energy-rich' chemical compounds of more or less well-defined structures as intermediates between the electron-transport and ATP-synthesizing systems. Despite intensive efforts in many laboratories, however, no experimental evidence could be obtained for these hypotheses. In addition, these hypotheses did not provide a rational explanation for the need for a membrane in oxidative and photosynthetic phosphorylation.

At this stage, in 1961, Peter Mitchell put forward his chemiosmotic hypothesis. The basic idea of this hypothesis is that the enzymes of the electron-transport and ATP-synthesizing systems are localized in the membrane with a well-defined orientation and are functionally linked to a vectorial transfer of positively charged hydrogen ions, or protons, across the membrane. Thus, electron transport will give rise to an electrochemical proton gradient across the membrane which can serve as a driving force for ATP synthesis. A requisite for the establishment of a proton gradient is, of course, that the membrane itself is impermeable to protons, which explains the need for an intact membrane structure in oxidative and photosynthetic phosphorylation.

The chemiosmotic hypothesis was received with reservation by many workers in the field which is, in a way, understandable, since it was unorthodox, fairly provocative, and based on little experimental evidence. Perhaps due to just these features, however, the hypothesis stimulated a great deal of activity; and it can be stated without exaggeration that during the last decade the chemiosmotic hypothesis has been the dominating issue in the field of bioenergetics both in the literature, at scientific meetings and, not least, in laboratories all over the world. As a result, a great deal of experimental data has been accumulated, both from Mitchell's own laboratory—there mostly in collaboration with Dr. Jennifer Moyle—and from other places, which strongly supports the hypothesis. In fact, the basic postulates of the chemiosmotic hypothesis are today generally regarded as experimentally proven, thus making it a fundamental theory of cellular bioenergetics.

To understand the detailed mechanisms by which protons interact with and are translocated by the electron-transport and ATP-synthesizing systems will require further work. It is already clear, on the other hand, that the principle of power transmission by protonmotive force—or, as Mitchell has recently begun to call it, "proticity" (in analogy with electricity)—will be applicable to a wide range of biological processes beyond those involved in oxidative and photosynthetic phosphorylation. Uptake of neutrients by bacterial cells, intracellular transport of ions and metabolites, generation of reducing power for biosyntheses, biological heat production, bacterial motion and chemotaxis are examples of energy-requiring biological processes already known to be driven by proticity.

Finally, a practical aspect should be mentioned. The discovery that membrane-bound energy-transducing enzymes are so constructed that they can generate an electrochemical potential may be of great practical interest. Chloroplasts, mitochondria and bacteria may be regarded as naturally occurring solar cells and fuel cells, and may as such serve as models, and in the future perhaps also as tools, in energy technology. Obviously, once again, Nature has preceded man in inventiveness and may help him with her millions of years of experience in his daily struggle for life.

Dr. Mitchell,

With ingenuity, courage and persistence you have innovated one of the classical fields of biochemistry. Your chemiosmotic theory has meant a breakthrough that has opened up new insights into the fundamental problems of bioenergetics. The details may need completion and adjustment; but the edifice you have raised will stand.

It is my great pleasure and privilege to convey to you the congratulations of the Royal Swedish Academy of Sciences on your outstanding achievements and to ask you to receive the Nobel Prize for Chemistry of 1978 from the hands of His Majesty the King.

Peter Mitchell

PETER MITCHELL

Peter Mitchell was born in Mitcham, in the County of Surrey, England, on September 29, 1920. His parents, Christopher Gibbs Mitchell and Kate Beatrice Dorothy (née) Taplin, were very different from each other temperamentally. His mother was a shy and gentle person of very independent thought and action, with strong artistic perceptiveness. Being a rationalist and an atheist, she taught him that he must accept responsibility for his own destiny, and especially for his failings in life. That early influence may well have led him to adopt the religious atheistic personal philosophy to which he has adhered since the age of about fifteen. His father was a much more conventional person than his mother, and was awarded the O.B.E. for his success as a Civil Servant.

Peter Mitchell was educated at Queens College, Taunton, and at Jesus College, Cambridge. At Queens he benefited particularly from the influence of the Headmaster, C. L. Wiseman, who was an excellent mathematics teacher and an accomplished amateur musician. The result of the scholarship examination that he took to enter Jesus College Cambridge was so dismally bad that he was only admitted to the University at all on the strength of a personal letter written by C. L. Wiseman. He entered Jesus College just after the commencement of war with Germany in 1939. In Part I of the Natural Sciences Tripos he studied physics, chemistry, physiology, mathematics and biochemistry, and obtained a Class III result. In part II, he studied biochemistry, and obtained a II-I result for his Honours Degree.

He accepted a research post in the Department of Biochemistry, Cambridge, in 1942 at the invitation of J. F. Danielli. He was very fortunate to be Danielli's only Ph.D. student at that time, and greatly enjoyed and benefited from Danielli's friendly and unauthoritarian style of research supervision. Danielli introduced him to David Keilin, whom he came to love and respect more than any other scientist of his acquaintance.

He received the degree of Ph.D. in early 1951 for work on the mode of action of penicillin, and held the post of Demonstrator at the Department of Biochemistry, Cambridge, from 1950 to 1955. In 1955 he was invited by Professor Michael Swann to set up and direct a biochemical research unit, called the Chemical Biology Unit, in the Department of Zoology, Edinburgh University, where he was appointed to a Senior Lectureship in 1961, to a Readership in 1962, and where he remained until acute gastric ulcers led to his resignation after a period of leave in 1963.

From 1963 to 1965, he withdrew completely from scientific research, and acted as architect and master of works, directly supervising the restoration of an attractive Regency-fronted Mansion, known as Glynn House, in the beautiful wooded Glynn Valley, near Bodmin, Cornwall—adapting and furnishing a major part of it for use as a research laboratory. In this, he was lucky to receive the enthusiastic support of his former research colleague Jennifer Moyle. He and Jennifer Moyle founded a charitable company, known as Glynn Research Ltd., to promote fundamental biological research and finance the work of the Glynn Research Laboratories at Glynn House. The original endowment of about £250,000 was donated about equally by Peter Mitchell and his elder brother Christopher John Mitchell.

In 1965, Peter Mitchell and Jennifer Moyle, with the practical help of one technician, Roy Mitchell (unrelated to Peter Mitchell), and with the administrative help of their company secretary, embarked on the programme of research on chemiosmotic reactions and reaction systems for which the Glynn Research Institute has become known. Since its inception, the Glynn Research Institute has not had sufficient financial resources to employ more than three research workers, including the Research Director, on its permanent staff. He has continued to act as Director of Research at the Glynn Research Institute up to the present time. An acute lack of funds has recently led to the possibility that the Glynn Research Institute may have to close.

Beside his interest in communication between molecules, Peter Mitchell has become more and more interested in the problems of communication between individual people in civilised societies, especially in the context of the spread of violence in the increasingly collectivist societies in most parts of the world. His own experience of small and large organisations in the scientific world has led him to regard the small organisations as being, not only more alive and congenial, but also more effective, for many (although perhaps not all) purposes. He would therefore like to have the opportunity to become more deeply involved in studies of the ways in which sympathetic communication and cooperative activity between free and potentially independent people may be improved. One of his specific interests in this field of knowledge is the use of money as an instrument of personal responsibility and of choice in free societies, and the flagrant abuse and basically dishonest manipulation of the system of monetary units of value practised by the governments of most nations.

Awards and affiliations:
CIBA Medal and Prize, British Biochemical Society, 1973; Member, European Molecular Biology Organisation, 1973; Fellowship of the Royal Society, 1974; Warren Triennial Prize, jointly with Efraim Racker, U.S.A., 1974; Louis and Bert Freedman Foundation Award, New York Academy of Sciences, 1974; Honorary Member, American Society of Biological Chemists, 1975; Foreign Honorary Member, American Academy of Arts and Sciences, 1975; Wilhelm Feldberg (Anglo/German) Foundation Prize, 1976; Dr. rerum naturalium honoris causa of the Technische Universität, Berlin, 1976; Lewis S. Rosenstiel Award, U.S.A., 1977; Foreign Associate, National Academy of Sciences, U.S.A., 1977; Honorary Degree of Doctor of Science, Exeter University, U.K., 1977; Sir Hans Krebs Lecture and Medal of the Federation of European Biochemical Societies, 1978; Honorary Degree of Doctor of Science, University of Chicago, 1978.

Peter Mitchell died in 1992.

DAVID KEILIN'S RESPIRATORY CHAIN CONCEPT AND ITS CHEMIOSMOTIC CONSEQUENCES

Nobel Lecture, 8 December, 1978
by
PETER MITCHELL
Glynn Research Institute, Bodmin, Cornwall, U. K.

0. INTRODUCTION

It was obviously my hope that the chemiosmotic rationale of vectorial metabolism and biological energy transfer might one day come to be generally accepted, and I have done my best to argue in favour of that state of affairs for more than twenty years. But, it would have been much too presumptuous to have expected it to happen. Of course, I might have been wrong, and in any case, was it not the great Max Planck (1928, 1933) who remarked that a new scientific idea does not triumph by convincing its opponents, but rather because its opponents eventually die? The fact that what began as the chemiosmotic hypothesis has now been acclaimed as the chemiosmotic theory—at the physiological level, even if not at the biochemical level—has therefore aroused in me emotions of astonishment and delight in full and equal measure, which are all the more heartfelt because those who were formerly my most capable opponents are still in the prime of their scientific lives.

I shall presently explain the difference between the physiological and the biochemical levels at which the chemiosmotic theory has helped to promote useful experimental research. But let me first say that my immediate and deepest impulse is to celebrate the fruition of the creative work and benevolent influence of the late David Keilin, one of the greatest of biochemists and—to me, at least—the kindest of men, whose marvellously simple studies of the cytochrome system, in animals, plants and microorganisms (Keilin, 1925), led to the original fundamental idea of aerobic energy metabolism: the concept of the respiratory chain (Keilin, 1929; and see Nicholls, 1963; King, 1966). Perhaps the most fruitful (and surprising) outcome of the development of the notion of chemiosmotic reactions is the experimental stimulus and guidance it has provided in work designed to answer the following three elementary questions about respiratory chain systems and analogous photoredox chain systems: What is it? What does it do? How does it do it? The genius of David Keilin led to the revelation of the importance of these questions. In this lecture, I hope to show that, as a result of the painstaking work of many biochemists, we can now answer the first two in general principle, and that considerable progress is being made in answering the third.

Owing to the broad conceptual background, and the very wide range of practical application of the chemiosmotic theory, I have had to be rather

selective in choosing aspects to review. I have chosen to consider the
evolution of the chemiosmotic theory from earlier fundamental biochemical and physicochemical concepts in three perspectives: 1, a middle-dis-
tance physiological-cum-biochemical perspective; 2, a longer physicochemical view; and 3, a biochemical close-up. These perspectives involve general
considerations of biochemical theory and knowledge, and scope is not
available to discuss experimental procedures on this occasion.

1. PHYSIOLOGICAL-CUM-BIOCHEMICAL PERSPECTIVE

A. *Oxidative and Photosynthetic Phosphorylation*

During the two decades between 1940 and 1960, the mechanism of oxidative phosphorylation (by which some 95 % of the energy of aerobic organisms is obtained), and the basically similar mechanism of photosynthetic
phosphorylation (by which much of the energy available from plant products is initially harvested from the sun) was recognised as one of the great
unsolved problems of biochemistry. At the beginning of this period, the
work of David Keilin (1925, 1929) on the cytochrome system, and work by
Warburg, Wieland and others on the respiratory hydrogen carriers, had
already led to the concept of the respiratory chain: a water-insoluble
complex of redox carriers, operating serially between the reducing substrates or coenzymes and molecular oxygen (see Nicholls, 1963; King,
1966).

As indicated in Fig. 1, according to Keilin's chemically simple concept of

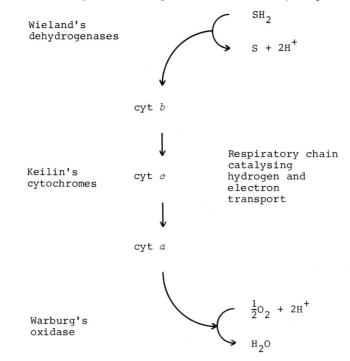

Fig. 1. David Keilin's chemically simple respiratory chain concept.

the respiratory chain, the respiratory-chain carriers (or their complexes of molecular dimensions) were involved chemically only in redox reactions. However, when, following the pioneer work of Kalckar (1937), Belitser & Tsybakova (1939), Ochoa (1940), Lipmann (1941, 1946), Friedkin & Lehninger (1948), and Arnon, Whatley & Allen (1954), attention was directed to the mechanism by which the redox process was coupled to the phosphorylation of ADP in respiratory and photoredox metabolism, it was natural for the metabolic enzymologists who were interested in this problem to use substrate-level phosphorylation as the biochemical model, and to assume that the mechanism of coupling between oxidation and phosphorylation in respiratory and photoredox chains would likewise be explained in terms of the classically scalar idiom of metabolic enzymology (see Slater, 1976).

In 1953, the general chemical coupling hypothesis, summarised in Fig. 2, was given formal expression in a historically important paper by Slater, which defined the reactions of the energy-rich intermediates at several coupling sites along the mitochondrial respiratory chain. Accordingly,

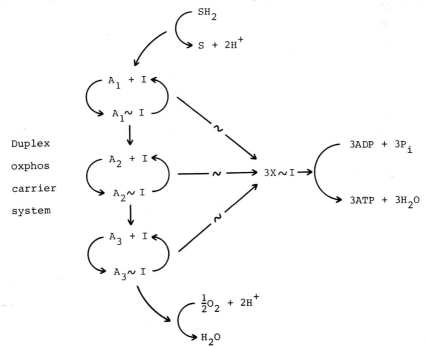

Fig. 2. Phosphorylating respiratory chain: Chemical coupling hypothesis. A~I and X~I represent hypothetical high-energy chemical intermediates; and the symbol ~ represents the so-called high-energy bond.

many expert metabolic enzymologists made it their prime objective to identify the energy-rich intermediates or other coupling factors supposed to be responsible for coupling oxidoreduction to phosphorylation in redox chain systems (Slater, 1953; Boyer et al., 1954; Chance & Williams, 1956;

Slater, 1958; Lehninger, 1959; Slater & Hulsmann, 1959; Chance, 1961; Racker, 1961; Lehninger & Wadkins, 1962; Williams, 1962; Boyer, 1963; Green et al., 1963; Hatefi, 1963; Ernster & Lee, 1964; Lardy et al., 1964; Griffiths, 1965; Racker, 1965; Sanadi, 1965; Slater, 1966; Chance et al., 1967). This development caused Keilin's chemically-simple concept of the respiratory chain to be almost universally rejected in favour of a chemically duplex concept according to which respiratory chain components participated directly, not only in the known redox changes, but also in other chemical changes involving the energy-rich intermediates—just as the phosphorylating glyceraldehyde-3-phosphate dehydrogenase is involved in both oxidative and phosphorylative reactions.

By the end of the two and a half decades between 1940 and 1965, the field of oxidative phosphorylation was littered with the smouldering conceptual remains of numerous exploded energy-rich chemical intermediates; the remarkable uncoupling action of 2,4-dinitrophenate and of other chemically unrelated reagents, and of physical membrane-lytic treatments, remained obscure; and the process of hypothesis-building, needed to keep faith with the chemical-coupling notion, reached such fantastic proportions as to be hardly intelligible to those outside the field (see Mitchell, 1961 c, 1966, 1967 a). Nevertheless, the quest for the energy-rich intermediates continued through the nineteensixties and persisted into the nineteenseventies with only a minor broadening of the conception of the type of coupling mechanism favoured by many of the metabolic enzymologists (Painter & Hunter, 1970; Storey, 1970, 1971; Slater, 1971, 1974, 1975; Chance, 1972, 1974; Hatefi & Hanstein, 1972; Wang, 1972; Cross & Boyer, 1973; Boyer et al., 1973, 1977; Ernster et al., 1973, 1974; Green, 1974; Weiner & Lardy, 1974; Griffiths et al., 1977; Nordenbrand et al., 1977). This conceptual broadening, which began to occur during the early nineteensixties, stemmed from ingenious suggestions by Boyer, Chance, Ernster, Green, Slater, Williams and others (see Boyer, 1965; Slater, 1971, 1974; Ernster et al., 1973). As indicated by Fig. 3, they assumed that coupling may be achieved through a direct conformational or other non-osmotic physical or chemical interaction—that might, for example, involve protons as a localised anhydrous chemical intermediate (Williams, 1962, 1970; and see Ernster, 1977 a; Mitchell, 1977 a), or might involve electrical interaction (Green, 1974)—between the redox carrier proteins and certain catalytic components associated with ATP synthesis in the supposedly duplex respiratory chain system, often described as the "phosphorylating respiratory chain".

Soon after 1950, it began to be recognised that the water-insoluble property of preparations of respiratory chain and photoredox chain complexes was related to the circumstance that, in their native state, these complexes were part of the lipid membrane system of bacteria, mitochondria and chloroplasts. But, such was the lack of liaison between the students of transport and the students of metabolism, that the significance of this fact for the field of oxidative and photosynthetic phosphorylation was

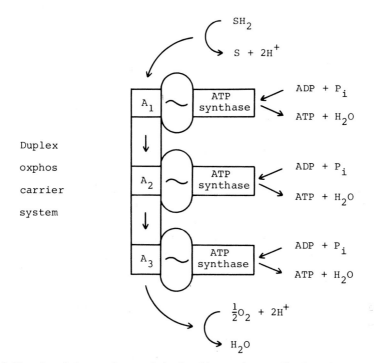

Fig. 3. Phosphorylating respiratory chain: Local interaction coupling hypothesis. The symbol ~ represents a localised "high-energy" chemical intermediary or physical state.

not appreciated, despite suggestive clues provided, for example, by Lunde-gardh (1945), Robertson & Wilkins (1948), Ussing (1949), Davies & Ogston (1950), Conway (1951) and me (Mitchell, 1954). These clues suggested that some osmotic type of protonic coupling mechanism might be feasible (see Robertson, 1960, 1968; Lehninger, 1962). It was in this context that I began to take an active outsider's interest in this fundamental problem of energy metabolism in the nineteenfifties (and occasionally talked to David Keilin about it), while I was mainly engaged in trying to develop general principles of coupling between metabolism and transport, by means of the biochemical concepts of chemiosmotic group-translocation reactions and vectorial metabolism (Mitchell, 1954, 1956, 1957, 1959, 1961 a, b; Mitchell & Moyle, 1956, 1958 a, b). I shall define these concepts more fully later. For the moment, suffice it to remark that it was these essentially biochemi-cal concepts (Mitchell, 1961 a, b, 1962, 1963, 1967 b, c, 1970 a, b, 1972 b, 1973 a, b, c, 1977 c), and not my relatively subsidiary interest in energy metabolism, that led me to formulate the coupling hypothesis, summarised in Fig. 4, which came to be known as the chemiosmotic hypothesis. As it happened, the main protonmotive ATPase principle of this hypothesis was first outlined at an international meeting held in Stockholm in 1960 (Mitchell, 1961 b). My motivation was simply a strategic conjectural one. There was a chance worth exploring that the chemiosmotic rationale might provide a generally acceptable conceptual framework in the field of membrane bioenergetics and oxidative phosphorylation, and that, if so, it

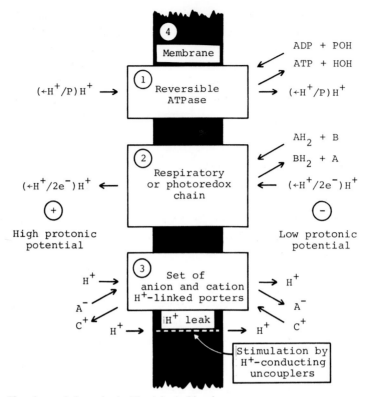

Fig. 4. Chemiosmotic hypothesis: Physiological level.

might encourage more adventurous and successful interdisciplinary research by improving communication and acting as a kind of navigational
aid (see Mitchell, 1961 c, 1963, 1976 a, 1977 b).

The two conceptual levels at which the chemiosmotic rationale has
helped to promote useful experimental research.

The level represented by Fig. 4 is essentially physiological. It aims to
answer the question: what does it do? At this conceptual level one makes
use of the general principle of coupling by proticity, the protonic analogue
of electricity. Separate protonmotive redox (or photoredox) and reversible
protonmotive ATPase complexes are conceived as being plugged through
a topologically closed insulating membrane between two proton-conducting aqueous media at different protonic potential. Thus, coupling may
occur, not by direct chemical or physical contact between the redox and
reversible ATPase systems, but by the flow of proticity around an aqueous
circuit connecting them. I use the word proticity for the force and flow of
the proton current by analogy with the word electricity, which describes
the force and flow of an electron current (Mitchell, 1972 a, 1976 a). However, the total protonic potential difference Δp has both electric ($\Delta\psi$) and
chemical activity (ΔpH) components, according to the equation:

$$\Delta p = \Delta\psi - Z\Delta pH \tag{1}$$

where Z is the conventional factor 2.303 RT/F, which is near 60 at 25°, when the potentials are expressed in mV (Mitchell, 1961 c, 1966, 1968).

To promote experimental research programmes designed to test, and if possible to falsify, the physiological-level chemiosmotic coupling concept, it was explicitly and unambiguously formulated (Mitchell, 1961 c, 1966) in terms of the following four fundamental postulates, corresponding to the four structural and functional systems represented in Fig. 4:

1. The ATP synthase is a chemiosmotic membrane-located reversible ATPase, having characteristic $\leftarrow H^+/P$ stoicheiometry.
2. Respiratory and photoredox chains are chemiosmotic membrane-located systems, having characteristic $\leftarrow H^+/2e^-$ stoicheiometry, and having the same polarity of proton translocation across the membrane for normal forward redox activity as the ATPase has for ATP hydrolysis.
3. There are proton-linked (or hydroxyl-ion-linked) solute porter systems for osmotic stabilisation and metabolite transport.
4. Systems 1 to 3 are plugged through a topologically closed insulating membrane, called the coupling membrane, that has a nonaqueous osmotic barrier phase of low permeability to solutes in general and to hydrogen ions and hydroxyl ions in particular. This is the cristae membrane of mitochondria, the thylakoid membrane of chloroplasts, and the plasma membrane of bacteria.

These postulates were almost entirely hypothetical and experimentally unexplored when they were given as the basis of the chemiosmotic hypothesis in 1961. My original guesses for the $\leftarrow H^+/P$ and $\leftarrow H^+/O$ stoicheiometries, and for the protonmotive polarity across the membrane (Mitchell, 1961c), required revision in the light of early experiments (Mitchell, 1963; Mitchell & Moyle, 1965, 1967, 1968). Fig. 5 represents the correct polarities, and what I think are probably the correct protonmotive stoicheiometries: A, for the mitochondrial oxidative phosphorylation system (not including the porters); and B, for the chloroplast non-cyclic photophosphorylation system. No change of fundamental principle was, however, involved; and the postulates, represented by Figs. 4 and 5, have now survived seventeen years of intensive scrutiny by many research workers, using a great variety of experimental methods in many different research laboratories, including my own (see: Racker, 1976; Papa, 1976; Boyer et al., 1977; Ernster, 1977b; Bendall, 1977; Harold, 1977a; Jagendorf, 1977; Kozlov & Skulachev, 1977; Skulachev, 1977; Witt, 1977; Dutton et al., eds, 1978; Hall et al., eds. 1978; Hinkle & McCarty, 1978; Junge et al., 1979a,b; Mitchell, 1976a, 1977a, 1979). The progress of research on oxidative and photosynthetic phosphorylation became much faster as soon as the molecular complexes represented by each of the four postulates began to be treated as biochemically separate systems (Fig. 4) that could best be studied individually – as in the laboratories of Jagendorf (1967), Racker (1967), Witt (1967), Chappell (1968) and Skulachev (1970) – instead of being mixed up in the so-called "phosphorylating respiratory chain" or its photosynthetic analogue.

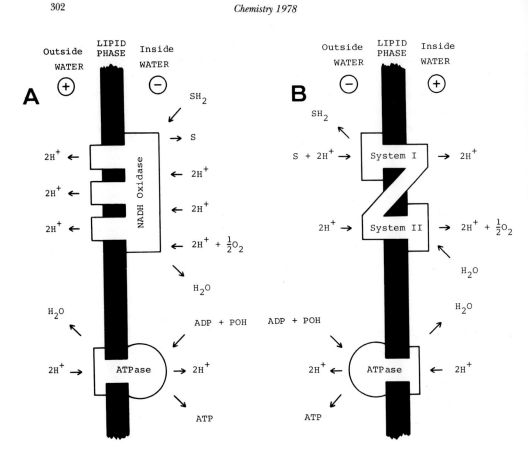

Fig. 5. Oxidative phosphorylation (A); and Photosynthetic phosphorylation (B) coupled by proticity (after Mitchell, 1967a).

The physiological-level concept of chemiosmotic coupling includes, not only the general principle of transformation of redox energy to phosphorylation energy, but also the principle of transmission of power from the redox complexes plugged through the membrane at one point to the reversible ATPase and other proticity-consuming complexes plugged through the membrane anywhere else (Mitchell, 1976a, 1977c). This circumstance has helped to place the mechanism of oxidative and photosynthetic phosphorylation in perspective in the broader field of membrane bioenergetics and general physiology — with particularly striking results for our understanding of microbial membrane transport and related processes (see Hamilton & Haddock, eds, 1977; Harold, 1977a,b; Rosen & Kashket, 1978).

The four postulates, representing the four systems with characteristic properties, are now widely regarded as experimentally established facts. Thus, we appear to have answered the question: What does it do? The plug-through respiratory chain and photoredox chain complexes generate proticity across the coupling membrane, energising the aqueous conductors on either side, so that power can be drawn off by other plug-through complexes, such as the reversible protonmotive ATPase. However, this acceptance of the chemiosmotic coupling concept at the physiological level is without prejudice to the possible biochemical mechanisms of the protonmotive ATPase and redox complexes, and fixes their relative but not their absolute protonmotive stoicheiometries (Boyer et al., 1977; Ernster, 1977b).

The other conceptual level at which the chemiosmotic rationale has helped to promote useful experimental research is essentially biochemical. It concerns the functional stoicheiometries, the molecular topologies and the molecular mechanisms of the protonmotive ATPase, the redox (or photoredox) complexes, and the porter complexes of postulates 1, 2, and 3.

In my opinion, the biochemical content and value of the chemiosmotic rationale depended from the outset on the feasibility of protonmotive chemiosmotic reaction mechanisms of the direct group-translocation type, exemplified by the redox loop and the hydrodehydration loop (Mitchell, 1966, 1967a,b,c), which are relatively orthodox biochemically, and require little more than the addition of a spatial dimension to Lipmann's concept of chemical group potential (Lipmann, 1941, 1946, 1960). Had this not been so, I would not have thought it worth fostering the chemiosmotic hypothesis, as I shall now proceed to explain.

B. *The Coupling of Metabolism and Transport: Vectorial Metabolism*
The photograph in Fig. 6 was taken in a tiny research room in the basement of the Department of Biochemistry in Cambridge, England, in 1942 or 1943, when I first began to do biophysical and biochemical research. It shows Jim Danielli, Joan Keilin (David Keilin's daughter), Mrs. Danielli (who was acting as Jim Danielli's technician) and me. Jim Danielli, who is seen there operating a surface balance on a Langmuir trough, was my research supervisor. He exposed me to the techniques and concepts of the membranologists and students of transport, while the general outlook of the Cambridge Biochemistry Department, in which we were working, was that of classical homogeneous solution metabolic enzymology. Intermediately was the position of David Keilin, at the Molteno Institute, with his studies of the insoluble cytochrome system and associated components, making up the respiratory chain.

I could not but be impressed by the great divergence of outlook, and even mutual antagonism, between the students of membranes and transport on the one hand, and the students of metabolic enzymology on the other hand — and I soon determined to try to understand both points of view in the hope that they might be brought together.

About seven years elapsed before I had accidentally become a microbi-

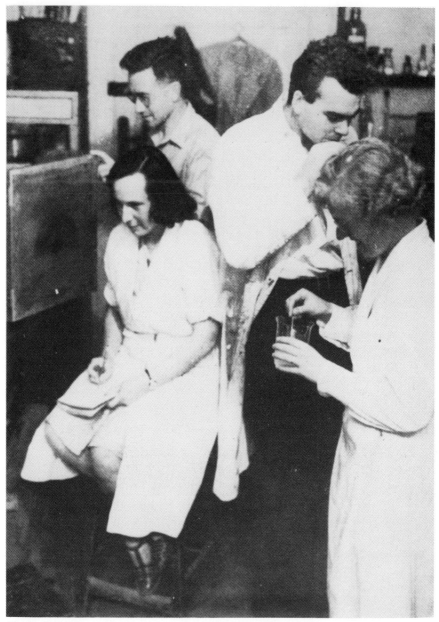

Fig. 6. Research at the bench in 1942 or 1943. Department of Biochemistry, Cambridge, England. From left to right: Joan Keilin, Jim Danielli, Peter Mitchell, Mary Danielli.

ologist and was involved: first, in studies of a functional aspect of the plasma membrane of bacteria, which I called the osmotic barrier (Mitchell, 1949); and soon after, in studies of the specific uptake and exchange of inorganic phosphate and arsenate through a catalytic system present in the osmotic barrier of staphylococci (Mitchell, 1954; and see Mitchell & Moyle, 1956). This enabled me to give my full attention to the functional and conceptual relationships between chemical and osmotic reactions. The remarkably high specificity of the phosphate translocation reaction in staphylococci, its susceptibility to specific inhibitors including SH-reactors, its high entropy of activation, which indicated a large conformational change in the translocator system, and the tight coupling of phosphate translocation against arsenate translocation that I observed – as in the phenomenon of exchange-diffusion previously described by Ussing (1947) – indicated how closely osmotic translocation reactions could resemble (or could be functionally related to) enzyme-catalysed group-transfer reactions (Mitchell, 1956, 1957, 1959). Further, my observation with Jennifer Moyle, that the plasma membrane material isolated from staphylococci contained the cytochrome system and associated enzyme activities (Mitchell, 1956;

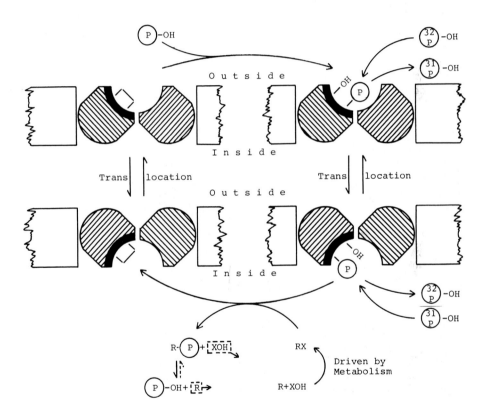

Fig. 7. Hypothetical enzyme system for phosphoryl group translocation (from Mitchell, 1957).

Mitchell & Moyle, 1956), suggested that one should generalise Lundegardh's idea of vectorial electron translocation through the cytochrome system (Lundegardh, 1945). Thus, from this work, and from related observations and ideas, notably of Lipmann (1941), Robertson & Wilkins

(1948), Rosenberg (1948), Pauling (1950), Davies & Ogston (1950) and Conway (1951), I surmised that certain of the group-transfer reactions, catalysed by the enzyme and catalytic carrier systems in the bacterial plasma membrane, might actually be vectorial group-translocation reactions, because of the anisotropic topological arrangement and specific conformational mobility of the catalytic systems. These, then were the circumstances that led me to remark at a symposium in 1953: ". . . in complex biochemical systems, such as those carrying out oxidative phosphorylation (e.g. Slater & Cleland, 1953) the osmotic and enzymic specificities appear to be equally important and may be practically synonymous" (Mitchell, 1954).

The general idea that I had in mind in the mid-nineteenfifties, illustrated by the hypothetical system for phosphoryl translocation reproduced in Fig. 7, was that of substrate-specific conformationally mobile enzyme and catalytic carrier systems catalysing the translocation, not only of solutes, but also of chemical groups. Thus, by the spatial extension of the group-potential concept of Lipmann (1941), transport could be conceived biochemically as being directly driven by the real vectorial forces of metabolic group-potential gradients. Hence the concepts of what I called chemiosmotic reactions and vectorial metabolism — bringing together transport and metabolism into one and the same chemiosmotic molecular-level biochemical process catalysed by group-conducting or conformationally mobile group-translocating enzyme systems (Mitchell, 1957, 1959). This led directly to the explicit concept of energetic coupling through enzyme-catalysed group translocation, as described in two papers by Jennifer Moyle and me in 1958. Fig. 8, from this work, represents a hypothetical example in which phosphoryl is conducted from ATP on the left to a substrate S on the right (Mitchell & Moyle, 1958a,b).

Mechanistically, the group translocation or conduction concept was a development of the idea, put forward by Pauling in 1950, that enzymic catalysis depends on tight binding of the transition-state complex rather than of reactants and resultants. As we pointed out (Mitchell & Moyle, 1958b), Pauling's idea required only a small adjustment to adapt it to the notion that the active centre regions of certain enzymes (and of certain catalytic carriers, such as cytochromes) may be conceived, not simply as specific group-binding centres, which would tend to lock in the transition state, but rather as specific group-conducting devices that facilitate the passage of chemical groups through a region of the catalytic complex between separate domains that specifically interact with the group-donor and group-acceptor species.

It was, of course, realised that the chemicomotive effect of group translocation — or group conduction, as I now prefer to call it — would not be manifested unless the enzyme or catalytic carrier molecules were inhomogeneously organised in space according to either of two main topological principles. According to one topological principle, the organisation could be at the macroscopic level in a membrane, thus giving rise to macroscopic

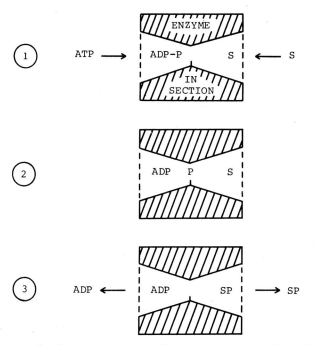

Fig. 8. Enzyme-catalysed group translocation illustrated by hypothetical phosphoryl translocation from ATP to a substrate S (from Mitchell & Moyle, 1958b).

chemiosmotic processes of which we gave some permutations and combinations for a phosphokinase system by way of example, as shown in Table
1. The two aqueous phases are represented to right and left of the central
line denoting the membrane containing the anisotropic phosphokinase;
and, in each example, the chemical (group-transfer) reaction is represented as progressing downwards, while the osmotic (group-translocation)
reaction is represented as progressing through the phosphokinase in the
membrane. The table illustrates that the overall chemiosmotic process
would depend as much on the osmotic translational (or conformational)
specificities as on the chemical specificities of the catalytic system. It is in
this respect that the vectorial chemiosmotic system differs fundamentally
from conventionally scalar chemical ones.

The second case in Table 1 is shown in diagram A of Fig. 9. It represents
the macroscopic chemiosmotic group-conduction principle applied to the
phosphorylative translocation of the substrate GH, which could be a sugar,
as in the phosphoenol pyruvate phosphotransferase system discovered in
bacteria by Kundig, Ghosh & Roseman in 1964.

Now, it was only a small, but nevertheless important, step from Table 1
to write heterolytic protonmotive ATPase reactions as shown in B and C of
Fig. 9. My original proposal (Mitchell, 1961b,c) for the protonmotive
ATPase, reproduced in Fig. 10, corresponds to the group-translocation
system of Fig. 9 B.

According to the other topological principle for manifesting the chemicomotive effect of group translocation, we suggested that the organisation

| EXAMPLE | CHEMI-OSMOTIC PROCESS | | | GROUPS TRANS-PORTED |
| | PHOSPHOKINASE | | | |
	LEFT PHASE	IN MEMBRANE	RIGHT PHASE	
1	ATP ADP		S SP	P– →
2	ATP ADP + SP		S	← S–
3	ATP + S ADP		SP	→ S– → P–
4	ATP		S ADP + SP	→ ADP– → P–
5	ADP		S + ATP SP	← ADP–
6	ATP SP		S ADP	→ ADP– ← S–
7	ATP + S		ADP + SP	→ ADP– → P– → S–

Table 1. Alternative translocation-specific chemiosmotic processes catalysed by a hypothetical phosphokinase in a membrane (from Mitchell & Moyle, 1958b).

could be at the microscopic level, by pairing and enclosure of a 'microscopic internal phase' between neighbouring catalytic units, thus giving rise to a chemical coupling effect. As illustrated in Fig. 11 A, we cited as possible examples the NADP-linked isocitrate dehydrogenase and the malic enzyme, which catalyse consecutive oxidation and decarboxylation reactions (with oxalosuccinate and oxaloacetate respectively as intermediates trapped in the microscopic internal phase). We pointed out that such pairing of catalytic units could be developed in three dimensions for branching or cycling reaction sequences in enzyme complexes (Mitchell &

PROTONMOTIVE ATPases

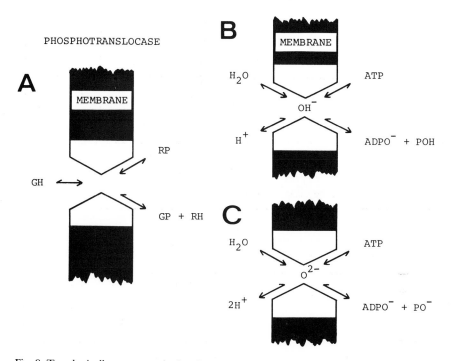

Fig. 9. Topologically macroscopic chemiosmotic group conduction: A, of group G, energised by phosphate transfer; B and C, of OH⁻ and O²⁻ respectively, energised by ATP hydrolysis.

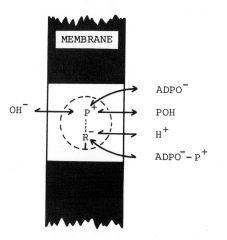

Fig. 10. Protonmotive ATPase translocating OH⁻ (from Mitchell, 1961c).

Moyle, 1958a,b). Diagrams B and C in Fig. 11 represent relevant hypothetical examples of the application of this microscopic coupling principle to redox (o/r) and hydrodehydration (h/d) complexes respectively. B represents consecutive hydrogen and electron transfer, via a flavoprotein (Fp) or ubiquinone (Q) and an ironsulphur protein (FeS) complex, as, perhaps, in NADH dehydrogenase; and C shows consecutive H_2O and O^{2-}

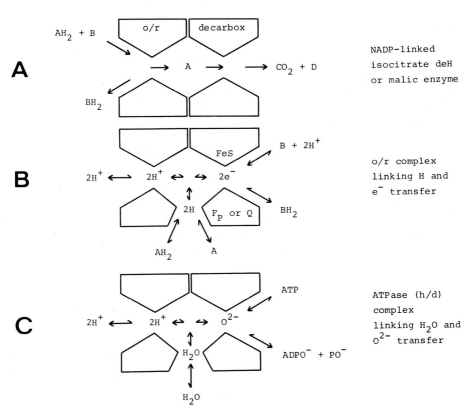

Fig. 11. Topologically microscopic chemiosmotic coupling systems: A, for oxidative decarboxylation; B, for successive H and e^- transfer; C, for successive H_2O and O^{2-} transfer (following Mitchell & Moyle, 1958a, b).

transfer, the latter via $(ADPO^- + PO^-)/ADPOP$ antiport in an ATPase complex.

The microscopic pairing and the macroscopic chemiosmotic principles of topological organisation have been developed together in chemiosmotic reaction systems. This is illustrated by Fig. 12, in which examples B and C of Fig. 11 have been plugged through a membrane to give a protonmotive redox loop complex and a protonmotive ATPase (hydrodehydration loop) complex respectively. Fig. 13 A and B represent the protonmotive redox loop in a more conventional nomenclature where X and Y are hydrogen and electron carrier respectively. It is, perhaps, more obvious from this cycling carrier style of representation that, in direct chemiosmotic mechanisms, the conformational movements or conduction processes of translocation overlap (in space-time) with the chemical processes of group transfer; and the chemicomotive stoicheiometry depends on the chemical properties, conformational articulations and/or conductive properties of the catalytic carriers, according to relatively conventional biochemical principles.

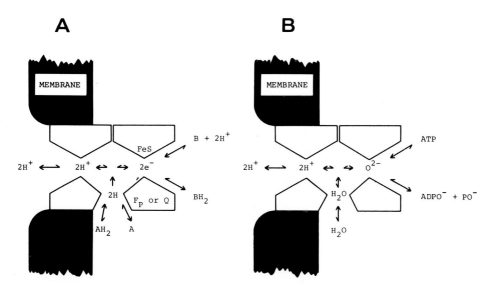

Fig. 12. Macroscopic and microscopic topological principles developed together in: A, proton-motive redox loop; B, protonmotive ATPase (hydrodehydration loop).

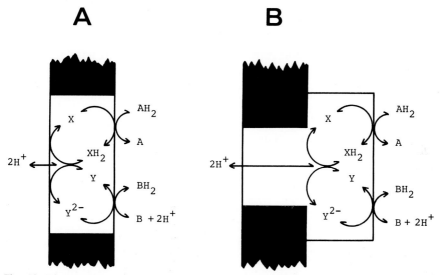

Fig. 13. Direct redox loop mechanism in cycling carrier idiom: A, plugged through the membrane; B, connected through a proton-conducting component (following Mitchell, 1966 and 1968).

C. *The Protonmotive Respiratory Chain and Photoredox Chain: What is it? How does it do it?*

Let us return to the theme of David Keilin's respiratory chain in the light of the essentially biochemical concept of direct group-translocating or group-conducting chemiosmotic mechanisms.

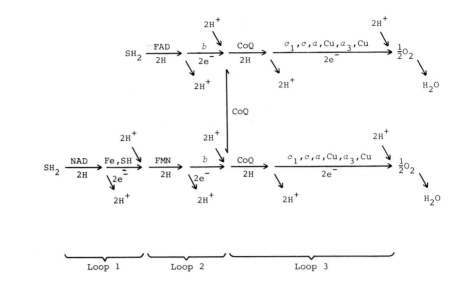

Fig. 14. Suggested alternation of H and e^- conductors in respiratory chain systems: A, for FAD-linked oxidations; B, for NAD-linked oxidations (after Mitchell, 1966).

As indicated in Fig. 14, the chemiosmotic hypothesis, at the biochemical level, permitted a return to David Keilin's notion of a chemically simple respiratory chain; but the protonmotive property would have to depend on a topological complexity in place of the chemical complexity favoured by the proponents of the chemical coupling type of hypothesis. This seemed to me to be an attractive notion because it was in accord with the evidence of the need for a special topological organisation of the components of the respiratory chain, as shown by David Keilin's early work on the reversible dissociability of cytochrome c, and by the important pioneering work of Keilin & King (see King, 1966) on the reversible dissociation of succinate dehydrogenase from the succinate oxidase complex of mitochondria – which, as we now know, set the scene for later topological resolution and reconstitution studies.

Fig. 15 shows how the alternation of hydrogen and electron carriers down the chain might translocate protons across the membrane, the chain being effectively looped across the osmotic barrier, forming three protonmotive redox loops that I called Loops 1, 2 and 3, corresponding to the three energy-transducing regions of the classical respiratory chain between NADH and oxygen (Mitchell, 1966). In this way, each loop would translocate two protons per bivalent reducing equivalent passing along – giving 6 in all for NADH oxidation, as observed experimentally (see: Mitchell, 1972a, 1976a, 1977b, 1979; Moyle & Mitchell, 1977, 1978a, b).

At equilibrium, the total protonmotive potential across the membrane would be equal to the total redox potential across each loop – that is, around 250 mV (Mitchell, 1966). Thus, we can relate quantitatively the

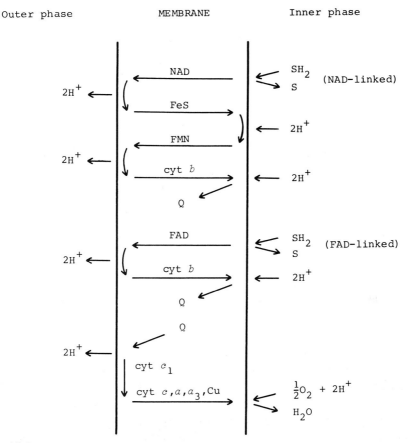

Fig. 15. Suggested looped configuration of respiratory chain systems (after Mitchell, 1966).

scalar group-potential differences of chemical reactions (i.e. of hydrogen and electron transfer reactions) to the real vectorial forces of transport (i.e. of oppositely directed H and e⁻ transport, adding up to net H^+ transloca-tion).

Fig. 16 illustrates my original guesses, (A) for the non-cyclic photoredox chain of chloroplasts, and (B) for the cyclic photoredox chain of certain photosynthetic bacteria, which employ the same direct group-conducting redox-loop principle as that applied to the respiratory chain (Mitchell, 1966). But a subtle difference of behaviour was expected because the orientation of the photosynthetic pigment systems, as represented in these diagrams, should cause a non-thermodynamic photoelectric effect across the membrane with a very short rise-time − as shown by Witt (1967) in elegant experiments on chloroplast thylakoids, and confirmed by Crofts and others in both chloroplasts and photosynthetic bacteria (see Junge, 1977; Witt, 1977; Hall et al., eds, 1978).

A great deal of ingenious experimental work in many laboratories over the last decade has shown that these schemes require some modification of detail, but their direct group-conducting redox-loop principle has been amply confirmed, as indicated in the diagrams of Fig. 17, which, as I shall

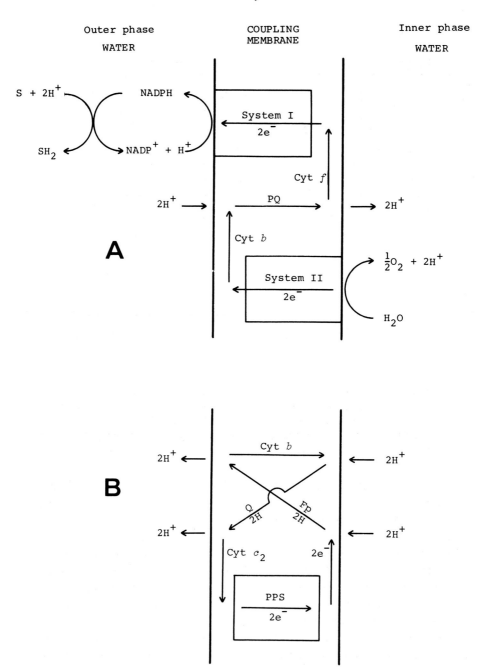

Fig. 16. Suggested protonmotive photoredox chain systems: A, for chloroplast non-cyclic photoredox activity; B, for bacterial cyclic photoredox activity. System I, System II and PPS stand for photosynthetic pigment systems (after Mitchell, 1966, 1967a).

discuss in a little more detail later, summarise recent knowledge about the mitochondrial respiratory chain (A), the chloroplast non-cyclic photoredox chain (B), and the reversible ATPases (F_0F_1 and CF_0CF_1), driven by these systems. It is noteworthy that these schemes show remarkable similarities.

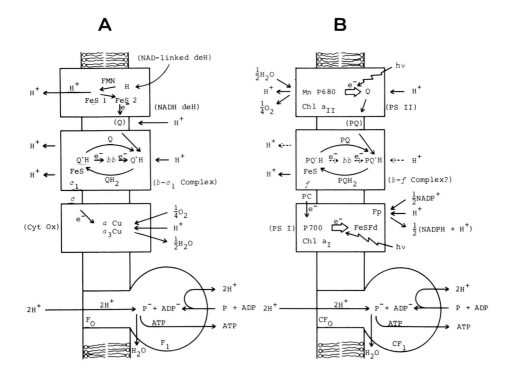

Fig. 17. Direct chemiosmotic mechanisms for: A, mitochondrial oxidative phosphorylation; B, chloroplast non-cyclic photosynthetic phosphorylation: Biochemical level chemiosmotic concept. Research information from many sources (see Dutton et al., eds., 1978; Hall et al., eds., 1978; Mitchell, 1977a, 1979; Trumpower, 1979). Symbols as conventionally used in references cited; and see text.

The photoredox chain system, shown in Fig. 17 B and Fig. 5 B, obviously has a real Z configuration that corresponds to the abstract Z scheme introduced by Hill & Bendall in 1960.

It may well be remarked that the protonmotive stoicheiometries of these schemes, which correspond to one proton per univalent reducing equivalent traversing each effective redox loop, represent a crucial datum. In my laboratory, and in most other laboratories where such stoicheiometric measurements have been made (see Mitchell, 1979), the $\leftarrow H^+/e^-$ ratio has been found to be near unity per effective redox loop. For reasons discussed elsewhere (Mitchell, 1972a, 1976a, 1977b, 1979; Moyle & Mitchell, 1977, 1978a, b; Jones et al., 1978), I do not think that recent dissent from the laboratories of Azzone (Azzone et al., 1977), Lehninger (Reynafarje & Lehninger, 1978) and Wikström (Wikström & Krab, 1978) constitutes a serious threat to this relatively hard experimental datum.

We seem, therefore, to have a partial answer to the questions: What is it; how does it do it? It is a system of specific hydrogen and electron conductors, which generates proticity by virtue of the fact that it is effectively looped across the coupling membrane, and catalyses the spontaneous diffusion of hydrogen atoms and electrons in opposite directions, adding up to net proton translocation across the coupling membrane.

2. A LONG PHYSICOCHEMICAL VIEW OF CHEMICOMOTIVE SYSTEMS

The first protonmotive device conceived by man was the electromotive hydrogen-burning fuel cell, invented by the remarkable William Grove in 1839. It is, perhaps, not self evident that such a fuel cell for generating electricity is also, potentially, a generator of proticity. This is illustrated by the diagrams of the hydrogen-burning fuel cell shown in Fig. 18. It simply depends where one opens the circuit to conduct away the power for external use. In A, the circuit is opened in the electron conductor to give electricity. In B, the circuit is opened in the proton conductor to give proticity (Mitchell, 1967a).

The fuel cell is a beautiful example of the truth of the principle, enunciated by Pierre Curie at the end of the last century (Curie, 1894), that effects cannot be less symmetric than their causes. The phenomena of transport in the fuel cell arise from the intrinsically vectorial disposition of the chemical reactions at the anisotropic metal/aqueous catalytic electrode interfaces (Liebhafsky & Cairns, 1968). Thus, the scalar group-potential differences of the chemical reactions are projected in space as vectorial chemical fields of force corresponding to the chemical group-potential gradients directed across the electrode interfaces. These simple considerations illustrate nicely the nonsensical character of the question asked by certain theoreticians in the context of the coupling between transport and metabolism around 1960, and still persisting in some circles: how can scalar chemical reactions drive vectorial transport processes? The answer is plainly: they can't (Mitchell, 1962, 1967b).

The idea of electrochemical cells and circuits was generalised by Guggenheim in 1933 to include the chemically motivated transport of any two species of chemical particle around a suitably conducting circuit. Guggenheim's rather abstract thermodynamic treatment effectively showed that chemical transport can be coupled reversibly to chemical transformation by splitting the chemical reaction spatially into two half reactions, connected internally by a specific conductor of one chemical species, and connected externally by a specific conductor of another chemical species needed to complete the overall reaction (see Mitchell, 1968). When we include the leading in and out of the reactants and resultants, as in the fuel cell of Fig. 18, we see that there have to be two internal specific ligand conductors arranged in a looped configuration between the interfaces where the chemical half reactions occur (Mitchell, 1967a). Obviously, the external specific ligand conduction process − for example, the flow of protons in Fig. 18 B − must be the sum of the internal specific ligand conduction processes − for example the flow of hydrogen atoms one way and of electrons the opposite way in Fig. 18 B.

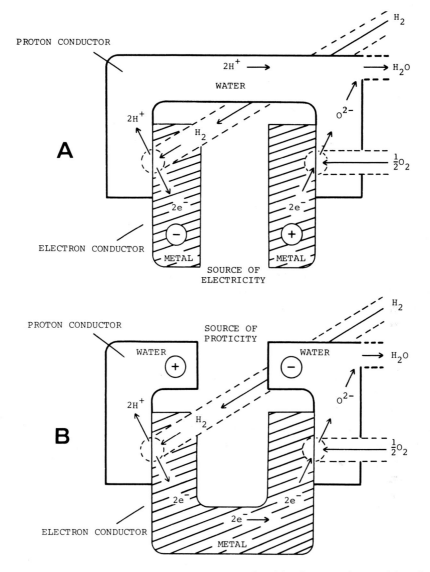

Fig. 18. Hydrogen-burning fuel cell: A, generating electricity; B, generating proticity (after Mitchell, 1967a).

The specification of chemical group conduction in an enzyme or catalytic carrier complex may usefully be considered to correspond to the specification of an internal ligand conduction reaction of a chemicomotive cell, the other internal and external circuit components of which may be determined by the topological arrangement of the specific group-conducting complex, relative to other osmotic or diffusion-regulating systems. Thus, as the name chemiosmotic implies, the intrinsic osmotic property of a

chemical group-translocation or group-conduction reaction in biology represents its chemicomotive potentiality, which may be exploited (through
natural selection) by appropriate topological organisation.

For example, the notion of the protonmotive redox loop is based on this
type of development of the specific ligand-conducting group-translocation
concept

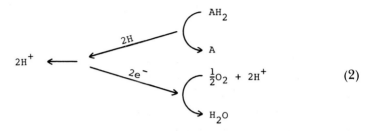

$$\tag{2}$$

As indicated in equation (2), the internal (trans osmotic barrier) ligand
conductors in the redox-loop complex are conceived as being specific for
hydrogen atoms that diffuse down their potential gradient one way, and
for electrons that diffuse down their (electrochemical) potential gradient
the opposite way — exactly as in the fuel cell of Fig. 18 B, and as illustrated
further in Fig. 19. The outer circuit consists of the aqueous proton conductors on either side of the insulating lipid membrane (Mitchell, 1966).

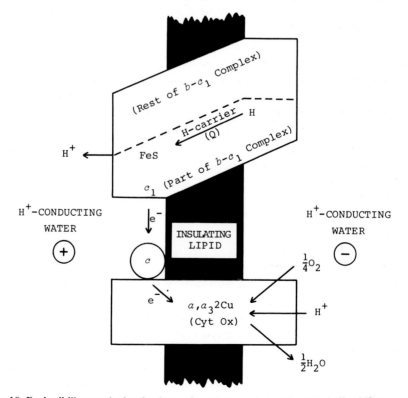

Fig. 19. Fuel cell-like terminal redox loop of respiratory chain (after Mitchell, 1967a).

The notion of the protonmotive hydrodehydration loop for reversible ATP hydrolysis depends on a similar principle (Mitchell, 1966).

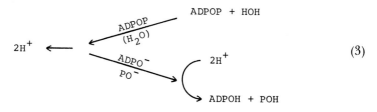

$$(3)$$

As illustrated formally in equation (3), protonmotive ATP hydrolysis may be conceived as the specific conduction of ATP (written ADPOP) and H_2O one way, and of $ADPO^- + PO^-$ the opposite way, giving, by difference, the net translocation of $2H^+$. In equation (3) the H_2O is bracketed to denote that there need be no specific H_2O conduction pathway, because lipid membranes generally have a high conductance to water (see Mitchell, 1977a).

The considerations of this section of my lecture look two ways. They show that the chemiosmotic rationale has its roots in physical and chemical theory going back more than a century. They also suggest that the general concept of specific vectorial ligand conduction has very powerful applications in physics, chemistry and biology, which are, as yet, by no means fully appreciated (Mitchell, 1977c).

The protonmotive redox loop and hydrodehydration loop, and other possible chemicomotive loops, as generally defined here, depend on a very simple specific ligand-conduction mechanism. I have called it the direct chemiosmotic mechanism in the biological systems, where it represents a spatial extension of the conventional biochemical concept of group transfer, but it is identical in general principle to the mechanism of man-made fuel cells. I find it remarkably paradoxical, therefore, that many physiologists and biochemists have tended to reject, or have found it difficult to accept intuitively, the feasibility of this direct and most biochemically conventional type of chemiosmotic mechanism (see Boyer, 1974; Skulachev, 1974; Boyer et al., 1977; DePierre & Ernster, 1977; Ernster, 1977a; Papa et al., 1978). Instead, as illustrated in Fig. 20, they have been inclined to adopt exclusively conformationally coupled mechanisms, in which the chemical and osmotic reaction centres are conceived as being separate in space-time, and interact exclusively via the conformational movements of an intervening polypeptide system, indicated by the squiggly line. Such essentially black-box mechanisms involving hypothetical translocators (T), unlike their direct counterparts, can be adjusted to have any stoicheiometry (n) to suit the experiments of the day, and they are of such low information-content biochemically as to be very difficult to disprove experimentally. By providing a blanket explanation, but without currently testable detail, it seems to me that the concept of exclusively conformational coupling in chemiosmotic reactions, which can be partly attributed to a spillover from studies of the Na^+/K^+-motive and Ca^{2+}-motive ATPases,

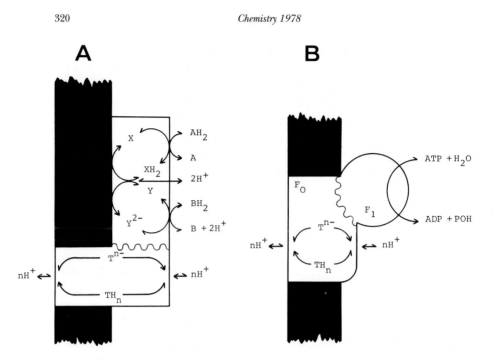

Fig. 20. Completely indirect or exclusively conformational type of chemiosmotic coupling concept: A, for redox proton pump; B, for ATPase proton pump (following Boyer, Chance, Ernster, Skulachev and others. See Mitchell, 1977a).

may actually inhibit productive research by acting as a palliative (see Mitchell, 1977a,c,1979). Conjectures about direct biochemically explicit chemiosmotic mechanisms, on the other hand, even if wrong, promote experimental activity and enthusiasm by suggesting crucial experiments for testing them. This, I think, has been an important strategic function of the biochemical conceptual aspect of the chemiosmotic theory in the recent past, and I would like to persuade more of my colleagues to make use of that function in the future—even, perhaps, in the field of the Na^+/K^+-motive and Ca^{2+}-motive ATPases (Mitchell, 1973c, 1979) where progress in understanding the general chemiosmotic principle of the mechanisms has so far been rather slow. The fact that the metal ions are not involved directly in stable covalent compounds does not militate against them being involved in stable electrostatic complexes with anionic phosphate groups that move along specific pathways down their group-potential gradients during ATP hydrolysis in the relatively non-aqueous environment of the enzyme active site domain.

Let me interject here that the protonic chemiosmotic theory has a much broader range of applicability than is encompassed by the central field of energy transduction in the classical oxidative and photosynthetic phosphorylation systems, treated in this lecture (see Mitchell, 1976a). For example: there is the protonmotive bacteriorhodopsin system of *Halobacterium halobium* (Stoeckenius, 1977; Schreckenbach & Oesterhelt, 1978;

Drachev et al., 1978; Schulten & Tavan, 1978); the protonmotive pyro-phosphatase of photosynthetic bacteria (Moyle et al., 1972); protic heating in fat cell mitochondria of hibernating animals (Heaton et al., 1978); the remarkable rotatory flagellar motor of bacteria, which is driven by proti-city (see Manson et al., 1977; Matsuura et al., 1977; Glagolev & Skulachev, 1978); the protonmotive ATPase and proton-coupled porter systems in chloroplast envelope membranes (Heldt, 1976; Edwards & Huber, 1978), in the plasma membranes of moulds, yeasts and higher plants (see Mitch-ell, 1970b, 1976a; Seaston et al., 1976; Bowman et al., 1978; Delhez et al., 1978), and also in the membranes of chromaffin granules (Njus & Radda, 1978) and synaptosomes (Toll & Howard, 1978); interesting and unusual redox chain systems, such as that of the acidophile *Thiobacillus ferro-oxidans* (Ingledew et al., 1977), which has incidentally helped to rule out the localised protonic anhydride coupling hypothesis; and probably other proticity producing and consuming systems yet to be discovered. More-over, it was never my wish or intention that the concepts of specific ligand conduction and chemial group translocation (Mitchell, 1956, 1957, 1961a, 1962, 1963), on which the chemiosmotic theory is based, should be con-fined to systems coupled by proticity. It is only the unique versatility of the uses of proticity (Mitchell, 1976a), and its importance for power transmis-sion in the main pathways of energy metabolism, that has accidentally tended to associate the use of the chemiosmotic theory with protonmotive and protonmotivated systems, rather than with other chemicomotive and chemicomotivated systems. In recent reviews (Mitchell, 1977c,1979), I have endeavoured to encourage the wider use of chemiosmotic theory, and the powerful biochemical concept of specific ligand conduction, in terms of the general idea of chemicalicity−an explicit extension of Lip-mann's marvellously precocious idea of metabolic process patterns (Lip-mann, 1946).

3. SOME QUESTIONS OF BIOCHEMICAL DETAIL IN THE PRO-TONMOTIVE RESPIRATORY CHAIN AND PHOTOREDOX CHAIN

The respiratory chain system summarised in Fig. 17A differs from my earlier suggestion of three linearly arranged redox loops (Figs. 14 and 15) in that loops 2 and 3 are coalesced into a cyclic Loop $2 + 3$ configuration, described as the Q cycle, catalysed by the cytochrome $b-c_1$ complex. In this way, many of the otherwise anomalous thermodynamic poising and kinetic characteristics of cytochromes b_{566} and b_{562} (represented by bb in the dia-gram) may be explained; and the presence of ubiquinone (Q) as the only hydrogen carrier in this redox region, and the site of action of the inhibi-tor antimycin, may also be rationalised (Mitchell, 1975,1976b; Rich & Moore, 1976; Rieske, 1976; Trumpower, 1976,1979; Konstantinov & Ruuge, 1977; King, 1978; Moore, 1978; Rich & Bonner, 1978; Yu et al., 1978; Ragan & Heron, 1978; Heron et al., 1978). As indicated by the broken symbols of Fig. 17B, it is not yet certain whether a similar plastoquinone (PQ) cycle may operate in chloroplasts (see Bendall, 1977). Another significant feature of the diagrams of Fig. 17 is the inclusion of ironsulphur centres (FeS), which, following the beautiful pioneer work of Helmut Beinert, are now thought to have as important a role in electron transport as the haem nuclei of cytochromes (Sands & Beinert, 1960; Beinert, 1977).

The representation of the respiratory and photoredox chains as a set of physically compact complexes (that may be partially resolved and reconstituted) stems from work by Keilin & King (1958), by Takemori & King (1962), by the Madison group, led by Green and Hatefi (see Hatefi, 1966), and by Efraim Racker's group (see Racker, 1976). They originally defined four complexes: NADH-Q reductase, succinate-Q reductase, QH_2-cytochrome c reductase (the cytochrome $b-c_1$ complex) and cytochrome oxidase—functionally linked by ubiquinone and cytochrome c. Racker's group added the F_0F_1 and CF_0CF_1 complexes, which are physically and chemically separate from the redox complexes (Racker, 1976,1977; Jagendorf, 1977; Kozlov& Skulachev, 1977; Kagawa, 1978; Senior, 1979).

The lipid coupling membrane through which the redox and ATPase complexes are plugged is now considered to be very mobile laterally (see Hackenbrock& Höchli, 1977; King, 1978; Heron et al., 1978), in accordance with the fluid membrane concept of Singer& Nicholson (1971).

There are about equal numbers of cytochrome $b-c_1$ and cytochrome oxidase complexes in mitochondrial respiratory chains. Counting all the different Q-linked dehydrogenases (NADH dehydrogenase, succinate dehydrogenase, electron transfer flavoprotein dehydrogenase, choline dehydrogenase, glycerol-1-phosphate dehydrogenase, etc.), there are about as many dehydrogenase complexes as there are cytochrome $b\text{-}c_1$ complexes. Thus, there is no special significance of the number four in the redox complexes of Green's group. There is, however, a special significance of two complexes: the cytochrome $b-c_1$ complex and the cytochrome oxidase complex, which are functionally linked by cytochrome c, and make up the protonmotive cytochrome system. This remarkably compact system serves all the Q-linked dehydrogenases, only one of which, the NADH dehydrogenase, is, so far, known to be protonmotive itself (Ragan, 1976).

There are generally at least ten Q molecules per cytochrome $b-c_1$ complex, so providing for the redox pool function of Q, identified by Kröger & Klingenberg (1973). However, recent work by Ragan and colleagues (Ragan & Heron, 1978; Heron et al., 1978) on functional interaction between NADH-Q reductase and cytochrome $b-c_1$ complexes in liposomal membranes confirms the thesis of King (1966,1978) that the most active redox-functional units are binary dehydrogenase-cytochrome $b-c_1$ complexes containing bound Q. Thus, it may be that the Q pool function arises more from rapid lateral lipid mobility, giving rise to a dynamic association-dissociation equilibrium of binary dehydrogenase-cytochrome $b-c_1$ complexes with associated Q, than to the lateral diffusion of free Q molecules between dehydrogenase and cytochrome $b\text{-}c_1$ complexes.

Hauska (1977a,b) and Lenaz et al. (1977) have argued that Q and PQ are sufficiently mobile across the lipid phase of liposomes to account for the observed rates of condution of H atoms across mitochondrial and chloroplast coupling membranes by the Q and PQ pools. It seems likely, however, that, in accordance with the ideas of King, and with the concept of the Q cycle (see Mitchell, 1975), the conduction of H atoms across the osmotic barrier may occur preferentially via specific ligand-conducting Q and PQ domains associated with Q-binding or PQ-binding proteins in the cytochrome $b-c_1$ (or $b-f$?) complexes, and in the neighbouring dehydrogenase (or PSII?) complexes (Trumpower, 1976,1978,1979; Gutman, 1977; Yu et al., 1977a,b,1978; King, 1978; Ragan & Heron, 1978; Heron et al., 1978; Hauska, 1977b).

The possible conduction of H atoms by flavin mononucleotide (FMN) in NADH dehydrogenase is based only on the known H-binding property of the flavin group (Mitchell, 1972a; Garland et al., 1972; Gutman et al., 1975). The very wide gap between the redox midpoint potentials of FeS1 and FeS2, and the effects of Δp in poising these centres, are difficult to reconcile with the arrangement shown in Fig. 17A (Ohnishi, 1979). As indicated in Fig. 21, I suggest rather speculatively that a protein-bound

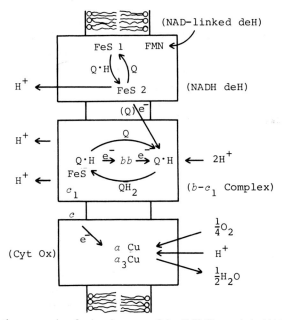

Fig. 21. Speculative suggestion for involvement of the $Q \cdot H/Q$ couple in NADH-Q reductase, and connection to cytochrome system.

$Q \cdot H/Q$ couple, acting between FeS1 and FeS2, might possibly account better for the general behaviour of the protonmotive NADH dehydrogenase, and for its requirement (Lenaz et al., 1975) for a specific homologue of Q.

The notion of the net conduction of O^{2-} groups by $(ADP^- + P^-)/ATP$ antiport in F_1 and CF_1 (Mitchell, 1972b, 1977a) is based on the precedent of the mitochondrial ADP/ATP antiporter, which is known to conduct ADP and ATP only in specific protonation states (Klingenberg, 1977). The protonmotive NAD(P) transhydrogenase, not shown in Fig. 17, may likewise translocate protons by the effective antiport of the phosphate groups of NAD and NADP, in different protonation states determined by the redox states of the nicotinamide groups (Mitchell, 1972b).

In summary, the bioenergetically efficient mechanisms, represented in outline by Fig. 17, depend on two main principles: 1, the semi-fluid bimolecular lipid membrane and the plug-through complexes form a condensed, continuous non-aqueous (protonically insulating) sheet that acts as the osmotic barrier and separates the aqueous proton conductors on either side; 2, components of the complexes plugged through the membrane catalyse the highly specific vectorially organised conduction of electrons, H atoms, H^+ ions and O^{2-} groups. As examples of specific ligand *binding*, we have the electron-accepting action of cytochromes or ironsulphur centres, the hydrogen-accepting action of flavoproteins or Q-proteins, and the O^{2-}−accepting action of the $ATP/(ADP^- + P^-)$ couple. But the action of specific ligand *conduction* in the plug-through chemiosmotic complexes requires additional dynamic topological, physical and chemical specifications that facilitate the diffusion of the ligands along uniquely articulated pathways down through-space or through-bond fields of force.

There is still much to be understood about the biochemical details of the specific ligand-conduction processes, even for electron conduction (King, 1978; Dockter et al., 1978). But, I think it is fair to say that the protonmotive property of the mitochondrial cytochrome system and the photosystems of chloroplasts can probably be correctly explained, in general principle, by the direct ligand-conduction type of chemiosmotic mechanism. The same may be said of the protonmotive property of the photosystems of bacterial chromatophores (Crofts & Bowyer, 1978; Dutton et al., 1978), and of certain bacterial redox chains (see Hamilton & Haddock, eds, 1977; Jones et al., 1978). The mechanism of the protonmotive ATPase is more controversial; but, at all events, mechanistic conjectures of the direct chemiosmotic type seem to me to be strategically valuable because they stimulate rational experimental research and thereby add to our biochemical knowledge, even if they ultimately turn out to be wrong.

4. CONCLUSION AND PROSPECT

The students of membrane biochemistry and bioenergetics have endured a long period of uncertainty and conceptual upheaval during the last thirty years − a time of great personal, as well as scientific, trauma for many of us.

The present position, in which, with comparatively few dissenters, we have successfully reached a consensus in favour of the chemiosmotic theory, augurs well for the future congeniality and effectiveness of experimental research in the field of membrane biochemistry and bioenergetics. At the time of the most intensive testing of the chemiosmotic hypothesis, in the nineteensixties and early nineteenseventies, it was not in the power of any of us to predict the outcome. The aspect of the present position of consensus that I find most remarkable and admirable, is the altruism and generosity with which former opponents of the chemiosmotic hypothesis have not only come to accept it, but have actively promoted it to the status of a theory. According to their classically Popperian view (see Mitchell, 1977b), the chemiosmotic theory is worth accepting, for the time being, as the best conceptual framework available (Slater, 1977). Thus to have falsified the pessimistic dictum of the great Max Planck is, I think, a singularly happy achievement.

Returning, finally, to the theme of the respiratory chain, it is especially noteworthy that David Keilin's chemically simple view of the respiratory chain appears now to have been right all along — and he deserves great credit for having been so reluctant to become involved when the energy-rich chemical intermediates began to be so fashionable. This reminds me of the aphorism: "The obscure we see eventually, the completely apparent takes longer".

ACKNOWLEDGEMENTS

I am especially indebted to my research associate, Jennifer Moyle, for constant discussion, criticism and help. Many colleagues have contributed to the evolution of ideas and knowledge traced in this lecture. In particular, I would like to mention the influence of Tsoo King, after the premaature death of David Keilin in 1963. I would also like to acknowledge that it was Bill Slater who persuaded me to become more deeply involved in practical research on oxidative phosphorylation in 1965, when Jennifer Moyle and I first commenced work at the Glynn Research Laboratories. I thank Jim Danielli for the photograph shown in Fig. 6; I thank Jack Dunitz and Ulrich Müller-Herold for help in researching the dictum of Max Planck (1928, 1933); and I thank Bernie Trumpower and Carol Edwards for informing me, prior to publication of their work, that factor OxF is probably a reconstitutively active form of the Rieske ironsulphur protein, thus influencing the representation of the Q cycle in Fig. 17A. I am grateful to Robert Harper and Stephanie Key for help in preparing the manuscript, and to Glynn Research Ltd for general financial support.

REFERENCES

Arnon, D. I., Whatley, F. R. and Allen, M. B. (1954) J. Am. Chem. Soc. 76, 6324–6328.

Azzone, G. F., Pozzan, T. and Bragadin, M. (1977) BBA Library 14, 107–116.

Beinert, H. (1977) BBA Library 14, 11–21.

Belitser, V. A. and Tsybakova, E. T. (1939) Biokhimia 4, 516–535.

Bendall, D. S. (1977) in *Internat. Rev. of Biochem. Plant Biochem. II* (Northcote, D. H., ed.) vol. 13, pp. 41–78, University Park Press, Baltimore.

Bowman, B. J., Mainzer, S. E., Allen, K. E., and Slayman, C. W. (1978) Biochim. Biophys. Acta 512, 13–28.

Boyer, P. D., Falcone, A. B. and Harrison, W. H. (1954) Nature 174, 401–404.

Boyer, P. D. (1963) Science 141, 1147–1153.

Boyer, P. D. (1965) in *Oxidases and Related Redox Systems* (King, T. E. et al., eds) vol. 2, pp. 994–1008, John Wiley, New York.

Boyer, P. D., Cross, R. L. and Momsen, W. (1973) Proc. Nat. Acad. Sci. U.S.A. 70, 2837–2839.

Boyer, P. D. (1974) BBA Library 13, 289–301.

Boyer, P. D., Chance, B., Ernster, L., Mitchell, P., Racker, E. and Slater, E. C. (1977) Ann. Rev. Biochem. 46, 955–1026.

Chance, B. and Williams, G. R. (1956) Adv. Enzymol. 17, 65–134.

Chance, B. (1961) J. Biol. Chem. 236, 1569–1576.

Chance, B., Lee, C. P. and Mela, L. (1967) Fed. Proc. 26, 1341–1354.

Chance, B. (1972) FEBS Lett. 23, 3–20.

Chance, B. (1974) Ann. N.Y. Acad. Sci. 227, 613–626.

Chappell, J. B. (1968) Brit. Med. Bull. 24, 150–157.

Conway, E. J. (1951) Science 113, 270–273.

Crofts, A. R. and Bowyer, J. (1978) in *The Proton and Calcium Pumps* (Azzone, G. F. et al., eds) pp. 55–64, Elsevier/North-Holland, Amsterdam.

Cross, R. L. and Boyer, P. D. (1973) in *Mechanisms in Bioenergetics* (Azzone, G. F. et al., eds) pp. 149–155, Academic Press, New York.

Curie, P. (1894) J. Phys., 3ème Ser. 393–415.

Davies, R. E. and Ogston, A. G. (1950) Biochem. J. 46, 324–333.

Delhez, J., Dufour, J.-P., Thines, D. and Goffeau, A. (1977) Eur. J. Biochem. 79, 319–328.

DePierre, J. W. and Ernster, L. (1977) Ann. Rev. Biochem. 46, 201–262.

Dockter, M. E., Steinmann, A. and Schatz, G. (1978) J. Biol. Chem. 253, 311–317.

Drachev, L. A., Kaulen, A. D. and Skulachev, V. P. (1978) FEBS Lett. 87, 161–167.

Dutton, P. L., Bashford, C. L., van den Bergh, W. H., Bonner, H. S., Chance, B., Jackson, J. B., Petty, K. M., Prince, R. C., Sorge, J. R. and Takamiya, K. (1978) in *Photosynthesis 77, Proc. Fourth Internat. Congr. Photosynth.* (Hall, D. O. et al., eds) pp. 159–171, Biochem. Soc., London.

Dutton, L., Leigh, J. and Scarpa, A., eds (1978) *Frontiers of Biological Energetics*, Academic Press, New York.

Ewards, G. E. and Huber, S. C. (1978) in *Photosynthesis 77, Proc. Fourth Internat. Congr. Photosynth.* (Hall, D. O. et al., eds) pp. 95–106, Biochem. Soc., London.

Ernster, L. and Lee, C. P. (1964) Ann. Rev. Biochem. 33, 729–788.

Ernster, L., Juntti, K. and Asami, K. (1973) J. Bioenerg. 4, 149–159.

Ernster, L., Nordenbrand, K., Chude, O. and Juntti, K. (1974) in *Membrane Proteins in Transport and Phosphorylation* (Azzone, G. F. et al., eds) pp. 29–41, North-Holland, Amsterdam.

Ernster, L. (1977a) in *Living Systems as Energy Converters* (Buvet, R. et al., eds) pp. 115–118, North-Holland, Amsterdam.

Ernster, L. (1977b) in *Bioenergetics of Membranes* (Packer, L. et al., eds) pp. 373–376, Elsevier/North-Holland, Amsterdam.

Friedkin, M. and Lehninger, A. L. (1948) J. Biol. Chem. 174, 757–758.

Garland, P. B., Clegg, R. A., Downie, J. A., Gray, T. A., Lawford, H. G. and Skyrme, J. (1972) FEBS Symp. *28*, 105–117.

Glagolev, A. N. and Skulachev, V. P. (1978) Nature *272*, 280–282.

Green, D. E., Beyer, R. E., Hansen, M., Smith, A. L. and Webster, G. (1963) Fed. Proc. *22*, 1460–1468.

Green, D. E. (1974) Biochim. Biophys. Acta *346*, 27–78.

Griffiths, D. E. (1965) in *Essays in Biochemistry* (Campbell, P. N. and Greville, G. D., eds) vol. 1, pp. 91–120, Academic Press, London.

Griffiths, D. E., Hyams, R. L. and Bertoli, E. (1977) FEBS Lett. *74*, 38–42.

Grove, W. R. (1839) Phil. Mag., *Ser. 3*, *14*, 127–130.

Guggenheim, E. A. (1933) *Modern Thermodynamics by the Methods of Willard Gibbs*, Methuen, London.

Gutman, M., Beinert, H. and Singer, T. P. (1975) in *Electron Transfer Chains and Oxidative Phosphorylation* (Quagliariello, E. et al., eds) pp. 55–62, North-Holland, Amsterdam.

Gutman, M. (1977) in *Bioenergetics of Membranes* (Packer, L. et al., eds) pp. 165–175, Elevier/North-Holland, Amsterdam.

Hackenbrock, C. R. and Höchli, M. (1977) FEBS Symp. *42*, 10–36.

Hall, D. O., Coombs, J. and Goodwin, T. W., eds (1978) *Photosynthesis 77, Proc. Fourth Internat. Congr. Photosynth.*, Biochem. Soc., London.

Hamilton, W. A. and Haddock, B. A., eds (1977) Symp. Soc. Gen. Microbiol. *27*.

Harold, F. M. (1977a) Curr. Top. Bioenerg. *6*, 83–149.

Harold, F. M. (1977b) Ann. Rev. Microbiol. *31*, 181–203.

Hatefi, Y. (1963) Adv. Enzymol. *25*, 275–328.

Hatefi, Y. (1966) Compr. Biochem. *14*, 199–231.

Hatefi, Y. and Hanstein, W. G. (1972) J. Bioenerg. *3*, 129–136.

Hauska, G. (1977a) in *Bioenergetics of Membranes* (Packer, L. et al., eds) pp. 177–186, Elsevier/North-Holland, Amsterdam.

Hauska, G. (1977b) in *Photosynthesis 77, Proc. Fourth Internat. Congr. Photosynth.* (Hall. D. O. et al., eds) pp. 185–196, Biochem. Soc., London.

Heaton, G. M., Wagenvoord, R. J., Kemp, A. and Nicholls, D. G. (1978) Eur. J. Biochem. *82*, 515–521.

Heldt, H. W. (1976) in *The Intact Chloroplast* (Barber, J., ed) pp. 215–234, Elsevier, Amsterdam.

Heron, C., Ragan, C. I. and Trumpower, B. L. (1978) Biochem. J. *174*, 791–800.

Hill. R. and Bendall, F. (1960) Nature *186*, 136–137.

Hinkle, P. C. and McCarty, R. E. (1978) Scientific American *238*, 104–123.

Ingledew, W. J., Cox, J. C. and Halling, P. J. (1977) FEBS Microbiology Letters *2*, 193–197.

Jagendorf, A. T. (1967) Fed. Proc. *26*, 1361–1369.

Jagendorf, A. T. (1977) in *Encyclopedia of Plant Physiol.* New Series (Trebst, A. and Avron, M., eds) vol. 5, pp. 307–377, Springer Verlag, Berlin.

Jones, C. W., Brice, J. M. and Edwards, C. (1978) FEBS Symp. *49*, 89–97.

Junge, W. (1977) Ann. Rev. Plant Physiol. *28*, 503–536.

Junge, W., Ausländer, W., McGeer, A. and Runge, T. (1979a) Biochim. Biophys. Acta, 546, 121–141.

Junge, W., McGeer, A. J., Ausländer, W. and Kolla, J. (1979b) in *29th Mosbach Colloq.* (Schäfer, G. and Klingenberg, M., eds) in press, Springer Verlag, Berlin.

Kagawa, Y. (1978) Biochim. Biophys. Acta *505*, 45–93.

Kalckar, H. (1937) Enzymologia *2*, 47–52.

Keilin, D. (1925) Proc. Roy. Soc. B *98*, 312–339.

Keilin, D. (1929) Proc. Roy. Soc. B *104*, 206–252.

Keilin, D. and King, T. E. (1958) Nature *181*, 1520–1522.

King, T. E. (1966) Adv. Enzymol. *28*, 155–236.

King, T. E. (1978) FEBS Symp. *45*, 17–31.

Klingenberg, M. (1977) BBA Library *14*, 275–282.

Konstantinov. A. A. and Ruuge, E. K. (1977) FEBS Lett. *81*, 137–141.

Kozlov, I. A. and Skulachev, V. P. (1977) Biochim. Biophys. Acta *463*, 29–89.

Kröger, A. and Klingenberg, M. (1973) Eur. J. Biochem. *34*, 358–368, and *39*, 313–323.

Kundig, W., Ghosh, S., and Roseman, F. D. (1964) Proc. Nat. Acad. Sci. U. S. A. *52*, 1067–1074.

Lardy, H. A., Connelly, J. L. and Johnson, D. (1964) Biochemistry *3*, 1961–1968.

Lehninger, A. L. (1959) Rev. Modern Physics *31*, 136–146.

Lehninger, A. L. (1962) Physiol. Rev. *42*, 467–517.

Lehninger, A. L. and Wadkins, C. L. (1962) Ann. Rev. Biochem. *31*, 47–78.

Lenaz, G., Pasquali, P., Bertoli, E., Parenti–Castelli, G. and Folkers, K. (1975) Archs. Biochem. Biophys. *169*, 217–226.

Lenaz, G., Mascarello, S., Laudi, L., Cabrini, L., Pasquali, P., Parenti-Castelli, G., Sechi, A. M. and Bertoli, E. (1977) in *Bioenergetics of Membranes* (Packer, L. et al., eds) pp. 189–198, Elsevier/North-Holland, Amsterdam.

Liebhafsky, H. A. and Cairns, E. J. (1968) *Fuel Cells and Fuel Batteries,* John Wiley, New York.

Lipmann, F. (1941) Adv. Enzymol. *1*, 99–162.

Lipmann, F. (1946) in *Currents in Biochemical Research* (Green, D. E., ed.) pp. 137–148, Interscience, New York.

Lipmann, F. (1960) in *Molecular Biology* (Nachmansohn, D., ed.) pp. 37–47, Academic Press, New York.

Lundegardh, H. (1945) Arkiv. Bot. 32 A *12*, 1–139.

Manson, M. D., Tedesco, P., Berg, H. C., Harold, F. M. and Van der Drift, C. (1977) Proc. Nat. Acad. Sci. U. S. A. *74*, 3060–3064.

Matsuura, S., Shioi, J. -i. and Imae, Y. (1977) FEBS Lett. *82*, 187–190.

Mitchell, P. (1949) in *The Nature of the Bacterial Surface* (Miles, A. A. and Pirie, N. W., eds) pp. 55–75, Blackwell, Oxford.

Mitchell, P. (1954) Symp. Soc. Exp. Biol. *8*, 254–261.

Mitchell, P. (1956) Discuss. Farad. Soc. *21*, 278–279 and 282–283.

Mitchell, P. and Moyle, J. (1956) Discuss. Farad. Soc. *21*, 258–265.

Mitchell, P. (1957) Nature *180*, 134–136.

Mitchell, P. and Moyle, J. (1958a) Nature *182*, 372–373.

Mitchell, P. and Moyle, J. (1958b) Proc. Roy. Phys. Soc., Edinburgh *27*, 61–72.

Mitchell, P. (1959) Biochem. Soc. Symp. *16*, 73–93.

Mitchell, P. (1961a) in *Membrane Transport and Metabolism* (Kleinzeller, A. and Kotyk, A., eds) pp. 22–34, Academic Press, New York.

Mitchell, P. (1961b) in *Biological Structure and Function, Proc. First IUB/IUBS Internat. Symp.,* Stockholm, 1960 (Goodwin, T. W. and Lindberg, O., eds) vol. 2, pp. 581–599, Academic Press, London.

Mitchell, P. (1961c) Nature *191*, 144–148.

Mitchell, P. (1962) J. Gen. Microbiol. *29*, 25–37.

Mitchell, P. (1963) Biochem. Soc. Symp. *22*, 142–168.

Mitchell, P. and Moyle, J. (1965) Nature *208*, 147–151.

Mitchell, P. (1966) *Chemiosmotic Coupling in Oxidative and Photosynthetic Phosphorylation,* Glynn Research, Bodmin, Cornwall, England.

Mitchell, P. (1967a) Fed. Proc. *26*, 1335–1340.

Mitchell, P. (1967b) Compr. Biochem. *22*, 167–197.

Mitchell, P. (1967c) Adv. Enzymol. *29*, 33–85.

Mitchell, P. and Moyle, J. (1967) Biochem. J. *105*, 1147–1162.

Mitchell, P. (1968) *Chemiosmotic Coupling and Energy Transduction,* Glynn Research, Bodmin, Cornwall, England.

Mitchell, P. and Moyle, J. (1968) Eur. J. Biochem. *4*, 530–539.

Mitchell, P. (1970a) Membr. Ion Transp. *1*, 192–256.

Mitchell, P. (1970b) Symp. Soc. Gen. Microbiol. *20*, 121–166.

Mitchell, P. (1972a) FEBS Symp. *28*, 353–370.

Mitchell, P. (1972b) J. Bioenerg. *3*, 5–24.

Mitchell, P. (1973a) J. Bioenerg. *4*, 63–91.

Mitchell, P. (1973b) in *Mechanisms in Bioenergetics* (Azzone, G. F. et al., eds) pp. 177–201, Academic Press, New York.

Mitchell, P. (1973c) FEBS Lett. *33*, 267–274.

Mitchell, P. (1975) in *Elecron Transfer Chains and Oxidative Phosphorylation* (Quagliariello, E. et al., eds) pp. 305–316, North-Holland, Amsterdam.

Mitchell, P. (1976a) Biochem. Soc. Trans. *4*, 399–430.

Mitchell, P. (1976b) J. Theoret. Biol. *62*, 327–367.

Mitchell, P. (1977a) FEBS Lett. *78*, 1–20.

Mitchell, P. (1977b) Ann. Rev. Biochem. *46*, 996–1005.

Mitchell, P. (1977c) Symp. Soc. Microbiol. *27*, 383–423.

Mitchell, P. (1979) Eur. J. Biochem. *95*, 1–20.

Moore, A. L. (1978) FEBS Symp. *49*, 141–147.

Moyle, J., Mitchell, R. and Mitchell, P. (1972) FEBS Lett. *23*, 233–236.

Moyle, J. and Mitchell, P. (1977) FEBS Lett. *84*, 135–140.

Moyle, J. and Mitchell, P. (1978a) FEBS Lett. *88*, 268–272.

Moyle, J. and Mitchell, P. (1978b) FEBS Lett. *90*, 361–365.

Nicholls, P. (1963) in *The Enzymes* (Boyer, P. D. et al., eds) vol. 8, part B, pp. 3–40, Academic Press, New York.

Njus, D. and Radda, G. K. (1978) Biochim. Biophys. Acta *463*, 219–244.

Nordenbrand, K., Hundal, T., Carlsson, C., Sandri, G. and Ernster, L. (1977) in *Bioenergetics of Membranes* (Packer, L. et al., eds) pp 435–446, Elsevier/North–Holland, Amsterdam.

Ochoa, S. (1940) Nature *146*, 267.

Ohnishi, T. (1979) in *Membrane Proteins in Energy Transduction* (Capaldi, R. A., ed.), 1–87, Marcel Dekker, New York.

Painter, A. A. and Hunter, F. E. (1970) Biochem. Biophys. Res. Commun. *40*, 360–395.

Papa, S. (1976) Biochim. Biophys. Acta *456*, 39–84.

Papa, S., Guerrieri, F., Lorusso, M., Izzo, G., Boffoli, D. and Stefanelli, R. (1978) FEBS Symp. *45*, 37–48.

Pauling, L. (1950) *Ann. Rep. Smithsonian Inst.* 225–241.

Planck, M. (1928) *Wissenschaftliche Autobiographie*, Leipzig. Planck's remark, as presented in my paper, is a paraphrase of the following passage on p. 22: "Eine neue wissenschaftliche Wahrheit pflegt sich nicht in der Weise durchzusetzen, dass ihre Gegner überzeugt werden und sich als belehrt erklären, sondern vielmehr dadurch, dass die Gegner allmählich aussterben und dass die heranwachsende Generation von vornherein mit der Wahrheit vertraut gemacht ist." But I have partly followed a later variant of this passage, given in Planck (1933), where he substituted the word Idee for Wahrheit.

Planck, M. (1933) *Ursprung und Auswirkung wissenschaftlicher Ideen*. A lecture reproduced in Planck's *Vorträge und Erinnerungen*, Darmstadt, 1975.

Racker, E. (1961) Adv. Enzymol. *23*, 323–399.

Racker, E. (1965) *Mechanisms in Bioenergetics*, Academic Press, New York.

Racker, E. (1967) Fed. Proc. *26*, 1335–1340.

Racker, E. (1976) *A New Look at Mechanisms in Bioenergetics*, Academic Press, New York.

Racker, E. (1977) Ann. Rev. Biochem. *46*, 1006–1014.

Ragan, C. I. (1976) Biochim. Biophys. Acta *456*, 249–290.

Ragan, C. I. and Heron, C. (1978) Biochem. J. *174*, 783–790.

Reynafarje, B. and Lehninger, A. L. (1978) J. Biol. Chem. *253*, 6331–6334.

Rich, P. R. and Moore, A. L. (1976) FEBS Lett. *65*, 339–344.

Rich, P. R. and Bonner, W. D. (1978) FEBS Symp. *49*, 149–158.

Rieske, J. S. (1976) Biochim. Biophys. Acta *456*, 195–247.

Robertson, R. N. and Wilkins, M. J. (1948) Aust. J. Sci. Res. *1*, 17–37.

Robertson, R. N. (1960) Biol. Rev. *35*, 231–264.

Robertson, R. N. (1968) *Protons, Electrons, Phosphorylation and Active Transport*, Cambridge University Press, Cambridge.

Rosen, B. P. and Kashket, E. R. (1978) in *Bacterial Transport* (Rosen, B. P., ed.) pp. 559–620, Marcel Dekker, New York.

Rosenberg, T. (1948) Acta Chem. Scand. *2*, 14–33.

Sanadi, D. R. (1965) Ann. Rev. Biochem. *34*, 21–48.

Sands, R. H. and Beinert, H. (1960) Biochem. Biophys. Res. Commun. *3*, 47–52.

Schreckenbach, T. and Oesterhelt, D. (1978) FEBS Symp. *45*, 105–119.

Schulten, K. and Tavan, P. (1978) Nature *272*, 85–86.

Seaston, A., Carr, G. and Eddy, A. A. (1976) Biochem. J. *154*, 669–676.

Senior, A. E. (1979) in *Membrane Proteins in Energy Transduction* (Capaldi, R. A., ed.) in press, Marcel Dekker, New York.

Singer, S. J. and Nicholson, G. L. (1971) Science *175*, 720–731.

Skulachev, V. P. (1970) FEBS Lett. *11*, 301–308.

Skulachev, V. P. (1974) Ann. N. Y. Acad. Sci. *227*, 188–202.

Skulachev, V. P. (1977) FEBS Lett. *74*, 1–9.

Slater, E. C. (1953) Nature *172*, 975–978.

Slater, E. C. and Cleland, K. W. (1953) Biochem. J. *53*, 557–567.

Slater, E. C. (1958) Rev. Pure Appl. Chem. *8*, 221–264.

Slater, E. C. and Hulsmann, W. C. (1959) in *Ciba Foundation Symp., Regulation of Cell Metabolism* (Wolstenholme, G. E. W. and O'Connor, C. M., eds) pp. 58–83, Churchill, London.

Slater, E. C. (1966) Compr. Biochem. *14*, 327–396.

Slater, E. C. (1971) Quart. Rev. Biophys. *4*, 35–71.

Slater, E. C. (1974) Biochem. Soc. Trans. *2*, 39–1163.

Slater, E. C. (1975) in *Electron Transfer Chains and Oxidative Phosphorylation* (Quagliariello, E. et al., eds) pp. 3–14, North-Holland, Amsterdam.

Slater, E. C. (1976) in *Reflections on Biochemistry* (Kornberg, A. et al., eds) pp. 45–55, Pergamon Press, Oxford.

Slater, E. C. (1977) in *Living Systems as Energy Converters* (Buvet, R. et al., eds) pp. 221–227, North-Holland, Amsterdam.

Stoeckenius, W. (1977) Fed. Proc. *36*, 1797–1798.

Storey, B. T. (1970) J. Theoret. Biol. *28*, 233–259.

Storey, B. T. (1971) J. Theoret. Biol. *31*, 533–552.

Takemori, S. and King, T. E. (1962) Biochim. Biophys. Acta *64*, 192–194.

Toll, L. and Howard, B. D. (1978) Biochemistry *17*, 2517–2523.

Trumpower, B. L. (1976) Biochem. Biophys. Res. Commun. *70*, 73–80.

Trumpower, B. L. (1978) Biochem. Biophys. Res. Commun. *83*, 528–535.

Trumpower, B. L. and Katki, A. G. (1979) in *Membrane Proteins in Electron Transport* (Capaldi, R. A., ed.) in press, Marcel Dekker, New York.

Ussing, H. H. (1947) Nature *160*, 262–263.

Ussing, H. H. (1949) Physiol. Rev. *29*, 127–155.

Wang, J. H. (1972) J. Bioenerg. *3*, 105–114.

Weiner, M. W. and Lardy, H. A. (1974) Archs. Biochem. Biophys. *162*, 568–577.

Wikström, M. and Krab, K. (1978) FEBS Lett. *91*, 8-14.

Williams, R. J. P. (1962) J. Theoret. Biol. *3*, 209–229.

Williams, R. J. P. (1970) in *Electron Transport and Energy Conservation* (Tager, J. M. et al., eds) p. 381, Adriatica Editrice, Bari.

Witt, H. T. (1967) in *Fast Reactions and Primary Processes in Chemical Kinetics*, Nobel Symp. 5, (Claesson, S., ed.) pp. 261–316, Interscience, London.

Witt, H. T. (1977) in *Living Systems as Energy Converters* (Buvet, R. et al., eds) pp. 185–197, Elsevier/North-Holland, Amsterdam.

Yu, C. A., Yu, L. and King, T. E. (1977a) Biochem. Biophys. Res. Commun. *78*, 259–265.

Yu, C. A., Yu, L. and King, T. E. (1977b) Biochem. Biophys. Res. Commun. *79*, 939–946.

Yu, C. A., Nagaoka, S., Yu, L. and King, T. E. (1978) Biochem. Biophys. Res. Commun. *82*, 1070–1078.

Chemistry 1979

HERBERT C. BROWN and GEORG WITTIG

for their development of the use of boron- and phosphorus-containing compounds, respectively, into important reagents in organic synthesis

THE NOBEL PRIZE FOR CHEMISTRY

Speech by Professor BENGT LINDBERG of the Royal Academy of
Sciences. Translation from the Swedish text.

Your Majesties, Your Royal Highnesses, Ladies and Gentlemen,

Chemistry is a natural science which is not entirely devoted to the study
of natural objects. The art of chemistry also includes the ability on the part
of the chemists to prepare or synthesize various chemical compounds. This
is especially true for carbon compounds, the chemistry of which is called
organic chemistry. Carbon atoms may be linked to other carbon atoms and
there appears to be no limit to how many times this may be repeated in a
molecule. The possibilities for variation therefore are immense. Chemists
have synthesized more than two million organic compounds. These possibilities to synthesizing compounds have greatly enriched chemistry and
have had enormous practical consequences.

Through the use of synthetic pharmaceuticals, vitamins and pesticides
against microorganisms, insects and weeds, millions of lives have been
saved, much suffering has been alleviated and world famine has been
reduced. Further significant progress may be expected in a range of
practical, important areas, especially concerning the development of specific pesticides less disturbing to the environment than the present ones.

One of the most important tasks for the present-day organic chemists, in
basic as well as in applied research, is the synthesis of biologically active
compounds. In order to achieve this, we need methods of joining carbon
atoms to one another and to modify organic compounds in a variety of
ways. A great many chemists devote their time to developing such methods. A few times in the history of chemistry have new synthetic methods
been deemed so important that the originators have been awarded the
Nobel Prize. This has once more happened, this year Brown and Wittig
have been awarded the prize for chemistry for their development of boron
and phosphorus compounds, respectively, into important reagents in organic synthesis.

Herbert C. Brown has systematically studied various boron compounds
and their chemical reactions. He has shown how various specific reductions can be carried out using borohydrides. One of the simplest of these,
sodium borohydride, has become one of the most used chemical reagents.
The organoboranes, which he discovered, have become the most versatile
reagents in organic synthesis. The exploitation of their chemistry has led
to new methods for rearrangements, for addition to double bonds and for
joining carbon atoms to one another.

Georg Wittig has provided many significant contributions in organic
chemistry. The most important of these is the discovery of the synthetic
method which bears his name, the Wittig reaction. In this, phosphorus

ylides, a type of compound which he discovered, are allowed to react with carbonyl compounds. An exchange of groups takes place and the result is a compound in which two carbon atoms have been joined by a double bond. Since many natural products with biological activity contain such bonds, this elegant method has found wide-spread use, for example in the industrial synthesis of vitamin A.

Professor Brown,

Your discovery and exploration of borohydrides and organo boranes have given the chemists new and powerful tools for organic syntheses. May I convey to you the warmest congratulations of the Royal Swedish Academy of Sciences.

Professor Wittig,

Die Reaktion die alle, aber nicht Sie selber, die Wittig Reaktion nennen, ist eine der wichtigsten Reaktionen der organischen Chemie geworden. Sie ist besonders geeignet für die Synthese verschiedener biologisch aktiver Moleküle. Ich überbringe Ihnen die herzlichsten Glückwünsche der Königlichen Schwedischen Akademie der Wissenschaften.

Professor Brown, Professor Wittig, may I ask you to receive the Nobel Prize for Chemistry from the hands of His Majesty the King.

Herbert C. Brown

HERBERT C. BROWN

My parents, Charles Brovarnik and Pearl Gorinstein, were born in Zhito-mir in the Ukraine and came to London in 1908 as part of the vast Jewish immigration in the early part of this century. They were married in London. In 1909 my sister, Ann, was born. I arrived on May 22, 1912. In June 1914 my father decided to join his mother and father and other members of his family in Chicago, much to the dismay of my mother, whose own family largely remained in England. My grandfather's name had been anglicized to Brown, and that became our name. In the United States, my two sisters, Sophie and Riva, were born in 1916 and 1918.

My father had been trained as a cabinet maker, doing delicate inlaid work. However, he found little market for his skills in the U.S. and turned to carpentry.

The Depression of 1920 persuaded him to go into business and he opened a small hardware store in Chicago at 18th and State Street, largely a black neighborhood. We lived in an apartment above the store and I attended the Haven School at Wabash and 16th Street with predominantly black classmates.

I did well in school and was advanced several times, graduating at 12. Indeed, I was offered, but refused, further advancement since I did not want to be in the same class with my sister, Ann.

On graduation, I went to Englewood High School on the South Side of Chicago. Unfortunately, my father became ill of some sort of infection and died in 1926. I left school to work in our store. I am afraid that I was not really interested in the business and spent most of my time reading. My mother finally decided that she would attend the store and I should go back to school. Accordingly, I reentered Englewood in February 1929 and graduated in 1930.

At Englewood I ran the humor column of the school paper and won a national prize. I never recovered.

We sold the store at that time. I had no hope of going on to college. However, this was the beginning of the Depression and I could find no permanent job. Studying appealed to me much more than the odd jobs I could find. I decided to go to college. I entered college intending to major in electrical engineering. I had heard that one could make a good living in that area. However, I took chemistry and became fascinated with that subject, and remained with chemistry thereafter. I had just completed one semester at Crane Junior College when it was announced in 1933 that the school was to be closed for lack of funds. I then went to night school at the

Lewis Institute, taking one or two courses, financing myself by working as a parttime shoe clerk.

I then heard that one of the instructors at Crane, Dr. Nicholas Cheronis, had opened his laboratory to several students, so that they could continue their studies on their own. I went there and grew to know and love a fellow student, Sarah Baylen. Sarah had been the brightest student in chemistry at Crane prior to my arrival. She has described ("Remembering HCB") how she initially "hated my guts." But since she could not beat me, she later decided to join me, to my everlasting delight.

In 1934 Wright Junior College opened its doors. We went there and nine of us graduated in 1935 as the first graduating class. In my yearbook Sarah predicted that I would be a Nobel Laureate!

I had been advised to take the competitive examination for a scholarship at the University of Chicago. I did so. To my astonishment, little of the examination was devoted to the chemistry, physics, and mathematics that had constituted the major portion of my studies. Instead, the examination emphasized general subjects: history, art, music, literature, etc.—subjects I had never studied formally. I did the best I could and was pleasantly surprised when I received a half scholarship.

I entered the University of Chicago in the Fall of 1935, accompanied by my girlfriend, Sarah. This was the time when the President of the University, Robert Maynard Hutchins, was arguing for the principle that students should be permitted to proceed as rapidly as possible. Indeed, at that time it cost no more to take ten courses than it did the usual three. I did so, and completed my junior and senior year in three quarters, receiving the B.S. in 1936.

I did not apply for graduate work. I wanted to find a job and marry my girlfriend. However, a famous organic chemist, Julius Stieglitz—then Emeritus, but still teaching—called me into his office and urged me to reconsider my decision. He predicted a favorable future as a research chemist. I discussed the matter with Sarah and she agreed that marriage could wait. Accordingly, I began graduate work.

On my graduation, Sarah presented me with a gift—a copy of Alfred Stock's book, *The Hydrides of Boron and Silicon*. This book interested me in the hydrides of boron and I undertook to study with Professor H. I. Schlesinger, then active in that area of research.

Sarah and I were married "secretly" on February 6, 1937. We were such innocents that we did not realize that marriages are published in the daily newspapers. Consequently, our marriage was a secret for the weekend!

Once the news got out, I had to begin supporting her. But my income as a graduate assistant was only $400 per year, out of which had to come $300 for tuition. But Sarah obtained a position at Billings Hospital in Medical Chemistry and kept us solvent.

I received my Ph.D. in 1938. Unfortunately (perhaps fortunately), I could not find an industrial position. Professor M. S. Kharasch then offered me a position as a postdoctorate at a stipend of $1600 and my academic career was initiated. The following year Professor Schlesinger invited me to become his research assistant with the rank of Instructor, replacing Anton B. Burg, who was moving on to the University of Southern California. Consequently, I am an unusual example of a chemist who ended up in academic work because he could not find an industrial position.

At that time one did not achieve tenure until after ten years. I had seen a number of individuals who had remained at Chicago as Instructors for nine years without tenure and then had to find another position under severe pressure. I decided to avoid this situation. Accordingly, after four years I asked Professor Schlesinger for a decision as to my future in the Department. When he came back with the word that there was no future, I undertook to find another position.

Fortunately, Morris Kharasch had a good friend, Neil Gordon, who had just gone as Department Head to Wayne University in Detroit. (Neil Gordon, the originator of the Gordon Research Conferences, had given Morris Kharasch his first position at the University of Maryland back in 1920.) Neil Gordon was persuaded to give me a position at Wayne as Assistant Professor, preserving my academic career. I became Associate Professor in 1946, and was invited to Purdue in 1947 by the Head of the Chemistry Department, Henry B. Hass, as Professor of Inorganic Chemistry. In 1959 I became Wetherill Distinguished Professor and in 1960 Wetherill Research Professor. I became Emeritus in 1978, but continue to work with a large group of postdoctorates.

Originally my research covered physical, organic and inorganic chemistry and I took students in all three areas. However, as the Department became more organized into divisions, it became necessary to make a choice, and I elected to work primarily with coworkers in organic chemistry.

In addition to my research program in the borane-organoborane area, described in my Nobel Lecture, my research program has involved the study of steric effects, the development of quantitative methods to determine steric strains, the examination of the chemical effects of steric strains, the non-classical ion problem, the basic properties of aromatic hydrocarbons, a quantitative theory of aromatic substitution, and the development of a set of electrophilic substitution constants, σ^+, which correlate aromatic substitution data and a wide variety of electrophilic reactions.

Recognitions:
Professor Brown was the Harrison Howe Lecturer in 1953, the Centenary Lecturer of The Chemical Society (London) in 1955, and the Baker Lecturer in 1968. He was elected to the National Academy of Sciences in 1957, the American Academy of Arts and Sciences in 1966, received an honorary Doctorate of Science degree from the University of Chicago in 1968 and was elected Honorary Fellow of The Chemical Society and Foreign Member of the Indian National Academy of Sciences in 1978. Finally, he is

the recipient of the Nichols Medal for 1959, the ACS Award for Creative Research in Synthetic Organic Chemistry for 1960, the Linus Pauling Medal for 1968, the National Medal of Science for 1969, the Roger Adams Medal for 1971, the Charles Frederick Chandler Medal for 1973, the Madison Marshall Award for 1975, the CCNY Scientific Achievement Award Medal for 1976, the Allied Award for 1978, the Ingold Memorial Lecturer and Medal for 1978, the Elliott Cresson Medal for 1978, and the Nobel Prize for 1979.

FROM LITTLE ACORNS TO TALL OAKS—FROM BORANES THROUGH ORGANOBORANES

Nobel Lecture, 8 December, 1979
by
HERBERT C. BROWN
Department of Chemistry, Purdue University, West Lafayette, Indiana, USA

I. INTRODUCTION

This Nobel Lecture provides me with an exceptional opportunity to trace my research program in boranes from its inception in 1936, as an investigation initiated for my Ph.D. thesis, to the present time, when this program has been recognized by the award of the Nobel Prize for 1979 (shared with my good friend, Georg Wittig).

In 1936 diborane, B_2H_6, was a rare substance, prepared in less than gram quantities in only two laboratories, that of Alfred Stock at Karlsruhe, Germany, and of H. I. Schlesinger, at the University of Chicago. The existence of the simplest hydrogen compound of boron, not as BH_3, but as B_2H_6, was considered to constitute a serious problem for the electronic theory of G. N. Lewis.[1] The reactions of diborane were under study at the University of Chicago by Professor H. I. Schlesinger and his research assistant, Anton B. Burg, in the hope that a knowledge of the chemistry would aid in resolving the structural problem.

I received the Assoc. Sci. degree from Wright Junior College (Chicago) in 1935 and the B.S. degree from the University of Chicago in 1936. Why did I decide to undertake my doctorate research in the exotic field of boron hydrides? As it happened, my girl friend, Sarah Baylen, soon to become my wife, presented me with a graduation gift, Alfred Stock's book, *The Hydrides of Boron and Silicon*.[1] I read this book and became interested in the subject. How did it happen that she selected this particular book? This was the time of the Depression. None of us had much money. It appears that she selected as her gift the most economical chemistry book ($2.06) available in the University of Chicago bookstore. Such are the developments that can shape a career!

Shortly before I undertook research on my Doctorate, H. I. Schlesinger and Anton Burg had discovered that carbon monoxide reacts with diborane to produce a new substance, borane-carbonyl, H_3BCO[2]. There was considerable discussion as to whether the product was a simple addition compound, or whether the reaction had involved a migration of a hydride unit from boron to carbon.

It was thought that an understanding of the reaction of diborane with aldehydes and ketones might contribute to a resolution of this problem. Accordingly, I was encouraged to undertake such a study.

Once I mastered the high-vacuum techniques developed by Stock for work with diborane, it did not take me long to explore the reactions of diborane with aldehydes, ketones, esters, and acid chlorides. It was established that simple aldehydes and ketones react rapidly with diborane at $0°$ (even at $-78°$) to produce dialkoxyboranes (1).

$$2 R_2CO + 1/2(BH_3)_2 \rightarrow (R_2CHO)_2BH \qquad (1)$$

These dialkoxyboranes are rapidly hydrolyzed by water to give the corresponding alcohols (2).

$$(R_2CHO)_2BH + 3 H_2O \rightarrow 2 R_2CHOH + H_2 + B(OH)_3 \qquad (2)$$

The reactions with methyl formate and ethyl acetate were slower, but quantitative reductions were achieved. No appreciable reaction was observed with chloral, acetyl chloride, and carbonyl chloride.[3].

My Ph.D. thesis was completed in 1938 and the contents were published in 1939.[3] At the time the organic chemist had available no really satisfactory method for reducing the carbonyl group of aldehydes and ketones under such mild conditions. Yet little interest in this development was evinced. Why?

In 1939 diborane was a very rare substance, prepared in only minor amounts in two laboratories in the world, handled only by very specialized techniques. How could the synthetic organic chemist consider using such a rare substance as a reagent in his work?

It would be nice to report that one of the three authors had the foresight to recognize that the development of practical methods of preparing and handling diborane would make this reductive procedure of major interest to organic chemists throughout the world. But that was not the case. The problem was later solved, but primarily because of the requirements of research supporting the war effort, and not because of intelligent foresight.

II. THE ALKALI METAL HYDRIDE ROUTE TO DIBORANE AND BOROHYDRIDES

In 1939 Anton B. Burg transferred to the University of Southern California and I became research assistant to Professor Schlesinger. In the Fall of 1940 he was requested to undertake for the National Defense Research Committee a search for new volatile compounds of uranium of low molecular weight. As his research assistant, I became his lieutenant in this war research program.

Just prior to this development, aluminum borohydride,[4] $Al(BH_4)_3$, beryllium borohydride,[5] $Be(BH_4)_2$, and lithium borohydride,[6] $LiBH_4$, had been synthesized in our laboratories. The lithium derivative was a typical non-volatile, salt-like compound, but the aluminum and beryllium derivatives were volatile, the most volatile compounds known for these elements. Accordingly, we undertook to synthesize the unknown uranium (IV) borohydride (3).

$$UF_4 + 2\,Al(BH_4)_3 \rightarrow U(BH_4)_4\uparrow + 2\,AlF_2(BH_4)\downarrow \qquad (3)$$

The synthesis was successful.[7] Moreover, the product, $U(BH_4)_4$, had a low molecular weight (298) and adequate volatility. We were requested to supply relatively large amounts of the material for testing.

The bottle-neck was diborane. We had six diborane generators operated by six young men. Each generator could produce 0.5 g of diborane per 8-hour working day, a total production, when all went well, of 3 g per day, or 1 kg per year! Clearly, we had to find a more practical route to diborane.

We soon discovered that the reaction of lithium hydride with boron trifluoride in ethyl ether solution provided such a route[8] (4).

$$6\,LiH + 8\,BF_3{:}OEt_2 \xrightarrow{Et_2O} (BH_3)_2\uparrow + 6\,LiBF_4\downarrow \qquad (4)$$

We could now prepare diborane in quantity and transform it into uranium (IV) borohydride by simple reactions (5, 6).

$$LiH + 1/2(BH_3)_2 \xrightarrow{Et_2O} LiBH_4 \qquad (5)[9]$$

$$AlCl_3 + 3\,LiBH_4 \xrightarrow{\triangle} Al(BH_4)_3\uparrow + 3\,LiCl\downarrow \qquad (6)[10]$$

$$UF_4 + 2\,Al(BH_4)_3 \rightarrow U(BH_4)_4\uparrow + 2\,AlF_2(BH_4)\downarrow \qquad (3)$$

Unfortunately, we were informed that lithium hydride was in very short supply and could not be spared for this synthesis. We would have to use sodium hydride instead.

Unfortunately, with the solvents then available, the direct use of sodium hydride was not successful. However, a new compound, sodium trimethoxyborohydride,[11] readily synthesized from sodium hydride and methyl borate, solved the problem (7).

$$NaH + B(OCH_3)_3 \rightarrow NaBH(OCH_3)_3 \qquad (7)$$

It proved to be very active and provided the desired transformations previously achieved with lithium hydride (4—6).

At this stage we were informed that the problems of handling uranium hexafluoride had been overcome and there was no longer any need for uranium borohydride. We were on the point of disbanding our group when the Army Signal Corps informed us that the new chemical, sodium borohydride appeared promising for the field generation of hydrogen. However, a more economical means of manufacturing the chemical was required. Would we undertake a research program with this objective?

We soon discovered that the addition of methyl borate to sodium hydride maintained at 250° provided a mixture of sodium borohydride and sodium methoxide[12] (8).

$$4 \text{ NaH} + \text{B(OCH}_3)_3 \xrightarrow{250°} \text{NaBH}_4 + 3 \text{ NaOCH}_3 \tag{8}$$

This provides the basis for the present industrial process for the manufacture of sodium borohydride.

III. REDUCTIONS WITH COMPLEX HYDRIDES

In the course of search for a solvent to separate the two reaction products, acetone was tested. Rapid reduction of the acetone was observed[9] (9).

$$\text{NaBH}_4 + 4 \text{ R}_2\text{C=O} \rightarrow \text{NaB(OCHR}_2)_4 \tag{9}$$

$$\downarrow \text{H}_2\text{O}$$

$$\text{NaB(OH)}_4 + 4 \text{ R}_2\text{CHOH}$$

In this way it was discovered that sodium borohydride is a valuable reagent for the hydrogenation of organic molecules.

At this stage I departed the University of Chicago for Wayne University (Detroit). With the much smaller opportunities for research at this institution, I concentrated on my program dealing with steric strains.[13, 14]

At the University of Chicago the alkali metal hydride route was successfully extended for the synthesis of the corresponding aluminum derivatives. Thus lithium aluminum hydride was synthesized in 1945 by the reaction of lithium hydride and aluminum chloride in ether solution[15] (10).

$$4 \text{ LiH} + \text{AlCl}_3 \xrightarrow{\text{Et}_2\text{O}} \text{LiAlH}_4 + 3 \text{ LiCl} \tag{10}$$

The discovery of sodium borohydride[9] in 1942 and of lithium aluminum hydride[15] in 1945 brought about a revolutionary change in procedures for the reduction of functional groups in organic molecules.[16] As first described by W. G. Brown and his coworkers, there is a major difference in the behavior of these two reducing agents.[16] Lithium aluminum hydride is an exceedingly powerful reducing agent, capable of reducing practically all functional groups. On the other hand, sodium borohydride is a remarkably mild reducing agent, readily reducing only aldehydes, ketones, and acid chlorides.[16] Consequently, we had available two reagents which exhibited extremes in their reducing capabilities (Fig. 1).

In 1947 I came to Purdue University with the opportunity for markedly expanding my research program. I decided to explore means of increasing the reducing properties of sodium borohydride and of decreasing the reducing

	NaBH$_4$										LiAlH$_4$
Aldehyde	+										+
Ketone	+										+
Acid chloride	R										+
Lactone	−										+
Epoxide	−										+
Ester	−										+
Acid	−										+
Acid salt	−										+
tert-Amide	−										+
Nitrile	−										+
Nitro	−										+
Olefin	−										−

Fig. 1. Sodium borohydride and lithium aluminum hydride as extremes in a possible spectrum of hydridic reducing agents. R = Reduced in non-hydroxylic solvents, reacts with hydroxylic solvents.

properties of lithium aluminum hydride. In this way the organic chemist would have at his disposal a full spectrum of reducing agents—he could select that reagent which would be most favorable for the particular reduction required in a given situation.

We quickly established that changes in the metal ion from sodium to lithium, to magnesium, and to aluminum greatly increase the reducing power of the borohydride moiety[17] (11).

$$\underset{\text{increasing reducing power}}{\underline{NaBH_4 < LiBH_4 < Mg(BH_4)_2 < Al(BH_4)_3}} \longrightarrow \qquad (11)$$

On the other hand, the reducing power of lithium aluminum hydride could be diminished by the introduction of alkoxy substituents[18, 19] (12).

$$\longleftarrow \underset{\text{decreasing reducing power}}{\underline{LiAlH(Ot\text{-}Bu)_3 < LiAlH(OMe)_3 < LiAlH_4}} \qquad (12)$$

Indeed, it has proven possible to enhance the reducing power of a borohydride ($LiEt_3BH$)[20] until it exceeds that of lithium aluminum hydride and to diminish the reducing power ($K(i\text{-}PrO)_3BH$)[21] so that it is even less than that of the parent borohydride (Fig. 2).

Finally, we discovered major differences between reduction by electrophilic reagents, such as diborane[22] and aluminum hydride,[23] and by nucleophilic reagents, such as sodium borohydride[16] and lithium aluminum hydride[16] (Fig. 3). It is often possible now to reduce group A in the presence of B, or group B in the presence of A, by a careful choice of reagents. This is nicely illustrated by the synthesis of both (R)- and (S)-mevalonolactone from a common precursor[24, 25] (Fig. 4).

It should be pointed out that these studies were greatly facilitated by many exceptional coworkers, among whom I would like to mention especially, Nung Min Yoon and S. Krishnamurthy.

EFFECT OF SUBSTITUENTS

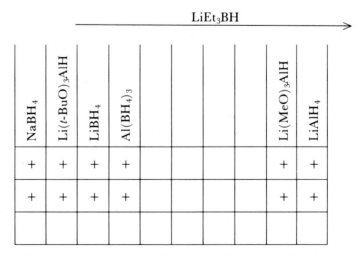

Fig. 2. Alteration of the reducing power of the two extreme reagents, sodium borohydride and lithium aluminum hydride.

	NaBH$_4$ in ethanol	Li(t-BuO)$_3$AlH in THF	LiBH$_4$ in THF	Al(BH$_4$)$_3$ in DG	B$_2$H$_6$ in THF	Sia$_2$BH in THF	9-BBN in THF	AlH$_3$ in THF	Li(MeO)$_3$AlH in THF	LiAlH$_4$ in THF
Aldehyde	+	+	+	+	+	+	+	+	+	+
Ketone	+	+	+	+	+	+	+	+	+	+
Acid chloride	(+)	+	+	+	−	−	+	+	+	+
Lactone	−	±	+	+	+	+	+	+	+	+
Epoxide	−	±	+	+	+	±	±	+	+	+
Ester	−	±	+	+	±	−	±	+	+	+
Acid	−	−	−	+	+	−	±	+	+	+
Acid salt	−	−	−	−	−	−	−	+	+	+
tert-Amide	−	−	−	−	+	+	+	+	+	+
Nitrile	−	−	−	−	+	−	±	+	+	+
Nitro	−	−	−	−	−	−	−	−	+	+
Olefin	−	−	−	−	+	+	+	−	−	−

Fig. 3. Variation in reduction characteristics with electrophilic and nucleophilic reagents.

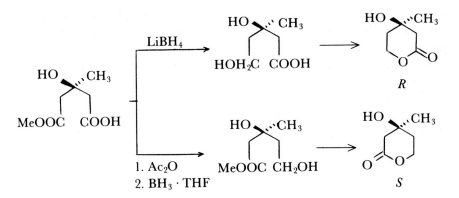

Fig. 4. Synthesis of both (*R*)- and (*S*)-mevalonolactone by selective reduction of a common precursor.

IV. HYDROBORATION

In the course of these studies of selective reductions, a minor anomaly resulted in the discovery of hydroboration. My coworker, Dr. B. C. Subba Rao, was examining the reducing characteristics of sodium borohydride in diglyme catalyzed by aluminum chloride[17]. He observed that the reduction of ethyl oleate under our standard conditions, 4 moles of hydride per mole of compound, one hour at 25°, took up 2.37 equivalents of hydride per mole of ester. This contrasted with a value of 2.00 for ethyl stearate. Investigation soon established

that the reagent was adding an H—B⟨ bond to the carbon-carbon double

bond to form the corresponding organoborane[26].

   Exploration of this reaction soon established improved procedures for carrying it out. Of special value was the discovery that the addition of diborane to alkenes was markedly catalyzed by ethers[27]. In the presence of such ethers, the reaction is practically instantaneous and quantitative (13).

$$C=C+H-B\big< \;\rightarrow\; H-C-C-B\big< \tag{13}$$

(My parents were far-seeing in giving me the initials H. C. B.)

   Dr. Subba Rao established that oxidation of such organoboranes, in situ, with alkaline hydrogen peroxide, proceeds quantitatively, producing alcohols with the precise structure of the organoborane[26, 27] (14).

$$H-C-C-B\big< \xrightarrow[\sim 40°]{H_2O_2} H-C-C-OH \tag{14}$$

At this stage in the development, Dr. B. C. Subba Rao returned to India, after spending five years with me. Fortunately, an equally competent and productive coworker, Dr. George Zweifel, soon joined my group. Although trained at the E. T. H. in Zurich as a carbohydrate chemist, he expressed a deep interest in the possibilities of the hydroboration reaction and progress was extraordinarily rapid[28].

It was soon established that the addition proceeds in an anti-Markovnikov manner (15).

$$\underset{CH_3CH_2C=CH_2}{\overset{CH_3}{\mid}} \xrightarrow{HB} \underset{\underset{H}{\mid}}{\overset{CH_3}{\mid}} CH_3CH_2CCH_2B\diagdown + \underset{\overset{\mid}{B}}{\overset{CH_3}{\mid}} CH_3CH_2CCH_3 \qquad (15)$$

99 %   1 %

The reaction involves a *cis*-addition of the H—B bond (16).

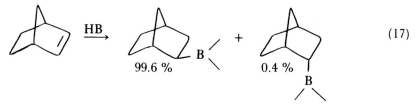

$$\text{(16)}$$

pure trans

The additon takes place preferentially from the less hindered side of the double bond (17).

99.6 %   0.4 %   (17)

No rearrangements of the carbon skeleton have been observed, even in molecules as labile as α-pinene (18).

(18)

Most functional groups can tolerate hydroboration (19).

$$CH_2=CHCH_2CO_2R \xrightarrow{HB} \diagup B-CH_2CH_2CH_2CO_2R \qquad (19)$$

Now the organic chemist can conveniently synthesize reactive intermediates containing functional groups and utilize those intermediates to form products with new carbon-carbon bonds.

Standardized procedures for hydroboration have been developed and are fully described, as well as the utilization of the organoborane products for organic syntheses[29]. (To conserve space, references will be given only to developments which have appeared since the publication of this book.)

V. NEW HYDROBORATING AGENTS

The hydroboration of simple olefins generally proceeds directly to the formation of the trialkylborane[28, 29] (20).

$$3 \text{ CH}_3\overset{\overset{\displaystyle CH_3}{|}}{C}=\text{CH}_2 + \text{BH}_3 \xrightarrow{\text{THF}} [(\text{CH}_3)_2\text{CHCH}_2]_3\text{B} \qquad (20)$$

However, in a number of instances it has been possible to control the hydroboration to achieve the synthesis of monoalkylboranes, dialkylboranes, and cyclic and bicyclic boranes (Fig. 5). Many of these reagents, such as thexylborane[30], disiamylborane[31], dipinylborane[32], and 9-borabicyclo-[3.3.1]nonane[33], have proven to be valuable in overcoming problems encountered with the use of diborane itself.

In a number of cases, hydroboration with heterosubstituted boranes has also proven valuable. Research in this area was greatly facilitated by exceptional contributions from S. K. Gupta, N. Ravindran, and S. U. Kulkarni. For example, catecholborane[34] and the chloroborane etherates[35] (Fig. 6) permit the synthesis of boronic and borinic acid esters, as well as the synthesis of the simple mono- and dialkylchloroboranes, RBCl_2 and R_2BCl. The corresponding haloborane-dimethyl sulfides are stabler and easier to work with[36] (Fig. 7).

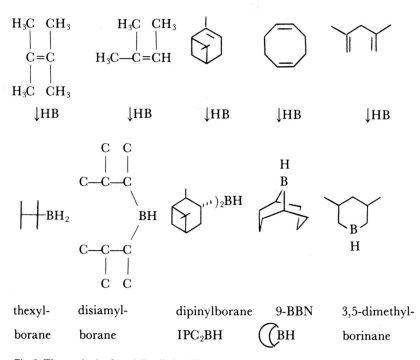

Fig. 5. The synthesis of partially alkylated boranes.

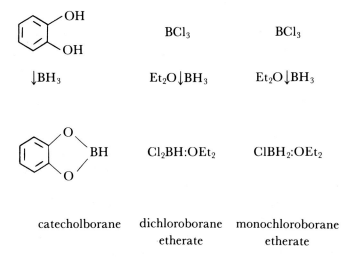

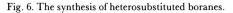

catecholborane   dichloroborane   monochloroborane
                        etherate   etherate

Fig. 6. The synthesis of heterosubstituted boranes.

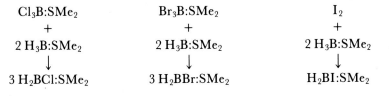

Fig. 7. The synthesis of heterosubstituted boranes.

These reagents often exhibit marked advantages in hydroboration over diborane itself. For example, disiamylborane yields far less of the minor isomer in the hydroboration of terminal olefins than does diborane (21).

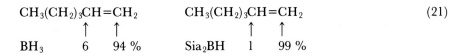

$$CH_3(CH_2)_3CH=CH_2 \qquad CH_3(CH_2)_3CH=CH_2 \qquad (21)$$

BH$_3$ 6   94 %   Sia$_2$BH   1   99 %

Disiamylborane also favors addition of the boron atom to the less substituted position of a 1,2-dialkylethylene (22).

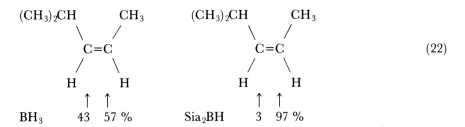

$$(CH_3)_2CH \qquad CH_3 \qquad\qquad (CH_3)_2CH \qquad CH_3$$

C=C   C=C   (22)

H   H   H   H

BH$_3$ 43   57 %   Sia$_2$BH   3   97 %

9-BBN exhibits an even greater selectivity (23).

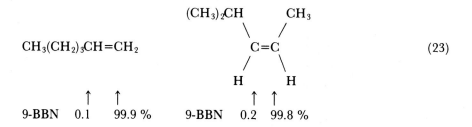

The use of optically active α-pinene yields dipinylborane, IPC$_2$BH, an asymmetric hydroborating agent. It achieves asymmetric syntheses with remarkable efficiency[32] (24).

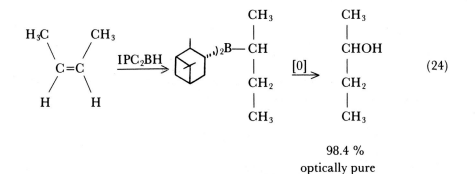

98.4 %
optically pure

VI. THE VERSATILE ORGANOBORANES

At the time we were exploring the hydroboration reaction, many individuals expressed skepticism to me as to the wisdom of devoting so much research effort to this reaction. After all, hydroboration produced organoboranes. Relatively little new chemistry of organoboranes had appeared since the original classic publication by Frankland in 1862. They took the position that the lack of published material in this area meant that there was little of value there.

In this case it is now clear that this position is not correct. After our exploration of the hydroboration reaction had proceeded to the place where we felt we understood the reaction and could apply it with confidence to new situations, we began a systematic exploration of the chemistry of organoboranes. This research, facilitated by a host of exceptionally capable coworkers, among whom may be mentioned M. M. Rogić, M. M. Midland, C. F. Lane, A. B. Levy, R. C. Larock, and Y. Yamamoto, made it clear that the organoboranes are among the most versatile chemical intermediates available to the chemist.

It is not possible here to give more than a taste of the rich chemistry. For a more complete treatment, the reader must go elsewhere[29, 38].

Simple treatment of the organoborane with halogen in the presence of a suitable base produces the desired organic halide[39, 40] (25, 26).

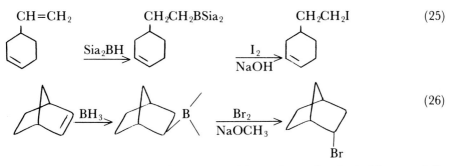

$$CH=CH_2 \qquad CH_2CH_2BSia_2 \qquad CH_2CH_2I \qquad (25)$$

$$\xrightarrow{Sia_2BH} \qquad \xrightarrow[NaOH]{I_2} \qquad (26)$$

$$\xrightarrow{BH_3} \qquad B \qquad \xrightarrow[NaOCH_3]{Br_2} \qquad$$

Oxidation with alkaline hydrogen peroxide produces the alcohol in essentially quantitative yield with complete retention of configuration[41] (27).

$$\xrightarrow{HB} \quad \cdots B \quad \xrightarrow[NaOH]{H_2O_2} \quad \cdots OH \qquad (27)$$
$$\text{Pure } trans$$

Either chloroamine or O-hydroxylaminesulfonic acid can be used to convert organoboranes into the corresponding amines[42] (28).

$$\xrightarrow{HB} \quad \cdots B \quad \xrightarrow{H_2NOSO_3H} \quad \cdots NH_2 \qquad (28)$$
$$\text{Pure } trans$$

The reaction of organoboranes with organic azides proceeds sluggishly with the more hindered organoboranes. Fortunately, this difficulty can be circumvented with the new hydroborating agents[43] (29).

$$\xrightarrow{HBCl_2} \quad \cdots BCl_2 \quad \xrightarrow[\text{2. } H_2O]{\text{1. } RN_3} \quad \overset{H}{\underset{R}{\cdots N}} \qquad (29)$$

Other organometallics can be synthesized from the organoboranes[44] (30).

$$CH=CH_2 \qquad CH_2CH_2B \qquad CH_2CH_2HgOAc$$
$$| \qquad\qquad \xrightarrow{HB} \quad | \qquad\qquad \xrightarrow{Hg(OAc)_2} \quad | \qquad (30)$$
$$(CH_2)_8CO_2R \qquad (CH_2)_8CO_2R \qquad (CH_2)_8CO_2R$$

The organoboranes can also be utilized to form carbon-carbon bonds[29]. One procedure utilizes transmetallation to the silver derivative, followed by the usual coupling reaction of such derivatives[45] (31).

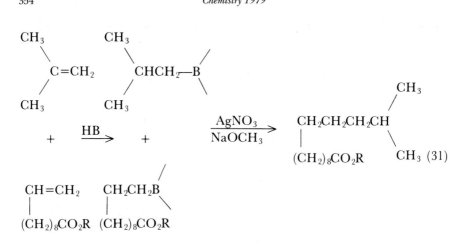

Cyclopropanes are readily synthesized[46] (32).

$$CH_2CHCH=CH_2 \xrightarrow{9\text{-BBN}} CH_2CHCH_2CH_2 \xrightarrow{NaOH} ClH_2C-\triangleleft \quad (32)$$

It is possible to achieve the alpha-alkylation and -arylation of esters, ketones, nitriles, etc.[47] (33, 34).

$$\xrightarrow{CH_2BrCO_2Et}{\text{base, 0°}} \qquad \qquad CH_2CO_2Et \qquad (33)$$

$$\xrightarrow{C_6H_5COCH_2Br}{\text{base, 0°}} \qquad CH_2COC_6H_5 \qquad (34)$$

α-Bromination provides still another route to achieve the synthesis of desired carbon structures[48] (35).

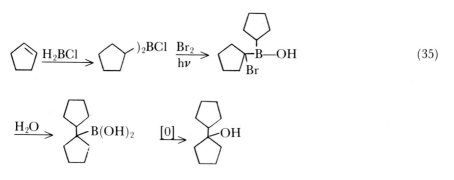

By means of this reaction, it is possible to combine three *sec*-butyl groups[49] (36).

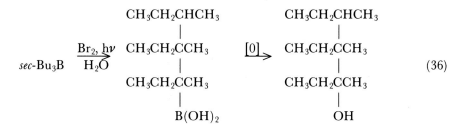

$$
\textit{sec-}Bu_3B \xrightarrow[H_2O]{Br_2, h\nu}
\begin{array}{c}
CH_3CH_2CHCH_3 \\ | \\ CH_3CH_2CCH_3 \\ | \\ CH_3CH_2CCH_3 \\ | \\ B(OH)_2
\end{array}
\xrightarrow{[O]}
\begin{array}{c}
CH_3CH_2CHCH_3 \\ | \\ CH_3CH_2CCH_3 \\ | \\ CH_3CH_2CCH_3 \\ | \\ OH
\end{array}
\qquad (36)
$$

Finally, it is possible to utilize this reaction to synthesize derivatives not realizable through the Grignard reaction[50] (37).

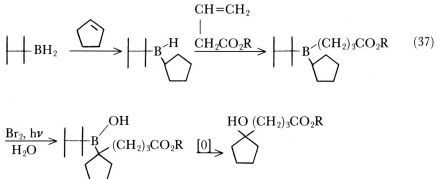

(37)

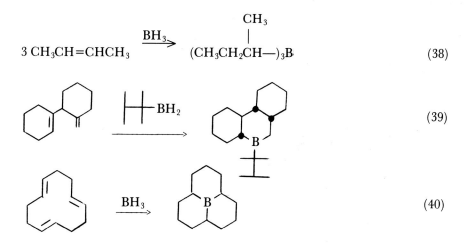

VII. STITCHING AND RIVETING

The hydroboration reaction allows the chemist to unite to boron under exceptionally mild conditions either three different olefins (38), or to cyclize dienes (39), or trienes (40).

$$
3\ CH_3CH{=}CHCH_3 \xrightarrow{BH_3}
\begin{array}{c}
CH_3 \\ | \\ (CH_3CH_2CH{-})_3B
\end{array}
\qquad (38)
$$

(39)

(40)

Thus, hydroboration allows us to use borane and its derivatives to "stitch" together with boron either the segments of individual molecules or the segments of a relatively open complex structure.

If we could replace boron by carbon, we would be in position to "rivet" these temporary structures into the desired carbon structure.

In fact, there are now three different procedures which can be used in this way: carbonylation[51] (41), cyanidation[52] (42), and the DCME reaction[53] (43).

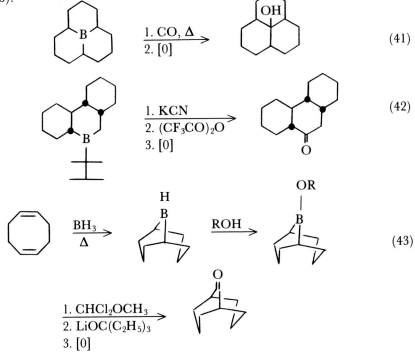

1. CO, Δ
2. [0]   (41)

1. KCN
2. (CF$_3$CO)$_2$O
3. [0]   (42)

BH$_3$
Δ

ROH

1. CHCl$_2$OCH$_3$
2. LiOC(C$_2$H$_5$)$_3$
3. [0]   (43)

Consequently, stitching and riveting provides an elegant new procedure for the synthesis of complex structures. Its versatility is indicated by the synthesis of an exceptionally hindered tertiary alcohol[54] (44) and by the annelation reaction[55] (Fig. 8).

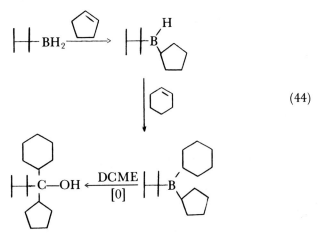

BH$_2$

DCME
[0]   (44)

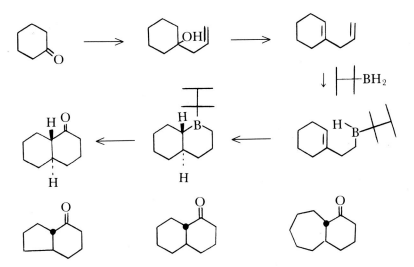

Fig. 8. General annelation reaction via hydroboration-carbonylation.

Again, these studies were greatly facilitated by a number of exceptional co-workers, among whom may be mentioned M. W. Rathke, Ei-ichi Negishi, J.-J. Katz, and B. A. Carlson.

VIII. HYDROBORATION OF ACETYLENES

Early attempts to hydroborate acetylenes with diborane led to complex mixtures[28]. Fortunately, the problem can be solved by use of borane derivatives[32, 33] (45, 46).

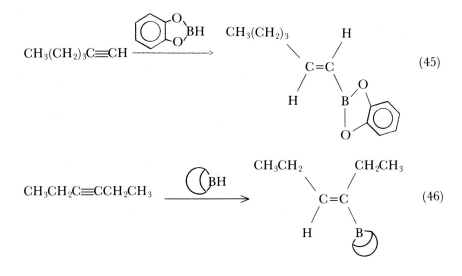

Chemistry 1979

Dibromoborane-dimethyl sulfide[56] appears to be especially valuable for such hydroboration of acetylenes[57]. The reaction readily stops at the mono-hydroboration step and it exhibits a valuable sensitivity to steric effects[56] (47).

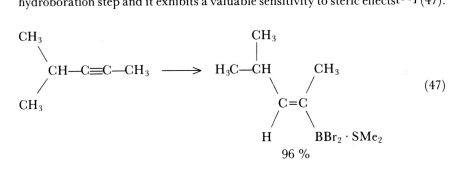

$$96\ \%$$

The different reagents exhibit very different selectivities toward double and triple bonds[56, 57]. Thus it is now possible to achieve the preferential hydro-boration in an appropriate enyne of the double bond in the presence of the triple bond[58] (48), or to hydroborate the triple bond selectively in the presence of the double bond[57] (49).

IX. VINYL BORANES

The monohydroboration of acetylenes makes the vinyl boranes readily available. These reveal an exceptionally rich chemistry.

For example, protonolysis proceeds readily and provides an excellent synthesis of *cis*-alkenes of high purity[59] (50).

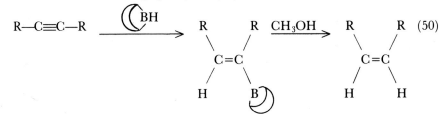

Oxidation produces the aldehyde or ketone[57] (51).

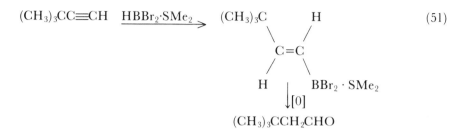

$$(CH_3)_3CCH_2CHO$$

The halogenation can be controlled to yield the halide either with retention[60] (52) or inversion (53) [61].

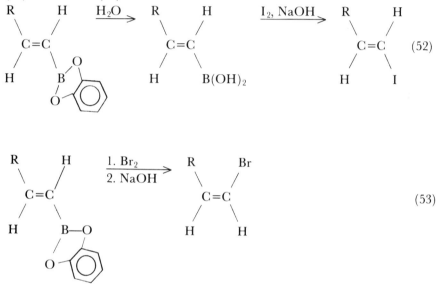

Mercuration readily yields the corresponding mercurial with complete retention of stereochemistry[62] (54).

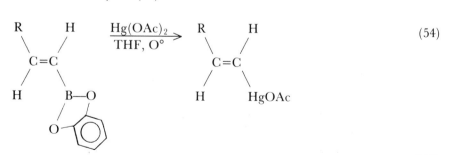

Pappo and Collins adopted this approach in their prostaglandin synthesis[63].

The ready conversion of these vinyl boranes into organomercurials suggested an exploration of their conversion into the organocopper intermediates. The research in this area was greatly facilitated by J. B. Campbell, Jr. Indeed, treatment of the 9-BBN adduct with sodium methoxide and the CuBr · SMe₂ complex at 0° gave the diene with complete retention of stereochemistry[64] (55).

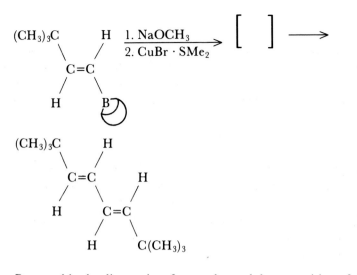

(55)

Presumably the diene arises from a thermal decomposition of the vinyl copper intermediate.

At $-15°$ the intermediate is sufficiently stable to be diverted along another reaction path by reaction with relatively reactive organic halides[65] (56).

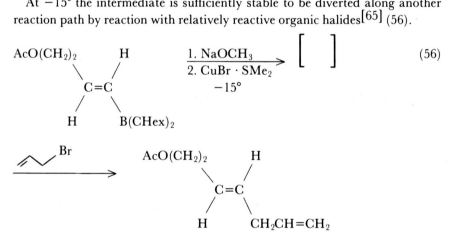

(56)

This gentle procedure for synthesizing vinyl copper intermediates can accommodate such functional groups as the acetoxy group utilized in the example shown.

Our research efforts in this area were greatly facilitated by exceptional contributions by a number of coworkers, including James B. Campbell, Jr. and Gary Molander.

Although time does not permit a detailed review here, attention is called to the elegant procedures developed by my former coworkers, George Zweifel and Ei-ichi Negishi, and their associates, for the synthesis of *cis-* and *trans-*olefins, and the synthesis of *cis, cis-, cis, trans-,* and *trans, trans-*dienes[66].

X. PHEROMONES

Pheromones offer a promising new means for controlling insect populations without the problems of some of the earlier methods[67]. The pheromones are chemicals of relatively simple structure emitted by insects as a means of communicating with other members of the same species. Typical examples are shown in Fig. 9.

Even though the structures are relatively simple, they must be very pure so that procedures utilized for their synthesis possess unusually severe requirements for high regio- and stereospecificity. It appeared that synthetic procedures based on organoborane chemistry should be especially favorable for this objective. Accordingly, we have undertaken a new program directed toward developing simple syntheses of such pheromones based upon organoborane chemistry. This research program has been greatly facilitated by Gary A. Molander and K. K. Wang.

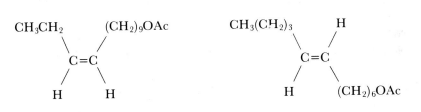

European Corn Borer

False Codling Moth

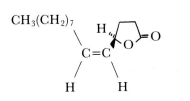

Japanese Beetle

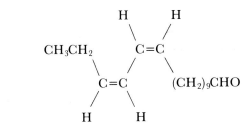

Navel Orangeworm

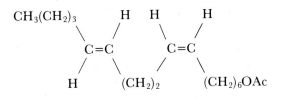

Pink Bollworm Moth

Fig. 9. Representative insect pheromones.

One example, the synthesis of the looper moth sex pheromone, will be presented[68].

Hydroboration of 6-acetoxy-1-hexene yields the corresponding organoborane (57).

$$3 \, AcOCH_2(CH_2)_3CH=CH_2 \xrightarrow{\text{BH}_3} (AcOCH_2(CH_2)_3CH_2CH_2)_3B \qquad (57)$$

Reaction with the lithium acetylide from 1-hexyne gives the ate complex (58).

$$[(AcOCH_2(CH_2)_3CH_2CH_2)_3BC{\equiv}C(CH_2)_3CH_3]Li \qquad (58)$$

Treatment of the ate complex with iodine at $-78°$ provides the acetylene[69] (59).

$$AcO(CH_2)_6C{\equiv}C(CH_2)_3CH_3 \qquad (59)$$

Hydroboration with 9-BBN, followed by protonolysis with methanol, gives the desired product in an isolated yield of 75 %, exhibiting a purity of > 98 % (60).

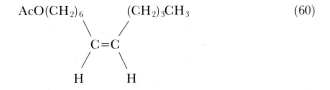

$$(60)$$

It should be noted that the entire synthetic procedure can be carried out in a single flask without isolation of any material until the final product.

IX. CONCLUSION

In this Nobel Lecture I elected to discuss the results of a research program over the past 43 years on the chemistry of borane and its derivatives. This was a deliberate choice. I felt that in this way I could transmit a valuable message to my younger colleagues.

In 1938, when I received my Ph. D. degree, I felt that organic chemistry was a relatively mature science, with essentially all of the important reactions and structures known. There appeared to be little new to be done except the working out of reaction mechanisms and the improvement of reaction products. I now recognize that I was wrong. I have seen major new reactions discovered. Numerous new reagents are available to us. Many new structures are known to us. We have at hand many valuable new techniques.

I know that many of the students of today feel the same way that I did in 1938. But I see no reason for believing that the next 40 years will not be as fruitful as in the past.

In my book, *Hydroboration* (ref. 28), I quoted the poet: "Tall oaks from little acorns grow."[70] But in this lecture I have started further back, to a time when the acorn was a mere grain of pollen. I have shown how that grain of pollen developed first into an acorn. Then the acorn became an oak. The oak tree became a forest. Now we are beginning to see the outlines of a continent.

We have been moving rapidly over that continent, scouting out the major mountain ranges, river valleys, lakes, and coasts. But it is evident that we have only scratched the surface. It will require another generation of chemists to settle that continent and to utilize it for the good of mankind.

But is there any reason to believe that this is the last continent of its kind? Surely not. It is entirely possible that all around us lie similar continents awaiting discovery by enthusiastic, optimistic explorers. I hope that one result of this lecture will be to inspire young chemists to search for such new continents.

Good luck!

REFERENCES

1. Stock, A. *Hydrides of Boron and Silicon*, Cornell University Press, Ithaca, New York, 1933.
2. Burg, A. B.; Schlesinger, H. I. J. Am. Chem. Soc. *59*, 780 (1937).
3. Brown, H. C.; Schlesinger, H. I.; Burg, A. B. J. Am. Chem. Soc. *61*, 673 (1939).
4. Schlesinger, H. I.; Sanderson, R. T.; Burg, A. B. J. Am. Chem. Soc. *62*, 3421 (1940).
5. Schlesinger, H. I.; Burg, A. B. J. Am. Chem. Soc. *62*, 3425 (1940).
6. Schlesinger, H. I., Brown, H. C. J. Am. Chem. Soc. *62*, 3429 (1940).
7. Schlesinger, H. I.; Brown, H. C. J. Am. Chem. Soc. *75*, 219 (1953).
8. Schlesinger, H. I., Brown, H. C.; Gilbreath, J. R.; Katz, J. J. J. Am. Chem. Soc. *75*, 195 (1953).
9. Schlesinger, H. I.; Brown, H. C.; Hoekstra, H. R.; Rapp, L. R. J. Am. Chem. Soc. *75*, 199 (1953).
10. Schlesinger, H. I.; Brown, H. C.; Hyde, E. K. J. Am. Chem. Soc. *75*, 209 (1953).
11. Brown, H. C.; Schlesinger, H. I., Sheft, I.; Ritter, D. M. J. Am. Chem. Soc. *75*, 192 (1953).
12. Schlesinger, H. I.; Brown, H. C.; Finholt, A. E. J. Am. Chem. Soc. *75*, 205 (1953).
13. Brown, H. C. J. Chem. Soc. 1248 (1956).
14. Brown, H. C. *Boranes in Organic Chemistry*, Cornell University Press, Ithaca, New York, 1972.
15. Finholt, A. E.; Bond, Jr., A. C.; Schlesinger, H. I. J. Am. Chem. Soc. *69*, 1199 (1947).
16. Brown, W. G. Org. React. *6*, 469 (1951).
17. Brown, H. C.; Subba Rao, B. C. J. Am. Chem. Soc. *78*, 2582 (1956).
18. Brown, H. C.; Weissman, P. M. Israel J. Chem. *1*, 430 (1963).
19. Brown, H. C.; Weissman, P. M. J. Am. Chem. Soc. *87*, 5614 (1964).
20. Brown, H. C.; Kim, S. C.; Krishnamurthy, S. J. Org. Chem. *45*, 1 (1980).
21. Brown. C. A.; Krishnamurthy, S.; Kim. S. C. J. C. S. Chem. Commun, 373 (1973).
22. Brown, H. C.; Heim, P.; Yoon, N. M. J. Am. Chem. Soc. *92*, 1637 (1970).
23. Brown, H. C.; Weissman, P. M.; Yoon, N. M. J. Am. Chem. Soc. *88*, 1458 (1966).
24. Huang, F. C.; Lee, L. F. H.; Mittal, R. S. D.; Ravikumar, P. R.; Chan, J. A.; Sih, C. J.; Caspi, E.; Eck, C. R. J. Am. Chem. Soc. *97*, 4144 (1975).
25. See Brown, H. C.; Krishnamurthy, S. Tetrahedron *35*, 567 (1979) for a detailed review of our program in selective reductions.
26. Brown, H. C.; Subba Rao, B. C. J. Am. Chem. Soc. *78*, 5694 (1956).
27. Brown, H. C.; Subba Rao, B. C. J. Org. Chem. *22*, 1136 (1957).
28. Brown, H. C. *Hydroboration*, W. A. Benjamin, New York, 1962.

29. Brown, H. C. (with techniques by Kramer, G. W.; Levy, A. B.; Midland, M. M.) *Organic Syntheses via Boranes,* Wiley-Interscience, New York, 1975.
30. Negishi, E.; Brown, H. C. Synthesis 77 (1974).
31. Chapter 13, ref. 28; Chapter 3, ref. 29.
32. Brown, H. C.; Yoon, N. M. Israel J. Chem *15,* 12 (1976/77).
33. Brown, H. C.; Lane, C. F. Heterocycles *7,* 454 (1977).
34. Brown, H. C.; Gupta, S. K. J. Am. Chem. Soc. *97,* 5249 (1975).
35. Brown, H. C.; Ravindran, N. J. Am. Chem. Soc. *98,* 1785, 1798 (1976).
36. Brown, H. C.; Ravindran, N.; Kulkarni, S. U. J. Org. Chem. *44,* 2417 (1979).
37. Frankland, E. J. Chem. Soc. *15,* 363 (1862).
38. Pelter, A.; Smith, K. in Comprehensive Organic Chemistry, eds. Sir Derek Barton and W. David Ollis, Vol. 3, Pergamon Press, Oxford, 1979.
39. Brown, H. C.; Rathke, M. W.; Rogić, M. M. J. Am. Chem. Soc. *90,* 5038 (1968).
40. Brown, H. C.; Lane, C. F. J. Am. Chem. Soc. *92,* 6660 (1970).
41. Zweifel, G.; Brown, H. C. Org. React. *13,* 1 (1963).
42. Rathke, M. W.; Inoue, N.; Varma, K. R.; Brown, H. C. J. Am. Chem. Soc. *88,* 2870 (1966).
43. Brown, H. C.; Midland, M. M.; Levy, A. B. J. Am. Chem. Soc. *95,* 2394 (1973).
44. Larock, R. C.; Brown, H. C. J. Am. Chem. Soc. *92,* 2467 (1970).
45. Brown, H. C.; Verbrugge, C.; Snyder, C. H. J. Am. Chem. Soc. *83,* 1002 (1961).
46. Brown, H. C.; Rhodes, S. P. J. Am. Chem. Soc. *91,* 2149 (1969).
47. Brown, H. C.; Rogić, M. M. Organomet. Chem. Syn. *1,* 305 (1972).
48. Brown, H. C.; Yamamoto, Y.; Lane, C. F. Synthesis, 303 (1972).
49. Lane, C. F.; Brown, H. C. J. Am. Chem. Soc. *93,* 1025 (1971).
50. Brown, H. C.; Yamamoto, Y.; Lane, C. F. Synthesis, 304 (1972).
51. Brown, H. C.; Negishi, E. J. Am. Chem. Soc. *89,* 5478 (1967).
52. Pelter, A.; Smith, K.; Hutchings, M. G.; Rowe, K. J. Chem. Soc., Perkin I, 129 (1975).
53. Carlson, B. A.; Brown, H. C. Synthesis, 776 (1973).
54. Brown, H. C.; Katz, J. -J.; Carlson, B. A. J. Org. Chem. *38,* 3968 (1973).
55. Brown, H. C.; Negishi, E. J. C. S. Chem. Commun. 594 (1968).
56. Brown, H. C.; Ravindran, N.; Kulkarni, S. U. J. Org. Chem. *45,* 384 (1980).
57. Brown, H. C.; Campbell, Jr., J. B. J. Org. Chem. *45,* 389 (1980).
58. Brown, C. A.; Coleman, R. A. J. Org. Chem. *44,* 2328 (1979).
59. Research in progress with G. A. Molander.
60. Brown, H. C.; Hamaoka, T.; Ravindran, N. J. Am. Chem. Soc. *95,* 5786 (1973).
61. Brown, H. C.; Hamaoka, T.; Ravindran, N. J. Am. Chem. Soc. *95,* 6456 (1973).
62. Larock, R. C.; Gupta, S. K.; Brown, H. C. J. Am. Chem. Soc. *94,* 4371 (1972).
63. Pappo, R.; Collins, P. W. Tetrahedron Lett., 2627 (1972).
64. Brown, H. C.; Campbell, Jr., J. B. J. Org. Chem. *45,* 549 (1980).
65. Brown, H. C.; Campbell, Jr., J. B. J. Org. Chem. *45,* 550 (1980).
66. For a review and leading references, see ref. 29.
67. Jacobson, M. Ed. *Insecticides of the Future,* Marcel Dekkar, Inc., New York, 1975.
68. Research in progress with K. K. Wang.
69. Suzuki, A.; Miyaura, N.; Abiko, S.; Itoh, M.; Brown, H. C.; Sinclair, J. A.; Midland, M. M. J. Am. Chem. Soc. *95,* 3080 (1973).
70. See Chapter 20, Epilog, ref. 28.

Georg Wittig

GEORG WITTIG

Translation from the German text

Born in Berlin in 1897. Doctorate and University Teaching Thesis in Marburg/Lahn from the Faculty of Chemistry. Head of Department at Braunschweig (Technical College) from 1932; Associate Professor at Freiburg/Brsg. from 1937; Professor and Faculty Director at the Institute of Chemistry, Tübingen, from 1944; turned down the same position as successor to H. Staudinger at Freiburg/Brsg.; accepted the same position at Heidelberg as successor to K. Freudenberg. Professor Emeritus since 1967.

Scientific Activities: Textbook on stereochemistry, 1930. Papers on the subject of ring tension and double bonds as well as valency tautomerism. Main research into organic reactions of alkali metals and elaboration of carbon-based chemistry. Discovery of the halogen–metal exchange reaction (simultaneously with H. Gilman). Development of ylide chemistry and, together with that, study of the Stevens and Sommelet rearrangements as well as intra-anionic ether isomerisation. Through the synthesis of the pentaaryl derivatives of the elements of group 5, the phosphorous ylides were discovered and also, in 1953, the carbonylolefins which have since proven to be crucial for the manufacture of synthetic fabrics and also important in other industrial processes. In 1942 dehydrobenzol was proven to be a short-lived by-product, a fact demonstrated by J. D. Roberts in 1953 and by me, only this time using different means, viz. control experiments on Diels–Alder adducts. More recently the concept of the "at"-complexes as a counterpart to the "onium" complexes has led to the development of a new chemistry from which have come the sodium tetra phenylborates.

Honours: Honorary Doctorate from the Sorbonne in 1957; Honorary Doctorate from the Universities of Tübingen and Hamburg in 1962; Adolf von Baeyer Memorial Medal from the German Chemical Society in 1953; Silver Medal from the University of Helsinki in 1957; Dannie Heinemann Award from the Göttingen Academy of Sciences in 1965; Otto Hahn Award for Chemistry and Physics in 1967; Silver Medal from the City of Paris in 1969; Paul Karrer Medal from the University of Zurich in 1972; Médaille de la Chaire Bruylants (University of Louvain) in 1972; Roger Adams Award from the American Chemical Society in 1973; Karl Ziegler Prize in 1975; Honorary Member of the Swiss Chemical Society in 1963; Honorary Member of the New York Academy of Sciences in 1965; Member of the Chemical Society of Peru, also in 1965; Honorary Fellow of the Chemical Society (London) in 1967; Member of the French Academy in 1971; Member of the Society of Medical Sciences, Córdoba (Argentina), in 1976. As well as these, member of several German academies: Bavarian Academy of Sciences, Heidelberg Academy of Sciences, German Academy of the Natural Scientist Leopoldina Halle.

Georg Wittig died in 1987.

FROM DIYLS TO YLIDES TO MY IDYLL

Nobel Lecture, 8 December 1979
by
GEORG WITTIG
Heidelberg, Federal Republic of Germany

Translation from the German text

Chemical research and mountaineering have much in common. If the goal or the summit is to be reached, both initiative and determination as well as perseverance are required. But after the hard work it is a great joy to be at the goal or the peak with its splendid panorama. However, especially in chemical research — as far as new territory is concerned — the results may sometimes be quite different: they may be disappointing or delightful. Looking back at my work in scientific research, I will confine this talk to the positive results (1).

Some 50 years ago I was fascinated by an idea which I investigated experimentally. The question was how ring strain acts on a ring if an accumulation of phenyl groups at two neighboring carbon atoms weakens the C–C linkage and predisposes to the formation of a diradical (for brevity called diyl) (Fig. 1). Among the many experimental results (2) I choose the synthesis of the hydrocarbons 1 and 4 (3), which we thought capable of diyl formation. Starting materials were appropriate dicarboxylic esters, which we transformed into the corresponding glycols. While these were obtained under the influence of phenylmagnesium halide only in modest yield, phenyllithium proved to be superior and was readily accessible by the method of K. Ziegler, using bromobenzene and lithium. The glycolates resulting from the reaction with potassium phenylisopropylide formed — on heating with methyl iodide — the corresponding dimethyl ethers, which supplied the equivalent hydrocarbons 1 and 4 by alkali metal splitting and demetalation with tetramethylethylene dibromide.

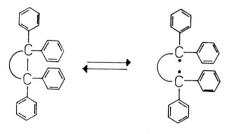

Fig. 1. Formation of a diradical (2).

The resulting tetraphenylbenzocyclobutane (**1**), however, rearranged to triphenyldihydroanthracene (**3**) (Fig. 2). In contrast, tetraphenyldihydrophenanthrene (**4**), which was prepared analogously, proved to be a stable hydrocarbon, even when substituents R were introduced that forced the biphenyl system to twist. While **4** did not decompose at 340°C and was stable in solution against oxygen, its aryl-weakened C–C bond could be observed since it split with potassium into the ring-opened dipotassium derivate. The results of these investigations on formation of radicals and ring strain seem to indicate that ring closure is more likely to contribute to stabilization of the ethane bond.

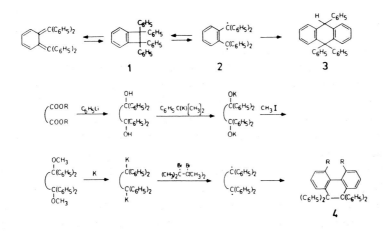

Fig. 2. Some reactions observed in the synthesis of **1** and **4** (*3*).

This stabilizing influence is documented impressively by the behavior of tris(biphenylene)ethane (**7**), which was also synthesized (*4*) (Fig. 3). The carbinol **6**, which was formed by the reaction of the ketone **5** with *o*-lithiobiphenyl, transformed into the desired hydrocarbon **7** by an acid-catalyzed twofold Wagner-Meerwein rearrangement.

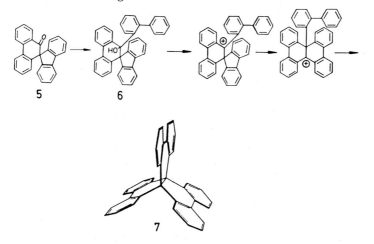

Fig. 3. Formation of tris(biphenylene)ethane (*4*).

This first aromatic propellane, which melted at 475°C without decomposition and whose structure agreed with the nuclear magnetic resonance spectrum, proved to be insensitive to ethane linkage-breaking sodium-potassium alloy. Evidently the close aryl packing prevents penetration of the metal into the interior of the molecule.

Since the tendency to form diradicals was not evident with the hydrocarbons mentioned, we intended to replace the phenyl groups by anisyls. Therefore, suitable dicarboxylic esters should be brought into reaction with *p*-lithioanisole (*5*). But, as we obtained unexpected smears, the functionally simple benzophenone was used to treat the mixture that resulted from the reaction of *p*-bromoanisole with lithium. Instead of the expected *p*-anisyldiphenylcarbinol, the bromine-containing compound **9** was isolated, whose structure could be proved by conversion into the well-characterized derivative by zinc dust distillation. Accordingly, *p*-lithioanisole, which was originally formed, metalates the *p*-bromoanisole that is still present into compound **8**, which then reacts with benzophenone to form the isolated compound **9** (Fig. 4).

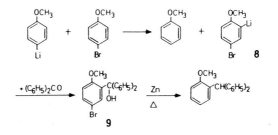

Fig. 4. Formation and characterization of compound **9** (*5*).

When it was noted that phenyllithium too can modify *p*-bromoanisole to form **9**, we decided to look closer at the lability of the aromatic proton as a function of the substituent. In the course of these studies we arrived at the surprising result that aryl iodide, bromide, and even chloride can exchange with the electropositive metal of phenyllithium (*6*). Later we called this principle of reaction *umpolung*, or reversal of polarity (*7*) (Fig. 5). Simultaneously and independently, H. Gilman found the same behavior when treating aryl halides with butyllithium.

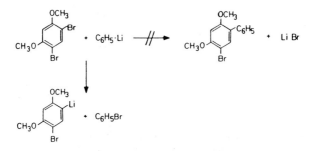

Fig. 5. Reversal of polarity (*umpolung*) (*6*).

Among the halogens of the various aromatic systems, fluorine proved to be not exchangeable with lithium (*8*). Here we found an unexpected reaction path. First we observed that in the formation of biphenyl by the reaction of monohalobenzene with phenyllithium, fluorobenzene acted rapidly, forming approximately 75 percent biphenyl, while the other halobenzenes produced only 5 to 7 percent. We interpreted this result as indicating that biphenyl formation was preceded by metalation of the halobenzene, which was stimulated by the inductive effect of the strongly electronegative fluorine. This explanation was supported by the finding that not biphenyl but *o*-lithiobiphenyl had been produced. In 1942 we further assumed that an elimination of metal and halogen results that leads to the occurrence of dehydrobenzene (*9*), and this is what changes phenyllithium into the *o*-lithiobiphenyl found experimentally (Fig. 6). Independent of our work, a proof for the intermediate occurrence of dehydrobenzene was given by Roberts *et al.* (*10*), who reacted [1-^{14}C]chlorobenzene with potassium amide in liquid ammonia and isolated the two expected anilines with approximately 50 percent yield (Fig. 7).

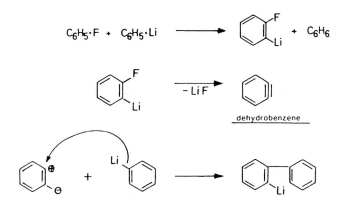

Fig. 6. Mechanism of formation of *o*-lithiobiphenyl (*8, 9*).

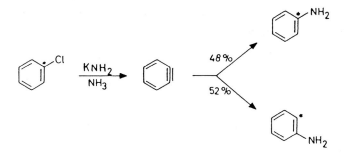

Fig. 7. Proof of the intermediate occurrence of dehydrobenzene in the reaction of ^{14}C-labeled chlorobenzene with potassium amide in liquid ammonia (*10, 24*).

Later we could prove the existence of the dehydrobenzene by expecting that it would react as dienophile (*11*). For the diene and solvent we chose furan, which, being an ether, should favor organometallic exchange while simultaneously serving as a trapping agent. In an exciting experiment we had *o*-fluorobromobenzene react with lithium amalgam in furan and isolated with good yield the endo-oxide **10**, which had been formed by a Diels-Alder addition (Fig. 8).

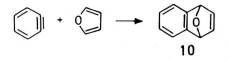

10

Fig. 8. Proof of the existence of dehydrobenzene through the formation of **10** (*11*).

We found the lifetime of dehydrobenzene in the gas phase (*12*) by the thermal decomposition of bis(*o*-iodophenyl)mercury as well as phthaloyl peroxide to biphenylene at 600°C in an argon atmosphere at reduced pressure. When furan was injected behind the decomposition zone naphthol was evolved as a by-product from the dihydronaphthalene endoxide that occurred first. Under the conditions used, the lifetime of dehydrobenzene was determined to be 20 milliseconds. See Fig. 9.

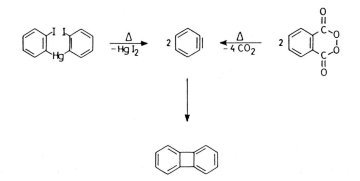

Fig. 9. Thermal decomposition reactions used to determine the lifetime of dehydrobenzene in the gas phase (*12*).

Phenyllithium, which had opened so many areas by acting as a sort of dowsing rod, was applied many times in the course of our research. Now our attention was drawn to the proton-labile substrates, to the aliphatic as well as the aromatic ones.

The process of proton–metal cation exchange appeared to us to be of fundamental importance, since the electron density at the carbon atom is enhanced after metalation. Thus the question arose of how carbanions, with their negative charge, would behave compared to carbonium ions, with their positively charged carbon atoms.

At the time we were not sure whether hydrogen bound to carbon would be proton-labile in quaternary ammonium salts. We came to this conclusion with an absurd experiment to prepare pentamethylnitrogen from tetramethylammonium salts by using the reaction of tetramethylammonium halide with methyllithium (*13*).

It was confirmed experimentally that the octet principle is strictly valid for the elements of the first eight-element period. The object of synthesizing compounds with a pentacoordinate central atom was reached only when we studied the higher elements of the fifth main group — that is, phosphorus, arsenic, antimony, and bismuth. It was easy to synthesize their pentaphenyl derivatives (*14*) and, in the case of antimony, also pentamethylantimony, which (as a nonpolar compound) is a liquid with a boiling point of 126°C (*15*).

Tetramethylammonium chloride reacting with methyl- or phenyllithium loses one proton and forms a product that we called trimethylammonium methylide (*11*) (Fig. 10). We gave the name *N*-ylides to this new class of substances since the bonding of the carbon to the neighboring nitrogen is homopolar (yl) and ionic (ide) at the same time. Trimethylammonium fluorenylide (*12*) could be isolated salt-free, thus its ylide structure is unambiguous (*16*). Subsequently, ylides as well as cryptoylides were studied more thoroughly (*17*).

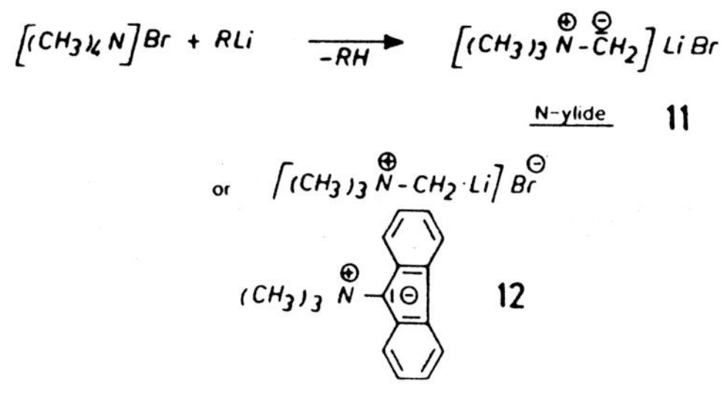

$$[(CH_3)_4 N]Br + RLi \xrightarrow{-RH} [(CH_3)_3 \overset{\oplus}{N} - \overset{\ominus}{CH_2}] Li Br$$

N-ylide   11

$$\text{or} \quad [(CH_3)_3 \overset{\oplus}{N} - CH_2 \cdot Li] \overset{\ominus}{Br}$$

$$(CH_3)_3 \overset{\oplus}{N} - \boxed{\ominus} \quad \textbf{12}$$

Fig. 10. Formation of N-ylides (13).

When we extended this concept to the phosphonium salts, we found that they could be converted into the corresponding *P*-ylides even more readily than the analogous ammonium salts when treated with organolithium compounds (Fig. 11). The reason for the greater proton mobility is that phosphorus, unlike nitrogen, can expand its outer electron shell to a decet. This allows an energy-lowering resonance stability between the ylide and ylene forms.

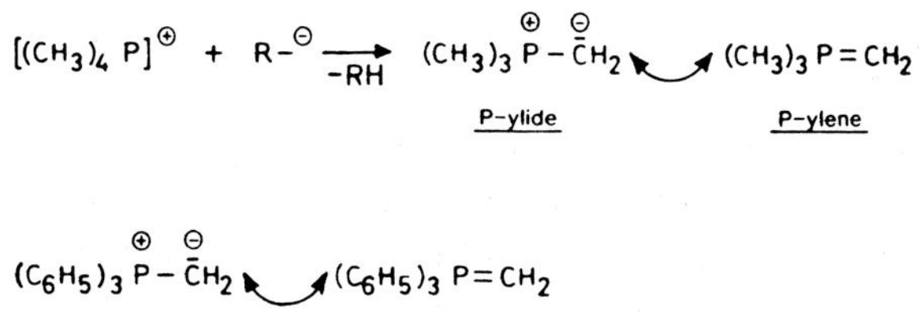

$$[(CH_3)_4 P]^{\oplus} + R^{\ominus} \xrightarrow{-RH} (CH_3)_3 \overset{\oplus}{P} - \overset{\ominus}{CH_2} \longleftrightarrow (CH_3)_3 P = CH_2$$

P-ylide   P-ylene

$$(C_6H_5)_3 \overset{\oplus}{P} - \overset{\ominus}{CH_2} \longleftrightarrow (C_6H_5)_3 P = CH_2$$

Fig. 11. Formation of P-ylides (17).

In the case of the *N*-ylide the semipolar nature of the N–C bond is demonstrated by its ability to add to benzophenone, forming the well-defined betaine (*18*) (Fig. 12). Now, if the same reaction was performed with triphenylphosphinemethylene, the expected betaine adduct was not obtained but, astonishingly, triphenylphosphine oxide and 1,1-diphenylethylene (*19*). Evidently the initially produced betaine **13** — due to the ability of the central atom to expand its electron shell — formed the four-membered ring **14**, which then decayed into two fragments as final products (*20*) and could not be isolated by itself (Fig. 13).

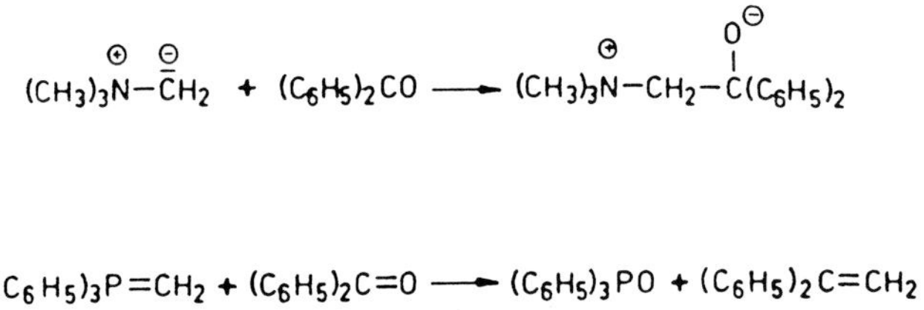

$$(CH_3)_3\overset{\oplus}{N}-\overset{\ominus}{C}H_2 \;+\; (C_6H_5)_2CO \longrightarrow (CH_3)_3\overset{\oplus}{N}-CH_2-\overset{O^{\ominus}}{\underset{}{C}}(C_6H_5)_2$$

$$(C_6H_5)_3P=CH_2 \;+\; (C_6H_5)_2C=O \longrightarrow (C_6H_5)_3PO \;+\; (C_6H_5)_2C=CH_2$$

Fig. 12. Reactions with benzophenone (*18–20*).

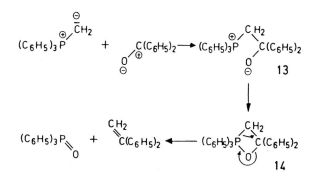

Fig. 13. Steps in the reaction of triphenylphosphinemethylene with benzophenone (*25*).

That the first step of the reaction is betaine formation was shown with the reaction of triphenylphosphinemethylene and benzaldehyde. In this case the betaine could be isolated as an intermediate product, and it decayed to triphenylphosphine oxide and styrene only on heating (*20*). This type of reaction (*21*) seemed to be of fundamental importance for preparative chemistry, and it also found industrial application (*22*). By these means it was possible to prepare vitamin A and β-carotene, among others. In the present context I restrict the discussion to the synthesis of vitamin A, which is produced industrially at BASF under the direction of Pommer (*22*). The phosphonium salt **16**, obtainable from vinyl-β-ionol **15**, triphenylphosphine, and acid, changes with HX splitting into the corresponding phosphinemethylene derivative, which reacts with γ-formyl-crotylacetate **17** to produce vitamin A in the form of its acetate **18** (Fig. 14).

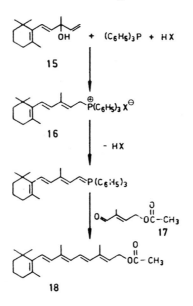

Fig. 14. Formation of vitamin A acetate (*22*).

With the addition of phenylsodium to triphenylboron, it could be demonstrated that boron can also act as tetracoordinate central atom (Fig. 15). Today this complex serves as an analytic reagent for the determination of potassium, rubidium, and cesium ions as well as for the quantitative determination and separation of ammonium and alkaloid salts.

Fig. 15. Addition of phenylsodium to triphenylboron (*26*).

We called the complex salts with negatively charged central atom "ate" complexes for understandable reasons (*23*). They can be compared with the "onium" complexes, which were already known, as shown in Fig. 16. Because

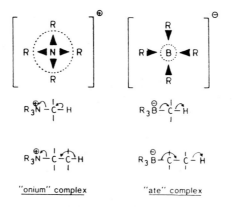

Fig. 16. Comparison of "onium" and "ate" complexes (*23*).

of the inductive effect of the central atom in onium complexes, all ligands R are cationically labilized and the hydrogen atoms at the neighboring carbon atoms are proton-mobile; however, in ate complexes all ligands at the central atom are anionically labilized and the hydrogen atoms at the neighboring carbon atoms are hydride-labile. This rule explains numerous reactions. I do not have time here to discuss its importance as a heuristic principle.

Thus I come to the end of my lecture. The excursion from diyls to ylides now ends at my idyll. With this I mean the conclusion of my research work as an emeritus, which allowed me to continue my work as a chemist free from the obligations of a teacher, and finally to devote myself completely to my interest in fine arts. I want to close my talk by offering cordial thanks to my collaborators. Without them my work could not have been accomplished.

REFERENCES AND NOTES

1. Compare G. Wittig, *Acc. Chem. Res.* **7**, 6 (1974).
2. _____ and M. Leo, *Ber. Dtsch. Chem. Ges.* **61**, 854 (1928); *ibid.* **62**, 1405 (1929).
3. _____, *ibid.* **64**, 2395 (1931); G. Wittig and H. Petri, *Justus Liebigs Ann. Chem.* **505**, 17 (1933).
4. G. Wittig and W. Schoch, *Justus Liebigs Ann. Chem.* **749**, 38 (1971).
5. G. Wittig, U. Pockels, H. Dröge, *Ber. Dtsch. Chem. Ges.* **71**, 1903 (1938).
6. G. Wittig and U. Pockels, *ibid.* **72,** 89 (1939).
7. G. Wittig, P. Davis, G. Koenig, *Chem. Ber.* **84**, 627 (1951).
8. G. Wittig, G. Pieper, F. Fuhrmann, *Ber. Dtsch. Chem. Ges.* **73**, 1193 (1940).
9. G. Wittig, *Naturwissenschaften* **30**, 696 (1942).
10. J. D. Roberts, H. E. Simmons, Jr., L. A. Carlsmith, C. W. Vaughan, *J. Am. Chem. Soc.* **75**, 3290 (1953).
11. G. Wittig and L. Pohmer, *Chem. Ber.* **89**, 1334 (1956).
12. G. Wittig and H. F. Ebel, *Justus Liebigs Ann. Chem.* **650**, 20 (1961); H. F. Ebel and R. W. Hoffmann, *ibid.* **673**, 1 (1964).
13. G. Wittig and M. Rieber, *ibid.* **562**, 187 (1949).
14. G. Wittig and K. Clauss, *ibid.* **577**, 26 (1952); *ibid.* **578**, 136 (1952).
15. G. Wittig and K. Torssell, *Acta Chem. Scand.* **7**, 1293 (1953).
16. G. Wittig and G. Felletschin, *Justus Liebigs Ann. Chem.* **555**, 133 (1944).
17. Compare A. W. Johnson, *Ylid Chemistry* (Academic Press, New York, 1966).
18. G. Wittig and M. Rieber, *Justus Liebigs Ann. Chem.* **562**, 177 (1949).
19. G. Wittig and G. Geissler, *ibid.* **580**, 44 (1953).
20. G. Wittig and U. Schöllkopf, *Chem. Ber.* **87**, 1318 (1954).
21. On a variant of synthesis of olefins with carbonyl compounds, see J. Boutagy and R. Thomas, *Chem. Rev.* **74**, 87 (1974).
22. H. Pommer, *Angew. Chem.* **89**, 437 (1977).
23. G. Wittig, *ibid.* **62**, 231 (1950); *ibid.* **70**, 65 (1958).
24. R. W. Hoffmann, *Dehydrobenzene and Cycloalkynes* (Academic Press, New York, 1967).
25. G. Wittig, *Angew. Chem.* **68**, 505 (1956); *Festschrift Arthur Stoll* (Birkhäuser, Basel, 1957).
26. _____, G. Keicher, A. Rückert, P. Raff, *Justus Liebigs Ann. Chem.* **563**, 110 (1949).

Chemistry 1980

PAUL BERG

for his fundamental studies of the biochemistry of nucleic acids, with particular regard to recombinant-DNA

WALTER GILBERT and FREDERICK SANGER

for their contributions concerning the determination of base sequences in nucleic acids

THE NOBEL PRIZE FOR CHEMISTRY

Speech by Professor BO G. MALMSTRÖM of the Royal Academy of
Sciences. Translation from the Swedish text.

Your Majesty, Your Royal Highnesses, Ladies and Gentlemen,

The body and soul of man is the most complex and refined chemical machine
that we know. Even the simplest forms of life, for example bacteria, are almost
immeasurably intricate systems compared to the dead matter that we find on
our Earth and out in the rest of the Universe. Modern biology has taught us
that there is no vital force, and living organisms consist wholly of dead atoms.
This does not mean that it is desirable, or even possible, to try to reduce all
problems of biology to biochemistry. To understand our own place in the
Universe we also need the softer data provided by the social sciences and, not
least, by literature. But the intimate relationship, seen in this century, between
fundamental research in biochemistry and medical progress has demonstrated
that it is not solely to satisfy the intellectual curiosity of the biochemist that we
ought to try as far as possible to reach a description of life processes in
chemical, molecular terms.

The machinery of life is made possible by a unique interplay between two
groups of biological giant molecules, nucleic acids and the proteins, in the form
of enzymes, the orchestra which plays the various expressions of life, motility,
feeling, reproduction etc. DNA is the carrier of the genetic traits in the chromosomes of the cells, and it governs the chemical machinery by determining which
enzymes a cell shall manufacture. We know through investigations which have
earlier been awarded with Nobel Prizes, that the sequence of the building
blocks in DNA, called nucleotides, determines the structure of a particular
enzyme that the cell produces. The investigators who have been awarded this
year's Nobel Prize in chemistry, Paul Berg, Walter Gilbert and Frederick
Sanger, have through their methodological contributions made it possible to
penetrate into further depth in our understanding of the relationship between
the chemical structure and biological function of the genetic material. Berg was
the first one to construct a recombinant-DNA molecule, i. e. a molecule which
contains part of DNA from different species, e. g. genes from a human being
combined with part of a bacterial chromosome. Berg has also used his method
to analyze the chromosome of a virus in considerable details.

Gilbert and Sanger have independently developed separate methods for the
determination of the exact sequence of the building blocks in DNA. Gilbert has
in addition studied those parts of DNA in a bacterial chromosome which
control the transcription of the genetic message in the cell. Sanger has, for
example, determined the complete nucleotide sequence for a small virus, whose
DNA still consists of not less than 5,375 building blocks.

It was the very research to be awarded today, which made a Swedish newspaper a couple of years ago have the headline "God knows which monsters the scientists have in their test tubes". How can the Royal Swedish Academy of Sciences reconcile its choice with the will of Nobel, which states that awards should be given for contributions which have conferred the greatest benefit on mankind? The recombinant-DNA debate was started by the scientists themselves, after an initiative from Berg, warning for possible dangers with the new technique. Continued research has, however, shown that the concern for hypothetical risks has been unwarranted. The recombinant-DNA technique is instead, together with methods for the determination of nucleotide sequences, an extremely important tool to widen our understanding of the way in which the DNA molecule governs the chemical machinery of the cell. The results of the three investigators has already given benefit to mankind, not only in the form of new fundamental knowledge but also in the form of important technical applications, e. g. the production of human hormones with the aid of bacteria. In a longer perspective the methods of Berg, Gilbert and Sanger should become important means in our efforts to understand the nature of cancer, as in this disease there is a malfunction in the control by the genetic material of the growth and division of a cell.

Drs. Berg, Gilbert and Sanger,

I have tried to put your fundamental contributions to the chemistry of the genetic material in the perspective of a biochemical concept of life. In particular, I have mentioned that you, Dr. Berg, constructed the first recombinant-DNA molecule and have applied your technique to the study of a viral chromosome. I have mentioned that you, Drs. Gilbert and Sanger, independently developed separate methods for the determination of the sequence of the nucleotide blocks in DNA, and have used these techniques, for example, to investigate viral and bacterial DNA. It is for these pioneering contributions that the Royal Academy of Sciences has decided to award this year's Nobel Prize for Chemistry to you together with Dr. Berg.

Drs. Berg, Gilbert and Sanger,
On behalf of the Royal Academy of Sciences I wish to convey to you our warmest congratulations, and I now ask you to receive your Prizes from the hand of His Majesty the King.

Paul Berg

PAUL BERG

Born June 30, 1926 to Harry and Sarah (Brodsky) Berg in Brooklyn, New York, U.S.A. Siblings—Jack (aged 53) and Irving (deceased).

Married September 13, 1947 to Mildred Levy. One son—John Alexander born September 30, 1958.

Attended public grade- and high-school (Abraham Lincoln) in New York, graduating early in 1943. Studied biochemistry at Pennsylvania State College from 1943 until 1948 (B. S. in Biochemistry). Served in U.S. Navy (1944—46).

Graduate studies in biochemistry at Western Reserve University (Ph. D. 1952). Postdoctoral training with Herman Kalckar, Institute of Cytophysiology, Copenhagen, Denmark (1952—53), and Arthur Kornberg, Washington University, St. Louis, MO (1953—54). Scholar in Cancer Research, Washington University (1954—57).

Professional positions: Assistant professor of microbiology, Washington University School of Medicine (1955—59); Associate professor and professor of biochemistry at Stanford University School of Medicine, Stanford, CA (1959—present); named Willson Professor of Biochemistry, Stanford University (1970). Chairman of Department of Biochemistry (1969—74). Non-resident Fellow of Salk Institute (1973—present).

Awards: Eli Lilly Prize in Biochemistry (1959); California Scientist of the Year (1963); V. D. Mattia Award of the Roche Institute for Molecular Biology (1974); Sarasota Medical Award (1979); Gairdner Foundation Annual Award (1980); Albert Lasker Medical Research Award (1980); New York Academy of Sciences Award (1980).

Honors: Elected to U. S. National Academy of Sciences and American Academy of Sciences (1966); Distinguished Alumnus Award Pennsylvania State University (1972); President of American Society of Biological Chemists (1975); Honorary D.Sc. Yale University and University of Rochester (1978); Foreign member of Japan Biochemical Society (1978).

Selected Bibliography
1. Physical and Genetic Characterization of Deletion Mutants of Simian Virus 40 Constructed *In Vitro*. Charles Cole, Terry Landers, Stephen Goff, Simone Manteuil-Brutlag, and Paul Berg. J. Virol., *24*: 277—294 (1977).
2. A Biochemical Method for Inserting New Genetic Information into SV40 DNA: Circular SV40 DNA Molecules Containing Lambda Phage Genes and the Galactose Operon of *E. coli*. David A. Jackson, Robert H. Symons, and Paul Berg. Proc. Nat. Sci. USA, *69*, 2904 (1972).

3. Construction of Hybrid Viruses Containing SV40 and Lambda Phage DNA Segments and Their Propagation in Cultured Monkey Cells. Stephen P. Goff, and Paul Berg. Cell, *9*: 695 (1976).
4. Synthesis of Rabbit β-Globin in Cultured Monkey Kidney Cells Following Infection with a SV40 β-Globin Recombinant Genome. R. C. Mulligan, B. H. Howard, and Paul Berg. Nature, *277*, 108−114 (1979).
5. Expression of a Bacterial Gene in Mammalian Cells. R. C. Mulligan and Paul Berg, Science *209*, 1422−1427 (1980).

DISSECTIONS AND RECONSTRUCTIONS OF GENES AND CHROMOSOMES

Nobel lecture, 8 December, 1980

by

PAUL BERG

Department of Biochemistry, Stanford University School of Medicine, Stanford, California 94305

ACKNOWLEDGEMENTS

The Nobel lecture affords a welcome opportunity to express my gratitude and admiration to the numerous students and colleagues with whom I have worked and shared, alternately, the elation and disappointment of venturing into the unknown. Without their genius, perseverance and stimulation much of our work would not have flourished. Those who have worked with students and experienced the discomfort of their curiosity, the frustrations of their obstinacy and the exhilaration of their growth know first hand the magnitude of their contributions. Each in our common effort left a mark on the other and, I trust, each richer from the experience. I have also been fortunate to have two devoted research assistants, Ms. Marianne Dieckmann and June Hoshi, who have labored diligently and effectively, always with understanding and sympathy for my idiosyncracies.

I have also been blessed with an amazing group of colleagues at Stanford University who have created as stimulating and liberating an environment as one could long for. Their many achievements have been inspirational and without their help—intellectually and materially—my efforts would have been severely handicapped. I am particularly grateful to Arthur Kornberg and Charles Yanofsky, both longtime close personal friends, for their unstinting interest, encouragement, support and criticism of my work, all of which enabled me to grow and thrive. And finally, there is my wife, Millie, without whom the rare triumphs would have lost their lustre. Her strength, assent and encouragement freed me to immerse myself in research.

Certainly my work could not have taken place without the generous and enlightened support of the U. S. National Institutes of Health, the National Science Foundation, the American Cancer Society and numerous foundations and individuals who invested their wealth in our research.

INTRODUCTION

"Although we are sure not to know everything and rather likely not to know very much, we can know anything that is known to man, and may, with luck and sweat, even find out some things that have not before been known to man."

J. Robert Oppenheimer

Although the concept that genes transmit and control hereditary characteristics took hold early in this century, ignorance about the chemical nature of genes forestalled most inquiries into how they function. All of this changed as a result of several dramatic developments during the 1940's to 1960's. First, Beadle and Tatum's researches (1—3) lent strong support for earlier (4) and widespread speculations that genes control the formation of proteins (enzymes); indeed, the dictum, "one gene—one protein", intensified the search for the chemical definition of a gene. The discovery by Avery and his colleagues (5) and subsequently by Hershey and Chase (6) that genetic information is encoded in the chemical structure of deoxyribonucleic acid (DNA) provided the first clue. Watson and Crick's solution (7) of the molecular structure of DNA— the three dimensional arrangement of the polymerized nucleotide subunits— not only revealed the basic design of gene structure but also the outlines of how genes are replicated and function. Suddenly, genes shed their purely conceptual and statistical characterizations and acquired defined chemical identities. Genetic chemistry, or molecular biology as it has frequently been called, was born.

Until a few years ago, much of what was known about the molecular details of gene structure, organization and function had been learned in studies with prokaryote microorganisms and the viruses that inhabit them, particularly, the bacterium *Escherichia coli* and the T and lambdoid bacteriophages. These organisms were the favorites of molecular biologists because they can be propagated readily and rapidly under controllable laboratory conditions. More significantly, utilizing several means of natural genetic exchanges characteristic of these organisms and phages, the mapping and manipulation of their relatively small genomes became routine. As a consequence, discrete DNA molecules, containing one or a few genes, were isolated in sufficient quantity and purity to permit extensive characterizations of their nucleotide sequences and chromosomal organisation. Moreover, such isolated genetic elements provided the models, substrates and reagents needed to investigate a wide range of basic questions: the chemical basis of the gentic code; mutagenesis; the mechanisms of DNA and chromosome replication, repair and recombination; the details of gene expression and regulation.

The astounding successes in defining the genetic chemistry of prokaryotes during the 1950's and 60's were both exhilarating and challenging. Not surprisingly, I and others wondered whether the more complex genetic structures of eukaryote organisms, particularly those of mammalian and human cells, were organized and functioned in analogous ways. Specifically, did the requirements of cellular differentiation and intercellular communication, distinctive characteristics of multicellular organisms, require new modes of genome structure, organisation, function and regulation? Were there just variations of the prokaryote theme or wholly new principles waiting to be discovered in explorations of the genetic chemistry of higher organisms? It seemed important to try to find out.

SV40's Minichromosome

Sometime during 1965–66 I became acquainted with Renato Dulbecco's work on the then newly discovered polyoma virus. The growing sophistication of animal cell culture methods had made it possible for Dulbecco's laboratory to monitor and quantify the virus' growth cycle *in vitro* (8). Particularly significant was the discovery that the entire virus genome resided in a single, relatively small, circular DNA molecule, one that could accomodate about five to eight genes (9). I was intrigued by the resemblance between polyoma's life styles and those of certain bacteriophages. On the one hand, polyoma resembled lytic bacteriophages in that the virus could multiply vegetatively, kill its host and produce large numbers of virus progeny (8). There was also a tantalizing similarity to lysogenic bacteriophages, since some infections yielded tumorigenic cells (10, 11). The acquisition of new morphologic and growth characteristics, as well as certain virus specific properties, suggested that tumorigenesis and cell transformation resulted from covalent integration of viral DNA into the cell's chromosomal DNA and the consequent perturbation of cell growth control by the expression of virus genes (12, 13).

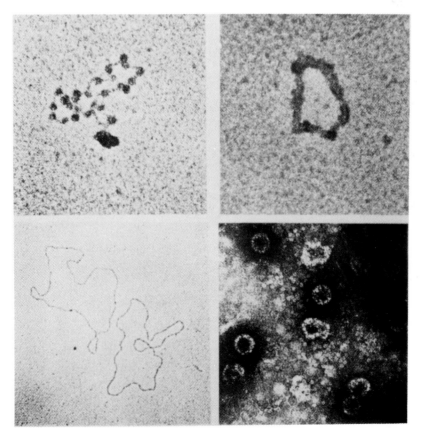

Fig. 1. Electron microgaphs of: SV 40 virions (u. l.); SV40 DNA (u. r.); "condensed" SV40 minichromosomes (1.1.); "relaxed-beaded" SV40 minichromosome (1. r.). Photo by J. Griffith.

These discoveries and provocative speculations, together with an eagerness
to find an experimental model with which to study the mechanisms of mammalian gene expression and regulation, prompted me to spend a year's sabbatical
leave (1967−68) in Dulbecco's laboratory at the Salk Institute. The work and
valuable discussions we carried on during that time (14) reinforced my conviction that the tumor virus system would reveal interesting features about mammalian genetic chemistry.

For somewhat technical reasons when I returned to Stanford, I adopted
SV40, a related virus, to begin our own research program. SV40 virions are
nearly spherical particles whose capsomers are organized in icosahedral symmetry (Fig. 1, upper left). The virions contain three viral coded polypeptides
and a single double-stranded circular DNA molecule (Fig. 1, upper right), that
is normally associated with four histones, H2a, H2b, H3 and H4 in the form of
condensed (Fig. 1, lower left) or beaded (Fig. 1, lower right) chromatin-like
structures. SV40 DNA contains 5243 nucleotide pairs (5.24 kbp), the entire
sequence of which is known from studies in S. M. Weissman's (15) and W.
Fier's (16) laboratories. Coding information for five (and possibly six) proteins
is contained in the DNA nucleotide sequence. Three of the proteins occur in
mature virions, possibly as structural components of the capsid shell, although
one might be associated with the DNA and have a regulatory function (17). Of
the two non-virion proteins encoded in the DNA sequence, one is localized in
the cell nucleus (large T antigen) and functions in Viral DNA replication and
cell transformation; the other, found in the cytoplasm (small t antigen), en-
hances the efficiency of cell transformation (18). Other proteins related in
structure to large T antigen have been speculated about, but their structures
and functions are unclear.

Restriction endonucleases have played a crucial role in defining the physical
and genetic organization of the SV40 genome (19, 20). The restriction or
cleavage sites serve as coordinates for a physical map of the viral DNA; the
availability of such map coordinates make it possible to locate, accurately,
particular physical features and genetic loci. In this system of map coordinates
the single *Eco*RI endonuclease cleavage site serves as the reference marker and
is assigned map position 0/1.0; other positions in the DNA are assigned
coordinates in DNA fractional length units measured clockwise from 0/1.0 (see
Fig. 2). At the present time knowledge of the entire nucleotide sequence has
made possible a more precise set of map coordinates: nucleotide pair number.
Thus, nucleotide 0/5243 is placed within ori, the site where DNA replication is
initiated, and the other nucleotide pairs are numbered consecutively in the
clockwise direction (see Fig. 2).

The SV40 minichromosome is expressed in a regulated temporal sequence
after it reaches the nucleus of infected primate cells. Initially, transcription in
the counter-clockwise direction of one strand (the E-strand) of about one half of
the DNA (the early region) yields the early mRNAs (Fig. 2). These mRNAs,
which encode the large T and small t antigen polypeptides (the stippled portion
of the mRNAs indicate the protein coding regions), have 5'-ends originating
from nucleotide sequences near the site marked ori and 3'-polyadenylated (poly

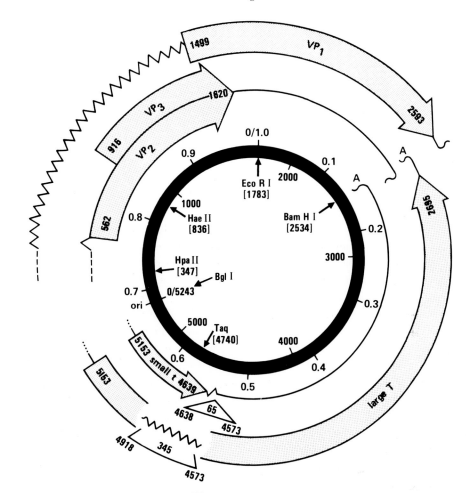

Fig. 2. A physical and genetic map of SV40 DNA.

The inner circle symbolizes the closed circular DNA molecule; indicated within the circle are the nucleotide-pair map coordinates starting and ending at 0/5243. Also shown by small arrows within the circle are the sites at which five restriction endonucleases cleave SV40 DNA once. Arrayed around the outside of the circle are the map coordinates, expressed in fractional lengths, beginning at the reference point 0/1.0 (the *Eco*RI endonuclease cleavage site) and proceeding clockwise around the circle. The coding regions for the early and late proteins are shown as stippled arrows extending from the nucleotide pair of the first codon to the nucleotide pair that specifies termination of the protein coding sequence. Each of the coding regions is embedded in a mRNA, the span of which is indicated by dotted or dashed 5'-ends and wavy poly A 3'-ends. The jagged or saw-toothed portions of each mRNA indicate the portions of the transcript that are spliced in forming the mature mRNAs.

A) ends from near map position 0.16. Synthesis of large T antigen triggers the initiation of viral DNA replication at ori, a specific site in the DNA (Fig. 2 identifies ori at map position 0.67 or nucleotide position 0/5243); replication then proceeds bi-directionally, terminating about 180° away near map position 0.17, yielding covalently closed circular progeny DNA. New viral mRNAs appear in the polyribosomes concommitant with DNA replication; these are synthesized in the clockwise direction from the L-strand of the other half of the virus DNA (the late region) and are referred to as late mRNAs. Transcription of the late mRNAs, which code for the virion proteins VP1, VP2, and VP3 (the stippled regions designate the protein coding regions of these proteins), begins from multiple positions between map positions 0.68−0.72 and terminates at about map position 0.16. Finally, the accumulation of progeny DNA molecules and virion proteins culminates in death of the cell and release of mature virion particles.

SV40 possesses an alternative life cycle when the virus infects rodent and other non-primate cells. The same early events take place−the E-strand mRNAs and large T and small t antigens are synthesized−but DNA replication does not occur and late strand mRNAs and virion proteins are not made. Frequently, replication of cell DNA and mitosis are induced after infection, and most infected cells survive with little evidence of prior infection. Generally, a small proportion of the cells (less than 10%) acquire the ability to multiply under culture conditions that restrict the growth of normal cells; moreover, these transformed cells can produce tumors after innoculation into appropriate animals. Invariably, the transformed cells contain all or part of the viral DNA covalently integrated into the cell's chromosomal DNA and produce the mRNAs and proteins coded by the early genes.

During the 1970's several different approaches, carried on in many laboratories including my own, clarified the arrangement of SV40 genes on the DNA and revealed how they function during the virus life cycle (21−23). Initially, viral genes were mapped on the DNA relative to restriction sites by localizing the regions from which early and late mRNAs were transcribed. Subsequently, more precise mapping was achieved by correlating the positions of discrete deletions and other alterations in the viral DNA with specific physiologic defects. But with the nucleotide sequence map, the boundaries of each SV40 gene and the nucleotide segments coding for each polypeptide can be specified with considerable precision (Figure 2). As expected, the availability of a precise genetic and physical map of SV40's minichromosome, has shifted the research emphasis to explorations of the molecular mechanisms governing each gene's expression and function, the replication and maturation of the viral minichromosome, recombination between the viral and host DNA, and how virus and host gene products interact to cause transformation of normal into tumorigenic cells. Excellent and more detailed summaries and analyses of the molecular biology of SV40 and polyoma containing acknowledgements to the important contributions made by many individuals can be found in several recent monographs (21−23).

SV40 as a Transducing Virus

The analysis of the organization, expression and regulation of bacterial genomes was greatly aided by the use of bacteriophage-mediated transfer of genes between cells. Indeed, specialized transducing phages of λ, Φ80, P22 and others, permitted the cloning and amplification of specific segments of bacterial DNA thereby, making it possible to construct cells with unusual and informative genotypes and to obtain valuable substrates and probes for exploring mechanisms of transcription, translation and regulation.

This background led me to consider, soon after beginning work with the tumor viruses, whether SV40 could be used to transduce new genes into mammalian cells. Initially, I had serious reservations about the success of such a venture because of the predictably low probability of generating specific recombinants between virus and cell DNA and the limited capability for selecting or screening animal cells that had acquired specific genetic properties. But it seemed that one possible way out of this difficulty, at least one worth trying, was to produce the desired SV40 transducing genomes synthetically. Consequently, in about 1970, I began to plan the construction, *in vitro*, of recombinant DNA molecules with SV40 and selected non-viral DNA segments. The goal was to propagate such recombinant genomes in suitable animal cells, either as autonomously replicating or integrated DNA molecules. At the time there were few if any animal genes available for recombination with SV40 DNA but I anticipated that a variety of suitable genes would eventually be isolated. Therefore, the first task was to devise a general way to join together, *in vitro*, any two different DNA molecules.

Hershey and his colleagues had already shown that λ-phage DNA could be circularized or joined end to end *in vitro* (24). This occurred because λ phage DNA has cohesive ends, i.e., single-stranded, overlapping, complementary DNA ends (25). So, it seemed that if cohesive ends could be synthesized onto the ends of DNA molecules, they could be covalently joined *in vitro* with DNA ligase.

During 1971−72, using then available enzymes and relatively straight-forward enzymologic procedures, David A. Jackson, Robert H. Symons and I (26) and, independently and concurrently, Peter E. Lobban and A. D. Kaiser (27), devised a way to synthesize synthetic cohesive termini on the ends of any DNA molecules, thereby paving the way for constructing recombinant DNAs *in vitro*. Our procedure (Fig. 3) was developed using as the model "foreign" DNA a bacterial plasmid that contained some bacteriophage λ DNA and three *E. coli* genes that specify enzymes required for galactose utilization (28). Circular SV40 DNA (5.24 kbp) and λdv gal plasmid DNA (about 10 kbp) were each cleaved with a specific endonuclease to convert them to linear molecules. Then, after a brief digestion with λ-exonuclease to remove about fifty nucleotides from the 5′-termini, it was possible for deoxynucleotidyl terminal transferase to add short "tails" of either deoxyadenylate or deoxythymidylate residues to the 3′-termini. After mixing and annealing under appropriate conditions the two DNAs were joined and cyclized via their complementary "tails" (Fig. 3). The gaps that occur where the two DNA molecules are held together, were filled in with DNA polymerase I and deoxynucleoside triphosphate substrates and the resulting molecules were covalently sealed with DNA ligase; exonuclease III was present to permit repair of nicks or gaps created during the manipulations.

The resulting hybrid DNA was approximately three times the size of SV40 DNA and, therefore, could not be propagated as an encapsidated virus. But we intended to test whether the *E. coli* galactose genes would be expressed after

Construction Of Hybrid Genome

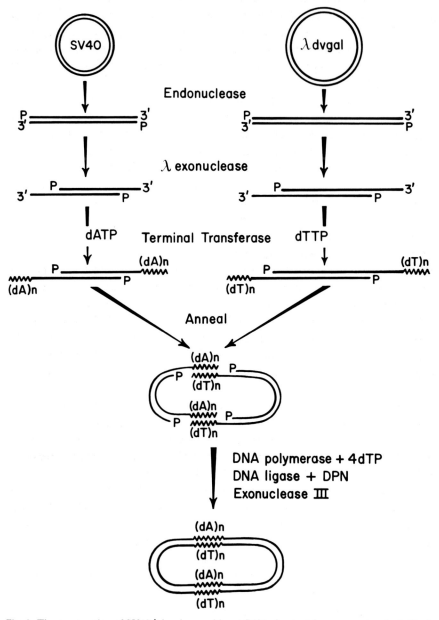

Fig. 3. The construction of SV40-λdvgal recombinant DNA. See text for comment on individual steps.

introduction into the chromosomes of cultured animal cells. Moreover, since the λdvgal plasmid could replicate autonomously in *E. coli* (28), we also planned to determine if SV40 DNA would be propagated in *E. coli* cells and if any SV40 genes would be expressed in the bacterial host. Although the SV40-λdv gal recombinant DNA shown in Fig. 3 could not have replicated in *E. coli*—

a gene needed for replication of the plasmid DNA in *E. coli* had been inactivated by the insertion of the SV40 DNA—a relatively simple modification of the procedure—the use of λdv gal dimeric DNA as acceptor for the SV40 DNA insert—could have circumvented this difficulty. Nevertheless, because many colleagues expressed concern about the potential risks of disseminating *E. coli* containing SV40 oncogenes, the experiments with this recombinant DNA were discontinued.

Since that time there has been an explosive growth in the application of recombinant DNA methods for a number of novel purposes and challenging problems. This impressive progress owes much of its impetus to the growing sophistication about the properties and use of restriction endonucleases, the development of easier ways of recombining different DNA molecules and, most importantly, the availability of plasmids and phages that made it possible to propagate and amplify recombinant DNAs in a variety of microbial hosts. (See 29, 30 for a collection of notable examples.)

By 1975, extensive cloning experiments had produced elaborate libraries of eukaryote DNA segments containing single genes or clusters of genes from many species of organisms. As expected, studies of their molecular anatomy and chromosomal arrangement have provided new insights about possible mechanisms of gene regulation in normal and developmentally interesting animal systems. But, it seemed likely from the beginning that ways would be needed to assay isolated genes for their biological activity *in vivo*. Consequently, I returned to the original goal of using SV40 to introduce cloned genes into cultured mammalian cells. But this time we explored a somewhat different approach.

During 1972—74 Janet Mertz and I (31) learned how to propagate SV40 deletion mutants by complementation using appropriate SV40 temperature-sensitive (ts) mutants as helpers. This advance made it feasible to consider propagating genomes containing exogenous DNA in place of specific regions of SV40 DNA. Accordingly, Stephen Goff and I devised a procedure to construct such recombinants by removing defined segments of SV40 DNA with appropriate restriction endonucleases and replacing them with "foreign" DNA segments using synthetic cohesive ends (32, 33) (Fig. 4). In this experimental design the recombinant genomes must contain the origin of SV40 DNA replication (ori) so that they can be propagated; also they must be smaller than 5.3 kbp, that is, not more than one mature viral DNA length, to be incorporated into virus particles. Furthermore, because the SV40 vector lacks genetic functions coded by the excised DNA segment, the recombinant genomes are defective and must be propagated with a helper virus that can supply the missing gene product or products. In our protocol the recombinant genome retains at least one functioning virus gene and, consequently, can complement a defective gene in the helper virus. For example, recombinants in which the inserted DNA replaces all or part of SV40's late region can be propagated with SV40 mutants that have a defective early region (e. g. at high temperature with ts early mutants); similarly, recombinants having exogenous DNA implants in place of

DNA segments in the early region can be propagated with a helper genome that is defective in its late region (in this instance, with ts late mutants).

Our initial attempts (32, 33) to obtain expression of cloned DNA segments as distinct mRNAs and proteins following introduction of the recombinant genomes into cultured cells were negative. But as soon as we recognized that expression of the new genetic information required that the transcript, originating from SV40 promoters and ending at SV40 specified poly A sites, be spliced, our fortunes changed. The initial success in obtaining expression of added genetic elements as mRNAs and proteins following transfection into cultured monkey cells was achieved with a DNA segment coding for rabbit β-globin (34). Soon afterward, a bacterial gene (Ecogpt) coding for xanthine-guanine phosphoribosyl transferase (XGPRT) (35), a mouse DNA specifying dihydrofolate reductase (DHFR) (36), and a bacterial gene (neo^R) for aminoglycoside phosphotransferase (37) were successfully transduced into mammalian cells via SV40 DNA based vectors. Generally, the transduced DNA segments are expressed at rates comparable to those of the SV40 genes they replace, but some anomalies in the RNA processing have been observed.

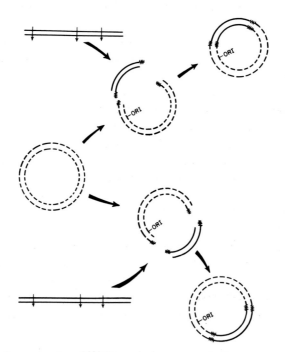

Fig. 4. A scheme for construction of SV40 transducing genomes *in vitro*.

Segments of the late (upward track) or early (downward track) regions of SV40 DNA (the dashed circle on the left) are removed by sequential cleavages with restriction endonucleases. Appropriate sized segments of any DNA, produced by restriction enzyme cleavages, enzymatic copying of mRNA or chemical synthesis, can be inserted in place of the resected SV40 DNA segment. Joining, via natural or synthetic cohesive ends (symbolized by the jagged lines), is mediated by a DNA ligase. Ori indicates the position of the origin of SV40 DNA replication.

Hamer and Leder have also constructed and propagated recombinants of SV40 DNA with cloned mouse genomic β-globin (38) or α-globin (39) genes. In certain of their recombinants the transduced genes are expressed from SV40 late promoter signals, but other constructions reveal that transcription can be initiated from the α-globin promoter as well (39). Their experiments also demonstrate that proper splicing of the globin intervening sequences and translation of the resulting mRNA's can occur in a heterologous host.

New Transducing Vectors for Mammalian Cells

In the experiments referred to above, our principal aim had been to exploit the ability of the recombinant genomes to replicate in the virus' permissive host. For example, following infection of monkey cells, the SV40 recombinant genomes are amplified about 10^4 to 10^5-fold, thereby ensuring a high yield of the products expressed from the transduced genes. This system has taught us a great deal about the necessity and mechanistic subtleties of RNA splicing (40), the rules governing expression of coding sequences inserted at different positions in SV40 DNA (41), and some novel features of SV40 gene expression itself (42). But this experimental design has several distinct shortcomings. During the course of the infection the cells are killed, precluding the opportunity to monitor the transduced gene's expression in continuously multiplying cell populations. Moreover, only cells which can replicate SV40 DNA are able to amplify the cotransduced genes. This constraint excludes many specialized and differentiated animal cells as hosts for the transduced genes.

To circumvent these disadvantages we have developed a new group of transducing vectors that can be used to introduce and maintain new genetic information in a variety of mammalian cells (Fig. 5). pSV2 (43), and its derivatives pSV3 and pSV5 (35, 44) contain a DNA segment (shown as the filled region in Fig. 5) from an *E. coli* plasmid (pBR322) that permits these DNAs to propagate in *E. coli* cells, thereby greatly simplifying the genetic manipulations involved in their use. Each of the vectors contains a marker gene (shown in Fig. 5 as the hatched segment) flanked at the 5'-end with a DNA segment containing the SV40 early promoter and origin of DNA replication (ori); another SV40 DNA segment that ensures splicing and polyadenylation of the transcript is located at the 3'-end of the marker segment (the SV40 derived DNA segments are shown stippled in Fig. 5). Additional DNA segments can also be inserted into the vector DNAs at any of several unique restriction sites; consequently, a single DNA molecule can transduce several genes of interest simultaneously.

pSV2 can not replicate in mammalian cells because it and the cell lack the means to initiate DNA replication at ori. This can be rectified by inserting, at pSV2's single *Bam*H1 cleavage site, DNA segments which contain either a complete early region from SV40 DNA (pSV3), or polyoma's early region (pSV5) (Fig. 5). The viral early regions inserted into pSV3 and pSV5 vectors code for proteins that promote DNA replications from their respective origins, therefore, pSV3 DNA can replicate in monkey cells and pSV5 DNA replicates in mouse cells (44).

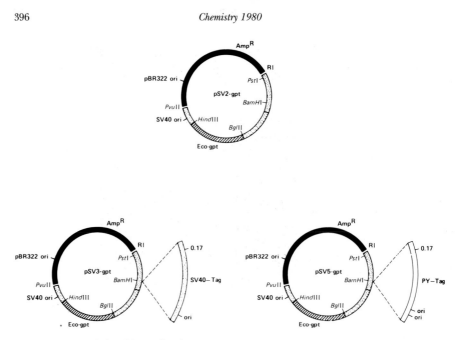

Fig. 5. Structure of plasmid transforming vectors.

The solid black segments in each diagram represent 2.3 kbp of pBR322 DNA sequence that contains the origin of pBR322 DNA replication and the ampicillinase gene. The stippled regions represent segments derived from SV40 DNA, the open region (in pSV5-gpt) is from polyoma DNA and the hatched segment represents Ecogpt DNA containing the gene coding for *E. coli* XGPRT.

To date three marker DNA segments have been used in conjunction with the pSV2, 3 and 5 vector DNAs: Ecogpt, an *E. coli* gene that codes for the enzyme XGPRT (35, 44); a mouse cDNA segment that specifies DHFR (36); neo[R], a bacterial plasmid gene specifying an aminoglycoside phosphotransferase that inactivates the antibacterial action of neomycin-kanamycin derivatives (37). Here, I shall only summarize several recent findings with Ecogpt as the marker gene but, suffice it to say, transfection of a variety of cells with any of the vector-dhfr or vector-neo[R] derivatives results in expression of the mouse DHFR, and neo[R] phenotype, respectively.

Richard Mulligan isolated the Ecogpt gene (35) to determine if it was useful for the detection and selective growth of mammalian cells that acquired that gene. The first priority was to establish that introduction of Ecogpt into mammalian cells would promote the production of *E. coli* XGPRT. We found that extracts from cultured monkey cells exposed to pSV2-gpt, pSV3-gpt and pSV5-gpt DNAs did contain two GPRT enzyme activities (Fig. 6A); one corresponds to the normal cellular enzyme, hypoxanthine-guanine phosphoribosyl transferase (HGPRT) by its electrophoretic behavior in a polyacrylamide gel and the other has the same electrophoretic mobility as the *E. coli* XGPRT activity. Other characterizations of XGPRT synthesized in monkey cells indicate that it is indistinguishable by several criteria from the same enzyme made in *E. coli*.

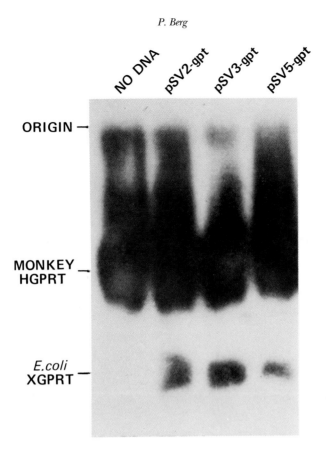

Fig. 6. Detection of GPRT activity following electrophoresis of cell extracts in polyacrylamide gels.

a) Extracts of about 5×10^6 CV1 cells harvested 3 days after transfection with 10 μg of pSV2-, -3- or -5-gpt DNAs were electrophoresed in polyacrylamide gels and assayed for GPRT activity *in situ* by incubation with ^{3}H-labeled guanine and detection of the labeled GMP product by fluorographic autoradiography (35). The black areas correspond to the location of guanine phosphoribosyl transferase activity on the gel and the arrows indicate the known positions of the monkey and *E. coli* GPRT enzymes.

E. coli XGPRT is analogous to the mammalian enzyme, HGPRT. Both use guanine as a substrate for purine nucleotide synthesis. But *E. coli* XGPRT differs from its mammalian counterpart in that it uses xanthine more efficiently then hypoxanthine as a substrate (45); the mammalian enzyme can use hypoxanthine but not xanthine as a precursor for purine nucleotide formation (46).

Since *E. coli* XGPRT is produced in monkey cells after introduction of the Ecogpt gene, it was important to learn if the bacterial enzyme could replace the cellular HGPRT function. This point could be tested with human Lesch-Nyhan cells because they lack HGPRT and, as a consequence, can not grow in a culture medium containing hypoxanthine, aminopterin and thymidine (HAT medium) (47). It seemed likely that transfection of Lesch-Nyhan cells with pSV-gpt DNAs would show whether the formation of *E. coli* XGPRT enabled these mutant cells to survive in HAT medium.

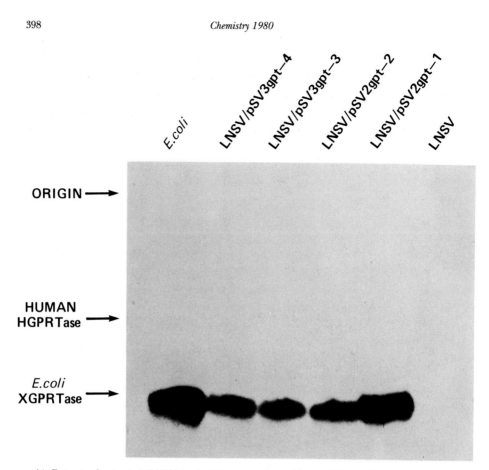

b) Extracts of untreated (LNSV) and pSV-gpt transformed Lesch-Nyhan cells were analyzed as described above. The track marked *E. coli* contained an extract of bacteria containing XGPRT.

As expected, Lesch-Nyhan cells did not survive in HAT medium if they had not received vector-gpt DNA. But when Lesch-Nyhan cell cultures received pSV2-gpt or pSV3-gpt DNA, surviving colonies were recovered at a frequency of about 10^{-4} (35). Representatives of the surviving clones, subcultured in HAT medium for 40 generations, were found to contain *E. coli* XGPRT (Fig. 6B). Thus, the acquisition of Ecogpt DNA and consequent expression of *E. coli* XGPRT rectified the defect caused by the absence of the cellular HGPRT in these cells.

The expression of Ecogpt provides a novel capability for mammalian cells; normal mammalian cells do not utilize xanthine for purine nucleotide formation but those that express Ecogpt and make *E. coli* XGPRT can do so. The acquisition and expression of Ecogpt can, therefore, be made the basis for an effective dominant selection for mammalian cells (48).

Purine nucleotides are synthesized *de novo* or by salvage pathways (Fig. 7). In the *de novo* pathway, inosinic acid (IMP), the first nucleotide intermediate, is converted to adenylic acid (AMP) via adenylosuccinate and to guanylic acid (GMP) via xanthylic acid (XMP). Salvage of free purines occurs by condensation with phosphoribosyl pyrophosphate (PRPP): adenine phosphoribosyl transferase (APRT) accounts for the formation of AMP from adenine (A) and hypoxanthine-guanine phosphoribosyl transferase (HGPRT) converts hypoxanthine (Hx) and guanine (G) to IMP and GMP, respectively. No mammalian enzyme comparable to the bacterial enzyme for converting xanthine (X) to XMP is known.

Mycophenolic acid (MPA), an inhibitor of IMP dehydrogenase (49), prevents the formation of XMP and, therefore, of GMP. The inhibition of cell growth by mycophenolic acid can be reversed by the addition of guanine to the medium, because guanine can be converted to its mononucleotide by HGPRT. Since normal mammalian cells do not convert xanthine to XMP, they do not grow if the medium containing mycophenolic acid is supplemented with xanthine. However, cells that contain *E. coli* XGPRT should grow under these conditions.

Transfection of monkey or mouse cells with pSV2, 3 or 5-gpt DNA and subsequent transfer to a medium containing mycophenolic acid plus xanthine, yields surviving colonies with frequencies ranging from 10^{-4} to 10^{-5} (48). Omission of the DNAs, or transfection with pSV2, 3 and 5 DNA containing other marker segments (e.g. β-globin or mouse DHFR DNA), or removal of the

PURINE NUCLEOTIDE SYNTHESIS

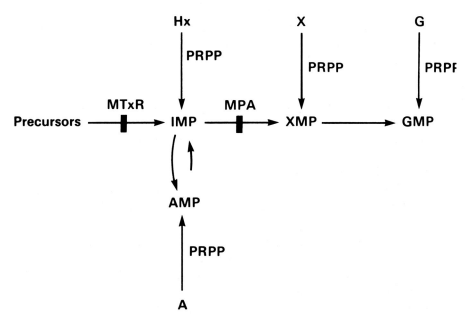

Fig. 7. Pathways of purine nucleotide synthesis.

Methotrexate (MtxR) inhibits *de novo* synthesis of purines and mycophenolic acid (MPA) specifically inhibits the conversion of IMP to XMP (49). The arrows between purine bases and their respective mononucleotides indicate the reactions of the bases with PRPP catalyzed by purine phosphoribosyl transferase. An arrow also indicates that AMP can be deaminated to IMP.

ori segment from pSV2-gpt DNA, results in fewer than 10^{-7} surviving colonies. Ecogpt transformed clones contain cells that produce both the normal HGPRT and the Ecogpt product, XGPRT. Analysis of the genetically transformed cells' DNA indicate that they have retained one to a few copies of the pSV-gpt DNA, probably integrated into their chromosomal DNA. The structure, organization and expression of these integrated genes and the relation of these parameters to their expression are presently being studied.

Our present studies suggest that the requirements for obtaining expression of prokaryote genes are not formidable. There is no reason to believe that the bacterial genes, Ecogpt and neo^R, are unique in their ability to be expressed in mammalian cells. Consequently, I foresee that bacteria, their viruses and simple eukaryotes will provide a rich source of genes for the modification of mammalian cells.

PROSPECTS

The development and application of recombinant DNA techniques has opened a new era of scientific discovery, one that promises to influence our future in myriad ways. It has already had a dramatic and far reaching impact in the field of genetics, indeed in all of molecular biology. Molecular cloning provides the means to solve the organization and detailed molecular structure of extended regions of chromosomes and eventually the entire genome of any organism including man's. Already, investigators have isolated a number of mammalian and human genes, and in some instances determined their chromosomal arrangement and even their detailed nucleotide sequence. Such detailed information has profound implications for the future of medicine. Just as our present knowledge and practice of medicine relies on a sophisticated knowledge of human anatomy, physiology and biochemistry, so will dealing with disease in the future demand a detailed understanding of the molecular anatomy, physiology and biochemistry of the human genome. There is no doubt that the development and application of recombinant DNA techniques has put us at the threshold of new forms of medicine. There are many who contemplate the treatment of crippling genetic diseases through replacement of defective genes by normal counterparts obtained by molecular cloning. Scenarios about how this could be done are rampant, only a few of which are plausible. Gene replacement as a therapeutic approach has many pitfalls and unknowns, amongst which are questions concerning the feasibility and desirability for any particular genetic disease, to say nothing about the risks. It seems to me that if we are ever to proceed along these lines we shall need a more detailed knowledge of how human genes are organized and how they function and are regulated. We shall also need physicians who are as conversant with the molecular anatomy, physiology and biochemistry of chromosomes and genes as the cardiac surgeon is with the structure and workings of the heart and circulatory tree. Gene therapy will have to be evaluated in terms of alternative and more conventional forms of treatment, just as is now done before undertaking heart valve replacements and renal transplants. Moreover, the ethical questions that have been raised in some quarters about such strategies will surely confound the scientific and medical issues that confront us.

The advent and widespread use of the recombinant DNA technology for basic and medical research and the implications for industrial and pharmaceutical application has also revealed, or perhaps created, an underlying apprehension, an apprehension about probing the nature of life itself, a questioning of whether certain inquiries at the edge of our knowledge and our ignorance should cease for fear of what we could discover or create. I prefer the more optimistic and uplifting view expressed by Sir Peter Medawar in his essay entitled "On The Effecting of All Things Possible" ("The Hope of Progress", McThuen and Co. Ltd, London, 1972).

"If we imagine the evolution of living organisms compressed into a year of cosmic time, then the evolution of man has occupied a day. Only during the past 10 to 15 minutes of the human day has our life been anything but precarious. We are still beginners and may hope to improve. To deride the hope of progress is the ultimate fatuity, the last word in the poverty of spirit and meanness of mind."

This passage speaks of the need to proceed. The recombinant DNA breakthrough has provided us with a new and powerful approach to the questions that have intrigued and plagued man for centuries. I, for one, would not shrink from that challenge.

REFERENCES

1. Beadle, G. W. and Tatum, E. L., Proc. Nat. Acad. Sci. USA *27*: 499 (1941).
2. Beadle, G. W., Chem. Revs. *37*: 15 (1945).
3. Beadle, G. W., The Harvey Lectures *40*: 179 (1945).
4. Garrod, A. E., Inborn Errors of Metabolism, Frowde, Hodder and Stoughton London (1941).
5. Avery, O. T., MacLeod, C. M., McCarty, M., J. Exp. Med. *79*: 137 (1944).
6. Hershey, A. D. and Chase, M. J., Gen. Physiol. *36*: 39 (1952).
7. Watson, J. D. and Crick, F. H. C., Nature *171*: 737, 964 (1953).
8. Dulbecco, R. and Freeman, G., Virology *8*: 396 (1959).
9. Dulbecco, R. and Vogt, M., Proc. Nat. Acad. Sci. USA *50*: 236 (1963).
10. Vogt, M. and Dulbecco, R., Proc. Nat. Acad. Sci. USA *46*: 365 (1960).
11. Dulbecco, R. and Vogt, M., Proc. Nat. Acad. Sci. USA *46*: 1617 (1960).
12. Sambrook, J. Westphal, H., Srinivasan, P. R., and Dulbecco, R., Proc. Nat. Acad. Sci. USA *60*: 1288 (1968).
13. Oda, K. and Dulbecco, R., Proc. Nat. Acad. Sci. USA *60*: 525 (1968).
14. Cuzin, F., Vogt, M., Dieckmann, M. and Berg, P., J. Mol. Biol. 47: 317 (1970).
15. Reddy, V. B., Thimappaya, B., Dhar, R., Subramanian, K. M., Zain, B. S., Pan, J., Ghosh, P. K., Celma, M. L. and Weismann, S., M. Science *200*: 494 (1978).
16. Fiers, W., Contreras, R., Haegeman, G., Rogiers, R., Van der Voorde, A., Van Heuverswyn, H., Van Herrewaghe, J., Volckgart, G., and Ysebaert, M., Nature *273*: 113 (1978).

17. Llopis, R. and Stark, G. R. A., J. Virol. 1981, submitted.
18. Bouck, N., Beales, N., Shenk, T., Berg, P. and DiMayorca, G., Proc. Nat. Acad. Sci. USA, 75: 2473 (1973).
19. Nathans, D., The Harvey Lectures 70: 111 (1976).
20. Nathans, D., Le Prix Nobel -1978 p. 198, Almquist and Wiksell International, Stockholm, Sweden.
21. Kelly, T. J. Jr. and Nathans, D., Adv. Virus Res. 21: 85 (1977).
22. Fried, M. and Griffin, B. E., Adv. Can. Res. 24: 67 (1977).
23. J. Tooze, ed., DNA Tumor Viruses (Cold Spring Harbor Laboratory Cold Spring Harbor, NY 1980) chapters 2–5.
24. Hershey, A. D., Burgi, E. and Ingraham, L., Proc. Nat. Acad. Sci. USA 49: 748 (1963).
25. Wu, R. and Kaiser, A. D., J. Mol. Biol. 35: 523 (1968).
26. Jackson, D. A., Symons, R. H. and Berg, P., Proc. Nat. Acad. Sci. USA 69: 2904 (1972).
27. Lobban, P. E. and Kaiser, A. D., J. Mol. Biol. 78: 453 (1973).
28. Berg, D. E. Jackson, D. A. and Mertz, J. E., J. Virol. 14: 1063 (1974).
29. Science 196: No. 183 (1977).
30. Science 209: No. 4463 (1980).
31. Mertz, J. E. and Berg, P., Virology 62: 112 (1974).
32. Goff, S. P. and Berg, P., Cell. 9: 695 (1976).
33. Goff, S. P. and Berg, P., J. Mol. Biol. 133: 359 (1979).
34. Mulligan, R. C., Howard, B. H. and Berg, P., Nature (London) 277: 108 (1979).
35. Mulligan, R. C. and Berg, P., Science 209: 1422 (1980).
36. Subramini, S., Mulligan, R. C. and Berg, P., Mol. Cell. Biol. 9: 854–864 (1981).
37. Southern, P. J. and Berg, P., Mol. & Applied Gen. 1: 327–341 (1982).
38. Hamer, D. H. and Leder, P., Nature (London) 281: 35 (1979).
39. Hamer, D. H., Koehler, M. and Leder, P., Cell, 21: 697 (1980).
40. Buchman, A. and Berg, P., Mol. Cell. Biol. 8: 4395–4405 (1988).
41. Southern, P. J., Howard, B. H. and Berg, P., J. Mol. Appl. Gen. 1: 177–190 (1982).
42. White, R. T., Villarreal, L. P. and Berg, P., J. Virol. 42: 262–274 (1982).
43. Howard, B. H., Mulligan, R. C. and Berg, P., unpublished.
44. Mulligan, R. C., Howard, B. H., Southern, P. J. and Berg P., unpublished.
45. Miller, R. L., Ramsey, G. A., Krenitsky, T. A. and Elion, G. B., Biochemistry 11: 4723 (1979).
46. Krenitsky, T. A., Papaioannou, R., Elion, G. B., Biol. Chem. 244: 1263 (1969).
47. Szybalski, E. H. and Szybalski, W., Proc. Nat. Acad. Sci. USA 48: 2026 (1962).
48. Mulligan, R. C. and Berg, P., Proc. Nat. Acad. Sci. USA 78: 2072–2076 (1981).
49. Franklin, T. J. and Cook, J. M., Biochem. J. 113: 515 (1969).

WALTER GILBERT

I was born on March 21, 1932 in Boston, Massachusetts. My father, Richard V. Gilbert, an economist, was at that time at Harvard University. He worked for the Office of Price Administration during the second World War and later headed up a planning group advising the Pakistani government. My mother, Emma Cohen, was a child psychologist, who practiced giving intelligence tests to me and my younger sister. She educated us at home for the first few years, to keep my sister and me amused. We loved reading and raided the adult section of the public library. In 1939 my family moved to Washington D.C.; I was educated there in public schools, later at the Sidwell Friends high school.

I always had an interest in science, in those years minerology and astronomy (I was a member of a minerological society and an astronomical society as a child). I became interested in inorganic chemistry at high school. In my last year in high school, 1949, I was fascinated by nuclear physics and would skip school for long periods to go down to the Library of Congress to read about Van de Graaf generators and simple atom smashers. I went to Harvard and majored in chemistry and physics. I became interested in theoretical physics and, as a graduate student, worked in the theory of elementary particles, the quantum theory of fields. I spent my first graduate year at Harvard, then went to the University of Cambridge for two years, where I received my doctorate degree in 1957. My thesis supervisor was Abdus Salam; I worked on dispersion relations for elementary particle scattering: an effort to use a notion of causality, formulated as a mathematical property of analyticity of the scattering amplitude, to predict some aspects of the interaction of elementary particles. I met Jim Watson during this period. I returned to Harvard and, after a postdoctoral year and a year as Julian Schwinger's assistant, became an assistant professor of Physics. During the late fifties and early sixties, I taught a wide range of courses in theoretical physics and worked with graduate students on problems in theory. However, after a few years my interests shifted from the mathematical formulations of theoretical physics to an experimental field.

In the summer of 1960, Jim Watson told me about an experiment that he and François Gros and his students were working on. I found the ideas exciting and joined in for the summer. We were trying to identify messenger RNA, a short-lived RNA copy of a DNA gene, which serves as a carrier of information from the genome to the ribosomes, the factories that make proteins. After each messenger is used a few times to dictate the structure of a protein, it is broken down and recycled to make other messenger RNA molecules. The experiments sought a fleeting new componant that we finally managed to pin down. I found the experimental work exciting and have continued research in molecular biology ever since.

After a year of work on messenger RNA, I returned briefly to physics then came back to biology to study how proteins are synthesized. I showed that a single messenger molecule can service many ribosomes at once and that the growing polypeptide chain always remains attached to a transfer RNA molecule. This last discovery illuminated the mechanism of protein synthesis: the protein chain is transferred in turn from one amino-acid-bearing transfer RNA to another as it grows, their order dictated by messenger RNA and ultimately by the genetic code on the DNA. In the middle sixties, Benno Müller-Hill and I isolated the lactose repressor: the first example of a genetic control element. A repressor is a protein product made by one gene in the bacterium in order to control a second gene by turning it off when its product is not wanted. This control function had been defined genetically by the work of Jacob and Monod, but a repressor is made in such small amounts that it was an extraordinarily elusive biochemical entity. We identified, characterized, and purified one. We developed bacterial strains which made several thousand fold more protein, and showed how that repressor functioned. In the late sixties, David Dressler and I invented the rolling circle model, which describes one of the two ways DNA molecules duplicate themselves. In the early seventies I isolated the DNA fragment to which the *lac* repressor bound and studied the interaction of the bacterial RNA polymerase and the lac repressor with DNA. In the middle seventies, Allan Maxam and I developed the rapid chemical DNA sequencing. At this time, I also became interested in and developed some of the recombinant DNA techniques, specifically showing that blunt end ligation was efficient in putting DNA fragments together. In the late seventies with Lydia Villa-Komaroff and Argiris Efstratiadis I worked on bacterial strains that expressed a mammalian gene product, insulin. Currently I am interested in, on one hand, the making of useful proteins in bacteria and, on the other hand, the structure of genes and the evolution of DNA sequences.

After my change from physics to molecular biology, I was promoted at Harvard in Biophysics and later in Biochemistry and Molecular Biology. Since 1974 I have been an American Cancer Society Professor of Molecular Biology.

I am married to Celia Gilbert, a poet, and have two children, John Richard and Kate.

Awards:

Westinghouse Science Talent Search, 1949.

United States Steel Foundation Award in Molecular Biology of the National Academy of Sciences, 1968

Elected American Academy of Arts and Sciences, May. 1968

Guggenheim Fellowship, 1968−69, Paris

Ledlie Prize, Harvard University, 1969 (with M. Ptashne)

Elected National Academy of Sciences, April 1976

V.D. Mattia Lectureship, Roche Institute of Molecular Biology, 1976

Smith, Kline, and French Lecturer, U.C, Berkeley, 1977

Warren Triennial Prize, Massachusetts General Hospital, 1977 (with S. Benzer)

Louis and Bert Freedman Award, New York Academy of Sciences, 1977

Prix Charles-Léopold Mayer, Académie des Sciences, Institut de France, 1977 (with M. Ptashne and E. Witkin)

Doctor of Science (Honorary), University of Chicago, June 1978

Doctor of Science (Honorary), Columbia University, October 1979

Harrison Howe Award of the Rochester Branch of the American Chemical Society, December 1978

Doctor of Science (Honorary), University of Rochester, May 1979

Louisa Gross Horwitz Prize, Columbia University, October 1979 (with F. Sanger)

Gairdner Foundation Annual Award, The Gairdner Foundation, November 1979

Albert Lasker Basic Medical Research Award, The Albert and Mary Lasker Foundation, November 1979 (with F. Sanger)

Prize Biochemical Analysis, The German Society for Clinical Chemistry, April 1980 (with A.M. Maxam, F. Sanger and A.R. Coulsen)

Sober Award, The American Society of Biological Chemists, June 1980

DNA SEQUENCING AND GENE STRUCTURE

Nobel lecture, 8 December, 1980

by

WALTER GILBERT

Harvard University, The Biological Laboratories, Cambridge, Massachusetts 02138, USA

When we work out the structure of DNA molecules, we examine the fundamental level that underlies all process in living cells. DNA is the information store that ultimately dictates the structure of every gene product, delineates every part of the organism. The order of the bases along DNA contains the complete set of instructions that make up the genetic inheritance. We do not know how to interpret those instructions; like a child, we can spell out the alphabet without understanding more than a few words on a page.

I came to the chemical DNA sequencing by accident. Since the middle sixties my work had focussed on the control of genes in bacteria, studying a specific gene product, a protein repressor made by the control gene for the *lac* operon (the cluster of genes that metabolize the sugar lactose. Benno Müller-Hill and I had isolated and characterized this molecule during the late sixties and demonstrated that this protein bound to bacterial DNA immediately at the beginning of the first gene of the three-gene cluster that this repressor controlled (1, 2). In the years since then, my laboratory had shown that this protein acted by preventing the RNA polymerase from copying the *lac* operon genes into RNA. I had used the fact that the *lac* repressor bound to DNA at a specific region, the operator, to isolate the DNA of this region by digesting all of the rest of the DNA with DNase to leave only a small fragment bound to the repressor, protected from the action of the enzyme. This isolated a twenty-five base-pair fragment of DNA out of the 3 million base pairs in the bacterial chromosome. In the early seventies, Allan Maxam and I worked out the sequence of this small fragment (3) by copying this DNA into short fragments of RNA and using on these RNA copies the sequencing methods that had been developed by Sanger and his colleagues in the late sixties. This was a laborious process that took several years. When a student, Nancy Maizels, then determined the sequence of the first 63 bases of the messenger RNA for the *lac* operon genes, we discovered that the *lac* repressor bound to DNA immediately after the start of the messenger RNA (4), in a region that lies under the RNA polymerase when it binds to DNA to initiate RNA synthesis. We continued to characterize the *lac* operator by sequencing a number of mutations (operator constitutive mutations) that damaged the ability of the repressor to bind to DNA. We wanted to determine more DNA sequence in the region to define the polymerase binding

site and other elements involved in *lac* gene control; however, that sequence
was worked out in another laboratory by Dickson, Abelson, Barnes and Reznikoff (5). Thus by the middle seventies I knew all the sequences that I had been
curious about, and my students (David Pribnow, and John Majors) and I were
trying to answer questions about the interaction of the RNA polymerase and
other control factors with DNA.

At this point, another line of experiments was opened up by a new suggestion. Andrei Mirzabekov came to visit me in early 1975. The purpose of his visit
was twofold: to describe experiments that he had been doing using dimethyl
sulfate to methylate the guanines and the adenines in DNA and to urge me to
do a similar experiment with the *lac* repressor. Dimethyl sulfate methylates the
guanines uniquely at the N7 position, which is exposed in major groove of the
DNA double helix, while it methylates the adenines at the N3 position which is
exposed in the minor groove (Fig. 1). Mirzabekov had used this property to
attempt to determine the disposition of histones and of certain antibiotics on
the DNA molecule by observing the blocking of the incorporation of radioactive
methyl groups onto the guanines and adenines of bulk DNA. He urged me to
use this groove specificity to learn something about the interaction of the *lac*
repressor with the *lac* operator. However, the amounts of *lac* operator available
were extremely small, and there was no obvious way of examining the protein
sitting on DNA to ask which bases in the sequence the protein would protect
against attack by the dimethyl sulfate reagent.

It was not until after a second visit by Mirzabekov that an idea finally
emerged. He and I and Allan Maxam and Jay Gralla had lunch together.

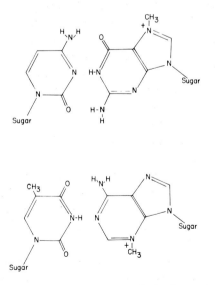

Figure 1. Methylated cytosine-guanine and thymine-adenine base pairs. The top of the figure
shows cytosine-guanine base pair methylated at the N7 position of guanine. The bottom of the
figure shows a thymine-adenine base pair methylated at the N3 position of adenine. The region
above each of the base pairs is exposed in the major groove of DNA. The region below each of the
base pairs lies between the sugar phosphate backbones in the minor groove of the DNA double
helix.

During our conversation I had an idea for an experiment, which ultimately underlies our sequencing method. We knew we could obtain a defined DNA fragment, 55 base-pairs long, which carried near its center the region to which the *lac* repressor bound. This fragment was made by cutting the DNA sequentially with two different restriction enzymes, each defining one end of the fragment (See fig. 2). Secondly, I knew that at every base along the DNA at which methylation occurred, that base could be removed by heat. Furthermore, once that had happened, only a sugar would be left holding the DNA chain together, and that sugar could be hydrolysed, in principle, in alkali to break the DNA chain. I put these ideas together by conjecturing that if we labelled one end of one strand of the DNA fragment with radioactive phosphate, we might determine the point of methylation by measuring the distance between the labelled end and the point of breakage. We could get such labelled DNA by isolating a DNA fragment (by length by electrophoresis through polyacrylamide gels) made by cutting with one restriction enzyme, labelling both ends of that fragment and then cutting it again with a second restriction enzyme to release two separable double-stranded fragments, each having a label at one end but not the other. Using polynucleotide kinase this procedure would introduce a radioactive label into the 5' end of one of the DNA strands of the fragment bearing the operator while leaving the other unlabelled (Fig. 2). If we then modified that DNA with dimethyl sulfate so that only an occasional adenine or guanine would be methylated, heated, and cleaved the DNA with alkali at the point of depurination, we would release among other fragments a labelled fragment extending from the unique point of labelling to the first point of breakage. Fig. 3 shows this idea. Any fragments from the other strand would be unlabelled, as would any fragments arising beyond the first point of breakage. If we could separate these fragments by size, as we could in principle by electrophoresis on a polyacrylamide gel, we might be able to associate the labelled fragments back to the known sequence and thus identify each guanine and adenine in the operator that had been modified by dimethyl sulfate. If we could do the modification in the presence of the *lac* repressor protein bound to the DNA fragment, then if the repressor lay close to the N7 of a guanine, we

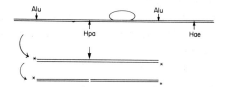

Figure 2. Procedure for obtaining a double-stranded DNA fragment uniquely labeled at one end of one strand. The figure shows the restriction cuts for the enzymes AluI (target AG/CT) and Hpa (target C/CGG producing an uneven end) in the neighborhood of the *lac* operator. The *lac* repressor is shown bound to the DNA. By cutting the DNA from this region first with the enzyme Alu, then labeling with radioactive P^{32} the 5' termini of both strands of the DNA with polynucleotide kinase, and then cutting in turn with the enzyme Hpa, we can isolate a DNA fragment that carries the binding site for the repressor uniquely labeled at one end of one strand.

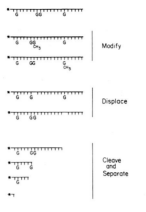

Figure 3. Outline of procedure to produce fragments of DNA by breaking the DNA at guanines. Consider an end-labeled strand of DNA. We modify an occasional guanine by methylation with dimethylsulfate. Heating the DNA will then displace that guanine from the DNA strand, leaving behind the bare sugar; cleaving the DNA with alkali will break the DNA at the missing guanine; the fragments are the separated by size, the actual size of the fragment (followed because it carries the radioactive label) determines the position of the modified guanine.

would not modify the DNA at that base, and the corresponding fragment would not appear in the analytical pattern.

I set out to do this experiment. Allan Maxam made the labelled DNA fragments, and I began to learn how to modify and to break the DNA. This involved analysing the release of the bases from DNA and the breakage steps separately. Finally the experiment was put together. Figure 4 shows the results: an autoradiogram of the electrophoretic pattern displays a series of bands extending downward in size from the full length fragment, each caused by the cleavage of the DNA at an adenine or a guanine. The same treatment of the DNA fragment with dimethyl sulfate, now carried out in the presence of the repressor produced a similar pattern, except that some of the bands were missing (lane one versus lane two in Figure 4). The experiment was clearly a success in that the presence of the repressor blocked the attack by dimethyl sulfate on some of the guanines and some of the adenines in the operator (6). I hoped that the size discrimination would be accurate enough to permit the assignment of each band in the pattern to a specific base in the sequence. This proved true because the spacing in the pattern, and the presence of light and dark bands, the dark bands corresponding to guanines and the light ones to adenines, were sufficiently characteristic to correlate the two. The guanines react about five times more rapdly with dimethyl sulfate while the methylated adenines are relesed from DNA more rapidly than the guanines during heating; the shift in intensities as a function of the time of heat treatment could by used to establish unambiguously which base was which. Furthermore, the gel pattern is so clear, that bands corresponding to fragments differing by one base were resolved. At this point it was evident that this technique could determine the adenines and guanines along DNA for distances of the order of 40 nucleo-

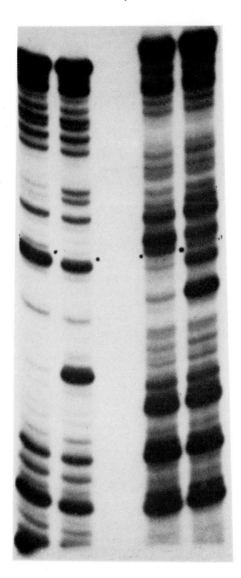

Figure 4. Methylation protection experiment with the *lac* repressor. The columns show the pattern of clevage along each strand of the 53–55 base long fragment bearing the *lac* operator. The second column from the left represents the DNA (labeled at the 5' end of the 53 base long strand) treated by dimethylsulfate, cleaved by heat and alkali. The dark bands correspond to breaks at guanines; the light bands are breaks at adenines. The second column of the figure reads from the bottom:
–G–GAAA––G––A–––G–––A–AA––––A–A–AA–A–A–. . . .
The first column shows this same double stranded piece of DNA treated with dimethylsulfate in the presence of the *lac* repressor. The repressor prevents the interaction of dimethylsulfate with the guanine a third of the way up the pattern and blocks the reaction with two adenines in the upper third of the pattern. The bands correspond to these two adenines represent fragments that differ in length by one base, 30 and 31 long. The right hand side of this pattern shows the same experiment done with the label at the end of the other DNA strand. The sequence at the far right would be:
–G–G–GGAA––G–GAG–GGA–AA–AA–––. . .

tides. By determining the purines on one strand and the purines on the complementary strand (as Fig 4 does) one has in principle a complete sequencing method.

Having in hand a reaction that will determine and distinguish adenines and guanines, could we find reactions that would distinguish cytosines and thymines? Allan Maxam and I turned our attention to this end. (First we examined a second binding site for the *lac* repressor that lies a few hundred bases further along the DNA, under the first gene of the operon. This binding site has no physiological function. We could locate this binding site on a restriction fragment by repeating the methylation-protection experiment and identifying bases protected by the *lac* repressor. I used the methylation pattern to attempt to predict the positions of the adenines and the guanines in the unknown DNA sequence; Allan Maxam then used the wandering spot sequencing method of Sanger and his coworkers to determine the DNA sequence of this region to verify that we had made a successful prediction.) Allan Maxam then went on to do the next part of the development. We knew that hydrazine would attack the cytosines and thymines in DNA and damage them sufficiently, or eliminate them to form a hydrazone, so that a further treatment of the DNA with benzaldehyde followed by alkali, (or a treatment with an amine), would cleave at the damaged base. This soon gave us a similar pattern, but broke the DNA at the cytosines and thymines without discrimination. Allan Maxam then discovered that salt, one molar salt in the 15 molar hydrazine, altered the reaction to suppress the reactivity of the thymines. The two reactions together then positioned and distinguished the thymines and the cytosines in a DNA sequence. This last discrimination conceptually completed the method. To improve the discrimination between the purines, and to provide redundant information which would serve to make the sequencing more secure against errors, we used the fact that the methylated adenosines depurinate more rapidly, especially in acid, to release the adenosines preferentially and thus to obtain four reactions: one for A's, one preferential for G's, one for C's and T's, and one for C's determining the T's by difference. This stage of the work was completed within a few months. As the range of resolution on the gels was extended toward 100 bases, the cleavage at the pyrimidines was not satisfactory, the result of the incomplete cleavage was that the longer fragments contained a variety of internal damages and the pattern blurred out. After many months of searching an answer was found. A primary amine, aniline, will displace the hydrazine products and produce a beta elimination that releases the phosphate from the 3' position on the sugar, but it will not release the other phosphate, and the mobility of a DNA fragment with a blocked 3' phosphate bearing a sugar-aniline residue is different from the free phosphate ended chains from the other reactions. A secondary amine, piperidine, is far more effective and triggers both beta eliminations as well as eliminating all the breakdown products of the hydrazine reaction from the sugar. This reagent completed the DNA sequencing techniques (7). Although the development of the techniques continued for another nine months, they were distributed freely to other groups that wanted to use them. Fig. 5 shows an actual sequencing

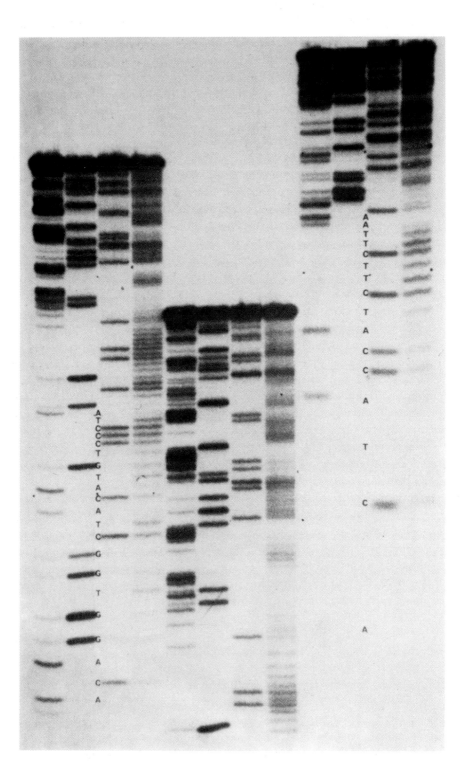

pattern from the 1978 period, used in the work described in (8). Fig. 6 shows two examples of the chemistry (9).

The logic behind the chemical method is to divide the attack into two steps. In the first we use a reagent that carries the specificity, but we limit the extent of that reaction—to only one base out of several hundred possible targets in each DNA fragment. This permits the reaction to be used in the domain of greatest specificity: only the very initial stages of a chemical reaction are involved. The second step, the cleavage of the DNA strand, must be complete. Since the target has already been distinguished from the other bases along the DNA chain by the preliminary damage, we can use vigorous, quantitative reaction conditions. The result is a clean break, releasing a fragment without hidden damages, which is required if the mobilities of the fragments are to be very closely correlated so that the bands will not blur. (The specificity need be only about a factor of ten for the sequence to be read unambiguously.)

Today, later developments of the technique (9) have modified the guanine reaction and replaced the dimethyl sulfate adenosine reaction with a direct depurination reaction that releases both the adenines and guanines equally. These changes, and the introduction of the very thin gels by Sanger's group (10), now make it possible to read sequences out between 200—400 bases from the point of labelling. The actual chemical workup, the analysis on gels, and the autoradiography is the short part of the process. The major time spent in DNA sequencing is spent in the preparation of the DNA fragments and on the elements of stratagy. The speed of the sequencing comes only in part from the ability to read off quickly several hundred bases of DNA—at a glance. The more important element is the linear presentation of the problem. Rather than sequence randomly, one can begin at one end of a restriction map and move rationally through a gene—or construct the restriction map as one goes.

The first long sequence was done by a graduate student, Phillip Farabaugh, who used the new techniques to sequence the gene for the *lac* repressor (11). The protein sequence of this gene product had been worked out in the early seventies by Beyreuther and his coworkers (12). Since the amino-acid sequence was known, he could quickly (a few months) establish the DNA sequence. However the DNA sequence showed that there were errors in the protein sequence, two amino acids dropped at one place and eleven at another. Since

Figure 5. Actual sequencing pattern from the 1978 period. Products of four different chemical reactions, applied to a DNA fragment about 150 bases long, are electrophoresed on a polyacryl-amide gel; three loadings produce sets of patterns that have moved different distances down the gel. The four columns correspond to reactions that break the DNA: 1) primarily at the adenines, 2) only at the guanines, 3) at the cytosines but not the thymines, and 4) at both the cytosines and thymines. The very shortest fragments are at the bottom right hand side of the picture and the sequence is read up the gel recognizing first the band in the left hand column corresponding to A, a band in the two! right hand columns corresponding to a G, a band in the far right hand column corresponding to T, a band in the left hand column corresponding to A and so forth. After reading up as far aspossible, the sequence continues in the sets of bands at the left hand side of the gel and then still further in the pattern in the center of the gel. From the original photograph the sequence of the entire fragment can be read. The fragment is from the genomic DNA corresponding to the variable region of the lambda light chain of mouse immunoglobulin (8).

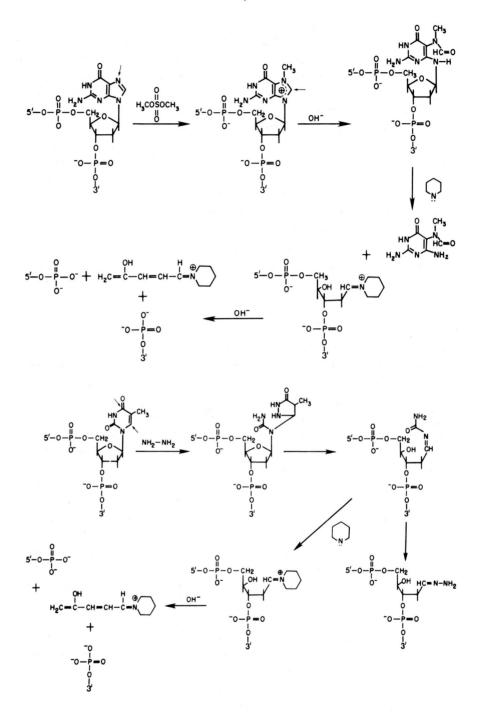

Figure 6. Examples of the detailed chemistry involved in breaking the DNA. Figure 6a, above, shows the guanine breakage. The guanines are first methylated with dimethylsulfate. The imidazole ring is opened by treatment with alkali (during the piperidine treatment). Piperidine displaces the base and then triggers two beta eliminations that release both phosphates from the sugar and cleave the DNA strand leaving a 3' and a 5' phosphate. Figure 6b, below, shows the hydrazine attack on a thymine that breaks the DNA at the pyrimidines.

the protein sequence contains 360 amino acids, he had to work out a gene of 1080 bases. DNA sequencing is faster and more accurate than protein sequencing. The reason for this is that DNA is a linear information store. Because the chemistry of each restriction fragment is like any other, they differ only in length, there is no particular reason for loosing track of them, except for the very smallest. By sequencing across the joins between the fragments, one established an unambiguous order. Proteins, on the other hand, are strings of amino acids used by Nature to create a wide variety of chemistries. When a protein is fragmented, the fragments can exhibit quite different properties, some of which may be unusually unfortunate in terms of solubility or loss. There is no simple way of keeping account of the total content of amino acids, or of the order of fragments, as there is for DNA, where the length of the restriction fragments can easily be measured.

Jeffrey Miller and his coworkers had done an extensive analysis of the appearance of mutations in the *lac* repressor gene. Three sites in the gene are hotspots, at which the mutation rate is some 10 times higher than at other sites. DNA sequencing showed that at each of these sites there was a modified base, a 5-methyl cytosine, in the sequence (13). (The chemical sequencing detects the presence of the 5-methyl cytosine directly, because the methyl group suppresses completely the reactivity of this base in the hydrazine reaction. A blank space appears in the sequence, but on the other strand is a guanine.) The high mutation rate is a transition to a thymine. 5-methyl cytosine occurs at a low frequency in DNA, this observation shows that it is a mutagen. What is the explaination? Deamination of cytosine to uracil occurs naturally. If this occurred in DNA it could lead to a transition; however it usually does not, since there is an enzyme that scans DNA examining it for deoxyuridine (14). When it finds this base in DNA, mismatched or not, it breaks the glycosidic bond and removes the uracil. This is then recognized as a defect in DNA, and another group of enzymes then repair the depyrimidinated spot. However, 5-methyl cytosine deaminates to thymine−a natural component of DNA. On repair or resynthesis a transition will ensue. This whole argument explains why thymine is used in DNA−the extra methyl group serves to suppress the effects of the natural rate of deamination.

To find out how easy and how accurate DNA sequencing was, I asked a student, Gregor Sutcliffe, to sequence the ampicillin resistance gene, the beta-lactamase gene, of *E. coli*. This gene is carried on a variety of plasmids, including a small constructed plasmid, pBR322, in *E. coli*. All that he knew about the protein was an approximate molecular weight, and that a certain restriction cut on the plasmid inactivated that gene. He had no previous experience with DNA sequencing when he set out to work out the structure of DNA for this gene. After seven months he had worked out about 1000 bases of double-stranded DNA, sequencing one strand and then sequencing the other for confirmation. The unique long reading frame determined the sequence of the protein product of this gene, a protein of 286 residues (15). We thought that the DNA sequence was unambiguous. Luckily there was available, from Ambler's laboratory, partial sequence information about the protein which had been obtained as a result of several years work attempting to develop a sequence for the beta-lactamase (16). This information, while not sufficient to determine the protein sequence directly, was adequate to confirm that the prediction of the DNA sequencing was correct. Sutcliffe then became very enthusiastic and sequenced the rest of the plasmid pBR322 during the next six months, to finish his thesis. He sequenced both strands of this 4362 base-pair long plasmid in order to confirm the sequence (17). The chemical sequencing is unambiguous, except for an occasional characteristic feature in the DNA fragment itself that causes it to move anomalously during the gel electrophoresis. As longer and longer strands are being analysed on the gel, a hairpin loop can form at one end of the fragment if the sequence is sufficiently self-complementary. As the fragmentation passes through this portion of the molecule, the mobilities on the gel do not decrease uniformly as a function of length, but some of the molecules move aberrantly, a feature called compression, because the bands on an autoradiograph become close together, or can overlap to conceal one or more bases. This rare feature occurs about once every thousand bases. It is resolved by sequencing the opposite strand in the other direction along the double stranded molecule (or the same strand in the opposite chemical direction) because the hairpin will form when a different region of the sequence is exposed and the compression feature will occur in a different place in the sequence. If both strands of the DNA helix are sequenced, the sequence can be unambiguous.

THE STRUCTURE OF GENES

The first genes to be sequenced, those in bacteria, yielded an expected structure: a contiguous series of codons lying upon the DNA between an initiation signal and one of the terminator signals. Before the position at which the RNA copy will start, there lies a site for the RNA polymerase, interacting with the Pribnow box, a region of sequence homology lying one turn of the helix before the initial base of the messenger RNA, and also with another region of homology, thirty-five bases before the start. Thus one could understand the bacterial gene in terms of a binding site for the RNA polymerase, and further binding sites for repressors and activator proteins around and under the polymerase. Alternatively, the control on transcription could be exercised by a control of the termination function: new proteins or an elegant translation control (18) could determine whether or not the polymerase would read past a stop signal into a new gene.

When the first genes from vertebrates were transferred into bacteria by the recombinant DNA techniques and sequenced an entirely different structure emerged. The coding sequences for globin (19,20), for immunoglobulin (21), and for ovalbumin (22) did not lie on the DNA as a continuous series of codons but rather were interrupted by long stretches of non-coding DNA. The discovery of RNA splicing in adenovirus by Sharp and his coworkers (23) and Broker and Roberts and their coworkers (24) paved the way for this new structure.

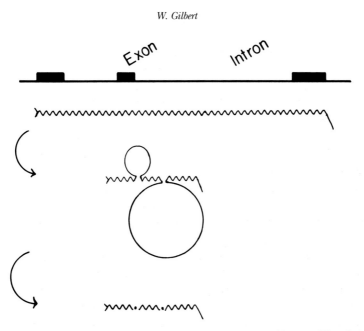

Figure 7. A transcription unit corresponding to alternating exons and introns. The whole gene, a transcription unit, is copied into RNA terminating in a poly (A) tail. The regions corresponding to introns are spliced out leaving a messenger RNA made up of the three exons, the regions that are expressed in the mature message.

They had shown that after the original transcription of DNA into a long RNA, regions of this RNA are spliced out: some stretches excised and the remaining portions fused together by an as yet undefined enzymatic process. The exons (25), regions of the DNA that will be expressed in mature message, are separated from each other by introns, regions of DNA that lie within the genetic element but whose transcripts will be spliced out of the message. Figure 7 shows this process: the original transcript of a gene (now thought of as a transcription unit) will undergo a series of splices before being able to function as a mature message in the cytoplasm. Figure 8 shows a few examples. Vertebrate genes can have many, eight, fifteen, even 50 exons (29,30), and the exons are for the most part short coding stretches separated by hundreds to several thousands of base pairs of intron DNA. The rapid sequencing has meant that we can work out the DNA sequence of any of these complex gene structures. But can we understand them?

The emerging generalization is that procaryotic genes have contiguous coding sequences while the genes for the highest eucaryotes are characterized by a complex exon-intron structure. As we move up from procaryotes, the simplest eucaryotes, such as yeasts, have few introns; further up the evolutionary ladder the genes are more broken up. (Yeast mitochondria have introns, are they an exception to this pattern?) Are we seeing the emergence of the intron-exon structure rising to ever greater degress of complexity as we move up to the vertebrates, or the loss of preexisting intron-exon structures as we move down to the simplest invertebrates and the procaryotes? One view considers the

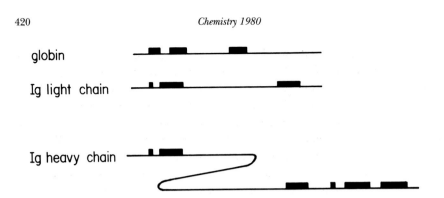

Figure 8. Examples of the intron-exon structure of a few genes. (1) The gene for globin is broken up by two introns into three exons (20). (2) The functional gene in a myeloma cell for the immunoglobulin lambda light chain is broken up into a short exon corresponding to the hydrophobic leader sequence, an exon corresponding to the V region, and then, after an intron of some thousand bases, an exon corresponding to the 112 amino acids of the constant region (26). (3) A typical gene for a gamma heavy chain of immunoglobulin (27, 28). The mature gene corresponds to a hydrophobic leader sequence, an exon corresponding to the variable region, and then, after a long intron, a series of exons; the first corresponding to the first domain of the constant region, the second corresponding to a 15 amino acid hinge region, the third corresponding to the second domain of the constant region, and the fourth exon corresponding to the third domain of the constant region.

splicing as an adaptation that becomes ever more necessary in more highly structured organisms. The other view considers the splicing as lost if the organism makes a choice to simplify and to replicate its DNA more rapidly, to go through more generations in a short time, and thus to be under a significant pressure to restrict its DNA content (31).

What role can this general intron-exon structure play in the genes of the higher organisms? Although most genes that have been studied have this structure, there are two notable exceptions: the genes for the histones and those for the interferons. This last demonstrates that there can be no absolutely essential role that the introns must play, there can be no absolute need for splicing in order to express a protein in mammalian cells. Although there is a line of experiments that shows that some messengers must have at least a single splice made before they can be expressed, there is no evidence that the great multiplicity of splices are needed. There is a pair of genes for insulin in the rat, that differ in the number of introns; both are expressed—which demonstrates that the intron that splits the coding region of one of them, has no essential role, in cis, in the expression of that gene (32). Although a common conjecture is that the splicing might have a regulatory role, so far there is no tissue dependent splicing pattern that could be interpreted as showing the existance of a gene (or tissue) specific splicing enzyme.

The introns are much longer than the exons. Their DNA sequence drifts rapidly by point mutation and small additions and deletions (accumulating changes as rapidly as possible, at the same rate as the silent changes in codons). This suggests that it is not their sequence that is relevant, but their length. Their function is to move the exons apart along the chromosome.

A consequence of the separation of exons by long introns is that the recombination frequency, both illegitimate and legitimate, between exons will be higher (25). This will increase the rate, over evolutionary time, at which the exons, representing parts of the protein structures, will be shuffled and reassorted to make new combinations. Consider the process by which a structural domain is duplicated to make the two domain structure of the light chain of the immunoglobulins (or duplicated again to make the four domain structure of the heavy chain, or combined to make the triple structure represented by ovomucoid (29)). Classically, this involved a precise unequal crossing over that fused the two copies of the original gene, in phase, to make a double length gene. As Figure 9 shows, this process involves an extremely rare, precise illegitimate event (a recombination event that leads to the fusing of two DNA sequences at a point where there is no matching of sequence) that has as its consequence the synthesis at a high level of the new, presumably more useful, double length gene product. Consider the same process against the background of a general splicing mechanism. Again, the process of forming the double gene must involve an illegitimate recombination event, but now that event can occur anywhere within a stretch of 1000 to 10,000 bases flanking the 3' side of one copy and the 5' side of the other to form an intron separating the genes for the two domains. From a long transcript across this region, even inefficient splicing may produce the new double-length gene product. This will happen some 10^6 to 10^8 times more rapidly than the classical process because of the many

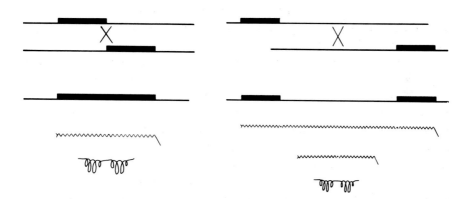

Figure 9. A double-length gene product arises through unequal crossing over. On the left, figure 9a, is the classical process by which a gene corresponding to a single polypeptide chain might have its length doubled by a crossing over. The top two lines indicate two coding regions, brought into accidental apposition by some act of illegitimate recombination which fuses the carboxy terminal region (the 3' end) of one copy of the gene to the amino terminal region (the 5' end) of the other. This rare illegitimate event (involving no sequence matching) would, if it occurs in phase, produce a double-length gene which could code for a double-length RNA which in turn translates into a double-length protein containing the reiteration of a basic domain. Figure 9b on the right illustrates the same process occurring in the presence of the splicing function. Now the unequal crossing over can occur anywhere to the 3' side of one copy of the gene and anywhere in front of the 5' end of the other copy of the gene to produce a gene containing two exons separated by long intron. I conjecture that the long transcript of this region now will be spliced at some low frequency to produce a mature message encoding the reiterated protein.

different combinations of sites at which the recombination can occur. If the long transcript can be spliced, even at a low frequency, some of the double-length product can be made. This is a faster way for evolution to form the final gene: proceeding through a rapid step to a structure that can produce a small amount of the useful gene product. Small mutational steps can be selected to produce better splicing signals and thus more of the gene product. If the splicing signals already exist, recombination within introns provides an immediate way to build polymeric structures out of simpler units. One would predict that polymeric structures, made up of simpler units, will be found to have genes in which the intron-exon structure of the primitive unit is repeated, separated again by introns. That is the case.

The rate of legitimate recombination between the exons of a gene will be increased by the introns. Consider two mutations to better functioning, arising in different parts of a gene and spreading, by selection, through the population. Classically, both mutations could end up in a single polypeptide chain, after both genes find their way into a single diploid individual, by homologous recombination within the gene. Figure 10 shows that this process also should be speeded some 10 to 100 fold by spreading the exons apart. This effect will be strongest if the exons can evolve separately—if they represent structures that can accumulate successful changes independantly.

Furthermore, one can change the pattern of exons by changing the initiation or termination of the RNA transcript, to add extra exons or to tie together exons from one region of the DNA to exons from another. This has been observed in adenovirus, and is found in notable examples in the immunoglobulins in which exons can be added or subtracted to the carboxy terminus of the heavy chain to modify the protein. Hood's laboratory has shown that this process is used to switch between two different forms of an IgM heavy chain (33). A membrane bound form is synthesized by a longer transcript, which splices on two additional exons and splices out part of the last exon of the shorter transcript. The shorter transcript synthesises a secreted form of the

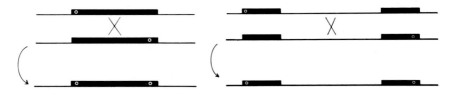

Figure 10. Introns speed legitimate recombination. Figure 10a, on the left, shows the classical pattern by which two mutations, one occurring in one copy of the gene, at the left end, and the other occurring in the other copy of the gene, at the right hand end, might get together by recombination happening in the homologous stretch of DNA that separates the two mutations. This recombination can create a single gene carrying both mutations. On the right, figure 10b, the same process happening in a gene in which the mutations occur in separate exons separated by an intron. Now the recombination can occur anywhere, either in the exon or within the intron, to produce a new gene carrying both mutations. Since the rate of recombination will be directly proportional to the distance along the DNA between the mutations, it will be faster.

protein. In a similar way the switch of the V region from IgM to an IgD constant region is probably the result of a different, still longer, transcript which splices across to attach the V region exons to the new constant region exons of the delta class. These combinations of genes have certainly been created by recombination events within the DNA that ultimately becomes the intron of the longer transcription unit.

The most striking prediction of this evolutionary view is that separate elements defined by the exons have some functional significance, that these elements have been assorted and put together in new combinations to make up the proteins that we know. Gene products are assembled out of prevously achieved solutions of the structure-function problem. Clear examples of this are still meager. The hydrophobic leader sequence which is involved in the transfer of proteins through membranes, and which is trimmed off after the secretion, is often on separate exons—most notably in the immunoglobulins (see Figure 8), but also in ovomucoid (29). In the pair of genes for insulin in the rat, a product of recent duplication (32), the two chains of insulin lie on separate in exons one gene, on a single exon in the other. The ancestral gene (the common structure in other species (34)) has the additional intron—suggesting that the gene was put together originally from separate pieces. The gene for lysozyme is broken up into four exons; the second one carries the critical amino acids of the active site and most of the substrate contacts (35). In the gene for globin the central exon encodes almost all of the heme contacts. Figure 11 shows a schematic disection of the molecule. A recent experiment (36) has shown that the poly-peptide that corresponds to the central exon in itself is a heme binding "minig-lobin"; the side exons have provided polypeptide material to stablize the protein.

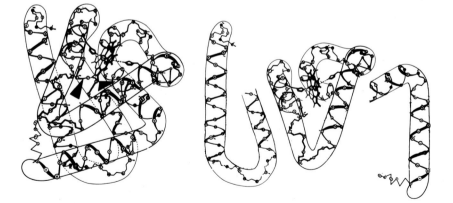

Figure 11. A schematic dissection of globin into the product of the separate exons. At the left the black arrows show the points at which the structure of a chain of globin is interrupted by the introns. (The structures of the chains of globin and of myoglobin are very similar, the schematic structure shown is myoglobin.) The introns interrupt the protein in the alpha-helical regions to break the protein into three portions shown on the right. The product of the central exon surrounds the heme; the products of the other two exons, I conjecture, wrap around and stablize the protein.

At the same moment that the rapid sequencing methods and the molecular cloning gave us the promise of being able to work out the structure of any gene, the ability to achieve a complete understanding of the genetic material, Nature revealed herself to be more complex than we had imagined. We can not read the gene product directly from the chromosome by DNA sequencing alone. We must appeal to the sequence of the actual protein, or at least the sequence of the mature messenger RNA, to learn the intron-exon structure of the gene. Nonetheless the hope exists, that as we look down on the sequence of DNA in the chromosome, we will not learn simply the primary structure of the gene products, but we will learn aspects of the functional structure of the proteins — put together over evolutionary time as exons linked through introns.

My interest in biology has always centered on two problems: how is the genetic information made manifest? and how is it controlled? We have learned much about the way in which a gene is translated into protein. The control of genes in prokaryotes is well understood, but for eukaryotes the critical mechanisms of control are still not known. The purpose of research is to explore the unknown. The desire for new knowledge calls forth the answers to new questions.

I owe a great debt to my students and collaborators over the years; the greatest to Jim Watson who stimulated my interest in molecular biology, to Benno Müller-Hill with whom I worked on the *lac* repressor, and to Allan Maxam with whom I developed the DNA sequencing.

BIBLIOGRAPHY

1 Gilbert, Walter and Müller-Hill, Benno "Isolation of the *Lac* Repressor" *Proc. Natl. Acad. Sci. USA 56*, 1891–1898 (1966).

2 Gilbert, Walter and Müller-Hill, Benno "The *Lac* Operator is DNA" *Proc. Natl. Acad. Sci. USA 58*, 2415–2421 (1967).

3 Gilbert, Walter and Maxam, Allan "The Nucleotide Sequence of the *lac* Operator" *Proc. Natl. Acad. Sci. USA 70*, 3581–3584 (1973).

4 Gilbert, W., Maizels, N. and Maxam, A. "Sequences of Controlling Regions of the Lactose Operon" Cold Spring Harbor Symposium on Quantitative Biology *38*, 845–855 (1973).

5 Dickson, R., Abelson, J., Barnes, W. and Reznikoff, W. "Genetic Regulation: the *Lac* Control Region" *Science 187*, 27–35 (1975).

6 Gilbert, Walter, Maxam, Allan and Mirzabekov, Andrei "Contacts Between the *lac* Repressor and DNA Revealed by Methylation" in Control of Ribosome Synthesis", 139–148, Alfred Benzon Symposium IX, Munksgaard 1976.

7 Maxam, Allan M. and Gilbert, Walter "A New Method for Sequencing DNA" *Proc. Natl. Acad. Sci. USA 74*, 560–564 (1977).

8 Tonegawa, S., Maxam, A. M., Tizard, R., Bernard, O. and Gilbert, W. "Sequence of a Mouse Germ-line Gene for a Variable Region of an Immunoglobulin Light Chain" *Proc. Natl. Acad. Sci. USA 75*, 1485–1489 (1978).

9 Maxam, Allan, M., and Gilbert, Walter "Sequencing End-Labelled DNA with Base-Specific Chemical Cleavages" *Methods In Enzymology 65*, 499–560 (editors K. Moldave and L. Grossman) (1980).

10 Sanger, F. and Coulson, A. R. "The Use of Thin Acrylamide Gels for DNA Sequencing" *FEBS Letters 87*, 107–110 (1978).

11 Farabaugh, Philip J. "Sequence of the *lacI* Gene" *Nature 274*, 765–769 (1978).

12 Beyreuther, K., Adler, K., Fanning, E., Murray, C., Klemm, A. and Geisler, N. "Amino-Acid Sequence of *lac* Repressor from *Escherichia coli*" *Eur. J. Biochem. 59*, 491–509 (1975).

13 Coulondre, Christine, Miller, Jeffrey H., Farabaugh, Philp J. and Gilbert, Walter "Molecular Basis of Base Substitution Hotspots in *Escherichia coli*" *Nature 274*, 775–780 (1978).

14 Lindahl, T., Ljungquist, S., Siegert, W., Nyberg, B. and Sperens, B. "DNA *N*-Glycosidases" *J. Biol. Chem. 252*, 3286–3294 (1977).

15 Sutcliffe, J. Gregor "Nucleotide Sequence of the Ampicillin Resistance Gene of *Escherichia coli* Plasmid pBR322" *Proc. Natl. Acad. Sci. USA 75*, 3737–3741 (1978).

16 Ambler, R. P. and Scott, G. K. "The Partial Amino Acid Sequence of the Penicillinase Coded by the *Escherichia coli* Plasmid R6K" *Proc. Natl. Acad. Sci. USA 75*, 3732–3736 (1978).

17 Sutcliffe, J. G. "Complete Nucleotide Sequence of the *Escherichia coli* Plasmid pBR322" Cold Spring Harbor Symposium *43*, 77–90 (1978).

18 For a review, see: Yanofsky, C. "Attenuation in the Control of Expression of Bacterial Operons" *Nature 289*, 751–758 (1981).

19 Tilghman, S. M., Tiemeister, D. C., Seidman, J. G., Peterlin, B. M., Sullivan, M., Maizel, J. V. and Leder, P. "Intervening Sequence of DNA Identified in the Structural Portion of a Mouse Beta-Globin Gene" *Proc. Natl. Sci. USA 75*, 725–729 (1978).

20 Konkel, D. A., Tilghman, S. M. and Leder, P. "The Sequence of the Chromosomal Mouse Beta-Globin Major Gene" *Cell 15*, 1125–1132.

21 Brack, C. and Tonegawa, S. "Variable and Constant Parts of the Immunoglobulin Light Chain of a Mouse Myeloma Cell are 1250 Nontranslated Bases Apart" *Proc. Natl. Acad. Sci. USA 74*, 5652–5656 (1977).

22 Breathnach, R., Mandel, J. L. and Chambon, P. "Ovalbumin Gene is Split in Chicken DNA" *Nature 270*, 314–319 (1977).

23 Berget, Susan M., Moore, Claire and Sharp, Phillip A. "Spliced Segments at the 5' Terminus of Adenovirus 2 Late mRNA" *Proc. Natl. Acad. Sci. USA 74*, 3171–3175 (1977).

24 Chow, L. T., Gelinas, R. E., Broker, T. R. and Roberts, R. J. "An Amazing Sequence Arrangement at the 5' Ends of Adenovirus 2 Messenger RNA" *Cell 12*, 1–8 (1977).

25 Gilbert, Walter "Why Genes in Pieces?" *Nature 271*, 501 (1978).

26 Bernard, O., Hozumi, N. and Tonegawa, S. "Sequence of Mouse Immunoglobulin Light Chain Genes Before and After Somatic Changes" *Cell 15*, 1133–1144 (1978).

27 Sakano, H., Rogers, J. H., Hüppi, K., Brack, C., Traunecker, A., Maki, R., Wall, R. and Tonegawa, S. "Domains and the Hinge Region of an Immunoglobulin Heavy Chain Are Encoded in Separate DNA Segments" *Nature 277*, 627–633 (1979).

28 Honjo, T., Obata, M., Yanawaki-Kataoka, Y., Kataoka, T., Kawakami, T., Takahashi, N. and Mano, Y. "Cloning and Complete Nucleotide Sequences of Mouse Gamma 1 Chain Gene" *Cell 18*, 559–568 (1979).

29 Stein, J. P., Catterall, J. F., Kristo, P., Means, A. R. and O'Malley, B. W. "Ovomucoid Intervening Sequences Specify Functional Domains and Generate Protein Polymorphisms" *Cell 21*, 681–687 (1980).

30 Yamado, Y., Avvedimento, V. E., Mudryj, M., Ohkubo, H., Vogeli, G., Irani, M., Pastan, I., and de Crombrugghe, B. "The Collagen Gene: Evidence for its Evolutionary Assembly by Amplification of a DNA segment Containing an Exon of 54 bp." *Cell 22*, 887–892 (1980).

31 Doolittle, W. F. "Genes in Pieces: Were They Ever Together?" *Nature 272*, 581 (1978).

32 Lomedico, P., Rosenthal, N., Efstratiadis, A., Gilbert, W., Kolodner, R., and Tizard, R. "The Structure and Evolution of the Two Non-Allelic Rat Preproinsulin Genes" *Cell 18*, 545–558 (1979).

33 Early, P., Rogers, F., Davis, M., Calami, K., Bond, M., Wall, R. and Hood, L. "Two mRNA's Can be Produced from a Single Immunoglobulin Gene by Alternative RNA Processing Pathways" *Cell 20*, 313–319 (1980).

34 Perler, F., Efstratiadis, A., Lomedico, P., Gilbert, W., Kolodner, R. and Dodgson, J. "The Evolution of Genes: The Chicken Preproinsulin Gene" *Cell 20*, 555–566 (1980).

35 Jung, Alexander, Sippel, Albrecht, E., Grez, Manuel and Schütz, Günther "Exons Encode Functional and Structural Units of Chicken Lysozyme *Proc. Natl. Acad. Sci. USA 77*, 5759–5763 (1980).

36 Craik, Charles, S., Buchman, Steven R. and Beychok, Sherman "Characterization of Globin Domains: Heme Binding to the Central Exon Product" *Proc. Natl. Acad. Sci. USA 77*, 1384–1388 (1980).

FREDERICK SANGER

I was born on 13th August 1918 in the village of Rendcombe in Gloucestershire, where my father, also Frederick Sanger, was a medical practitioner. Influenced by him, and probably even more so by my brother Theodore (a year older than me), I soon became interested in biology and developed a respect for the importance of science and the scientific method. At Bryanston School and St John's College, Cambridge, I was probably above average but not an outstanding scholar. Initially I had intended to study medicine, but before going to University I had decided that I would be better suited to a career in which I could concentrate my activities and interests more on a single goal than appeared to be possible in my father's profession. So I decided to study science and, on arrival at Cambridge, became extremely excited and interested in biochemistry when I first heard about it, principally through Ernest Baldwin and also other members of the relatively young and enthusiastic Biochemistry Department that had been founded by F. G. Hopkins. It seemed to me that here was a way to really understand living matter and to develop a more scientific basis to many medical problems.

After taking my B. A. degree in 1939 I remained at the University for a further year to take an advanced course in Biochemistry, and surprised myself and my teachers by obtaining a first class examination result. I was a conscientious objector during the war and was allowed to study for a Ph. D. degree, which I did in the Biochemistry Department with A. Neuberger, on lysine metabolism and a more practical problem concerning the nitrogen of potatoes. It was Neuberger who first taught me how to do research, both technically and as a way of life, and I owe much to him. In 1943 A. C. Chibnall succeeded F. G. Hopkins as Professor of Biochemistry at Cambridge and I joined his research group working on proteins and, in particular, insulin. This was an especially exciting time in protein chemistry. New fractionation techniques had been developed, particularly by A. J. P. Martin and his colleagues, and there seemed to be a real possibility of determining the exact chemical structure of these fundamental components of living matter. I succeeded in developing new methods for amino acid sequencing and used them to deduce the complete sequence of insulin, for which I was awarded the Nobel Prize for Chemistry in 1958. This award had an important and stimulating effect on my subsequent career. I had remained in Cambridge concentrating only on basic research and avoiding as far as possible teaching or administrative responsibilities. This recognition of my work gave me renewed confidence and enthusiasm to continue in this way of life, which I enjoyed. It also enabled me to obtain better research facilities and, even more important, to attract excellent colleagues.

Until 1943 I received no stipend. I was able to support myself as my mother was the daughter of a relatively wealthy cotton manufacturer. From 1944 to 1951 I held a Beit Memorial Fellowship for Medical Research, and since 1951 I have been on the staff of the Medical Research Council. In 1962 I moved to their newly built Laboratory of Molecular Biology in Cambridge, together with M. F. Perutz's unit from the Cavendish Laboratory which included F. H. C. Crick, J. C. Kendrew, H. E. Huxley and A. Klug. In this atmosphere I soon became interested in nucleic acids. Although at the time it seemed to be a major change from proteins to nucleic acids, the concern with the basic problem of "sequencing" remained the same. And indeed this theme has been at the centre of all my research since 1943, both because of its intrinsic fascination and my conviction that a knowledge of sequences could contribute much to our understanding of living matter. My work on nucleic acids is summarised in my Nobel lecture. This work has not been done single-handed and it owes much to the excellent collaborators I have had. Most of these have been students and postdoctoral fellows spending a few years in the laboratory and bringing their experience and ideas with them, but I feel particularly indebted to my more permanent colleagues, B. G. Barrell, A. R. Coulson and G. G. Brownlee, who have contributed so much to the methods we have developed.

I was married to Margaret Joan Howe in 1940. Although not a scientist herself she has contributed more to my work than anyone else by providing a peaceful and happy home. We have two sons, Robin and Peter, born in 1943 and 1946, and a daughter, Sally Joan, born in 1960. Apart from my work my main interests are gardening and what can best be described as "messing about in boats".

DETERMINATION OF NUCLEOTIDE SEQUENCES IN DNA

Nobel lecture, 8 December 1980

by

FREDERICK SANGER

Medical Research Council Laboratory of Molecular Biology,
Cambridge, England

INTRODUCTION

In spite of the important role played by DNA sequences in living matter, it is only relatively recently that general methods for their determination have been developed. This is mainly because of the very large size of DNA molecules, the smallest being those of the simple bacteriophages such as φX174 (which contains about 5,000 nucleotides). It was therefore difficult to develop methods with such complicated systems. There are however some relatively small RNA molecules − notably the transfer RNAs of about 75 nucleotides, and these were used for the early studies on nucleic acid sequences (1).

Following my work on amino acid sequences in proteins (2) I turned my attention to RNA and, with G.G. Brownlee and B.G. Barrell, developed a relatively rapid small-scale method for the fractionation of ^{32}P-labelled oligonucleotides (3). This became the basis for most subsequent studies of RNA sequences. The general approach used in these studies, and in those on proteins, depended on the principle of partial degradation. The large molecules were broken down, usually by suitable enzymes, to give smaller products which were then separated from each other and their sequence determined. When sufficient results had been obtained they were fitted together by a process of deduction to give the complete sequence. This approach was necessarily rather slow and tedious, often involving successive digestions and fractionations, and it was not easy to apply it to the larger DNA molecules. When we first studied DNA some significant sequences of about 50 nucleotides in length were obtained with this method (4,5), but it seemed that to be able to sequence genetic material a new approach was desirable and we turned our attention to the use of copying procedures.

Abbreviations The abbreviations C, A and G are used to describe both the ribonucleotides and the deoxyribonucleotides, according to context.

COPYING PROCEDURES

In the RNA field these procedures had been pioneered by C. Weissmann and his colleagues (6) in their studies on the RNA sequence of the bacteriophage Qβ. Qβ contains a replicase that will synthesize a complementary copy of the single-stranded RNA chain, starting from its 3′ end. These workers devised elegant procedures involving pulse-labelling with radioactively labelled nucleotides, from which sequences could be deduced.

For DNA sequences we have used the enzyme DNA polymerase, which copies single-stranded DNA as shown in Fig. 1. The enzyme requires a primer, which is a single-stranded oligonucleotide having a sequence that is complementary to, and therefore able to hybridize with, a region on the DNA being sequenced (the template). Mononucleotide residues are added sequentially to the 3′ end of the primer from the corresponding deoxynucleoside triphosphates, making a complementary copy of the template DNA. By using triphosphates containing ^{32}P in the α position the newly synthesized DNA can be labelled. In the early experiments synthetic oligonucleotides were used as primers, but after

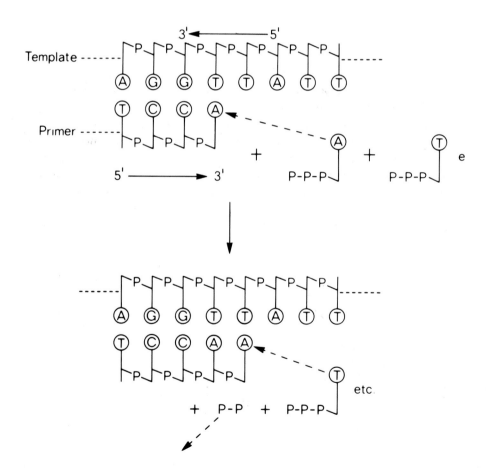

Fig. 1. Specificity requirements for DNA polymerase.

the discovery of restriction enzymes it was more convenient to use fragments resulting from their action as they were much more easily obtained.

The copying procedure was used initially to prepare a short specific region of labelled DNA which could then be subjected to partial digestion procedures. One of the difficulties of sequencing DNA was to find specific methods for breaking it down into small fragments. No suitable enzymes were known that would recognise only one nucelotide. However, Berg, Fancher & Chamberlin (7) had shown earlier that under certain conditions it was possible to incorporate ribonucleotides, in place of the normal deoxyribonucleotides, into DNA chains with DNA polymerase. Thus, for instance, if copying were carried out using riboCTP and the other three deoxynucleoside triphosphates, a chain could be built up in which the C residues were in the ribo form. Bonds involving ribonucleotides could be broken by alkali under conditions where those involving the deoxynucleotides were not, so that a specific splitting at C residues could be obtained. Using this method we were able to extend our sequencing studies to some extent (8). However extensive fractionations and analyses were still required.

THE 'PLUS AND MINUS' METHOD

In the course of these experiments we needed to prepare DNA copies of high specific radioactivity, and in order to do this the highly labelled substrates had to be present in low concentrations. Thus if $\alpha[^{32}P]$-dATP was used for labelling its concentration was much lower than that of the other three triphosphates and frequently when we analysed the newly synthesized DNA chains we found that they terminated at a position immediately before that at which an A should have been incorporated. Consequently a mixture of products was produced all having the same 5' end (the 5' end of the primer) and terminating at the 3' end at the position of the A residues. If these products could be fractionated on a system that separated only on the basis of chain length, the pattern of their distribution on fractionation would be proportional to the distribution of the A's along the DNA chain. And this, together with the distribution of the other three mononucleotides, is the information required for sequence determination. Initial experiments carried out with J.E. Donelson suggested that this approach could be the basis for a more rapid method, and it was found that good fractionations according to size could be obtained by ionophoresis on acrylamide gels.

The method described above met with only limited success but we were able to develop two modified techniques that depended on the same general principle and these provided a much more rapid and simpler method of DNA sequence determination than anything we had used before (9). This, which is known as the "plus and minus" technique, was used to determine the almost complete sequence of the DNA of bacteriophage φX174 which contains 5,386 nucleotides (10).

THE 'DIDEOXY' METHOD

More recently we have developed another similar method which uses specific
chain-terminating analogues of the normal deoxynucleoside triphosphates
(11). This method is both quicker and more accurate than the plus and minus
technique. It was used to complete the sequence of φX174 (12), to determine
the sequence of a related bacteriophage, G4 (13), and has now been applied to
mammalian mitochondrial DNA.

The analogues most widely used are the dideoxynucleoside triphosphates
(Fig. 2). They are the same as the normal deoxynucleoside triphosphates but
lack the 3′ hydroxyl group. They can be incorporated into a growing DNA
chain by DNA polymerase but act as terminators because, once they are
incorporated, the chain contains no 3′ hydroxyl group and so no other nucelotide can be added.

The principle of the method is summarised in Fig. 3. Primer and template
are denatured to separate the two strands of the primer, which is usually a
restriction enzyme fragment, and then annealed to form the primer-template
complex. The mixture is then divided into four samples. One (the T sample) is
incubated with DNA polymerase in the presence of a mixture of ddTTP
(dideoxy thymidine triphosphate) and a low concentration of TTP, together
with the other three deoxynucleoside triphosphates (one of which is labelled

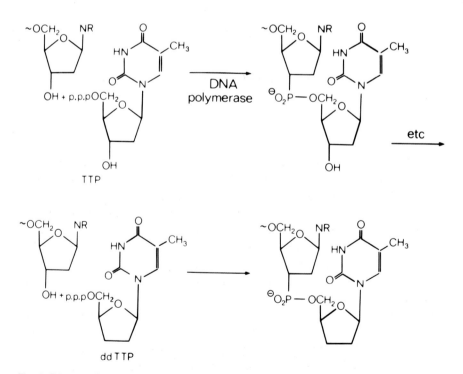

Fig. 2. Diagram showing chain termination with dideoxythymidine triphosphate (ddTTP). The
top line shows the DNA polymerase-catalysed reaction of the normal deoxynucleoside triphosphate
(TTP) with the 3′ terminal nucleotide of the primer: the bottom line the corresponding reaction
with ddTTP.

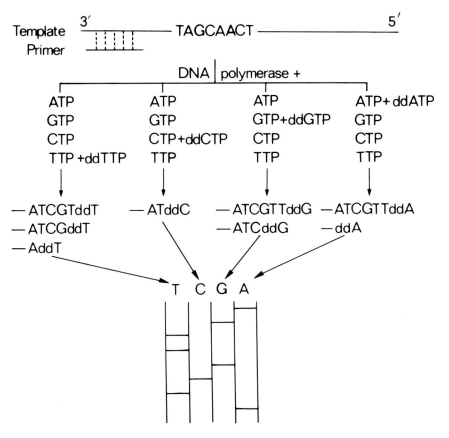

Fig. 3. Principle of the chain-terminating method.

with ^{32}P) at normal concentration. As the DNA chains are built up on the 3' end of the primer the position of the T's will be filled, in most cases by the normal substrate T and extended further, but occasionally by ddT and terminated. Thus at the end of incubation there remains a mixture of chains terminating with T at their 3' end but all having the same 5' end (the 5' end of the primer). Similar incubations are carried out in the presence of each of the other three dideoxy derivatives, giving mixtures terminating at the positions of C, A and G respectively, and the four mixtures are fractionated in parallel by electrophoresis on acrylamide gel under denaturing conditions. This system separates the chains according to size, the small ones moving quickly and the large ones slowly. As all the chains in the T mixture end at T the relative position of the T's in the chain will define the relative sizes of the chains, and therefore their relative positions on the gel after fractionation. The actual sequence can then simply be read off from an autoradiograph of the gel (Fig. 4). The method is comparatively rapid and accurate and sequences of up to about 300 nucleotides from the 3' end of the primer can usually be determined.

2·5 hr   5 hr
G A T C   G A T C

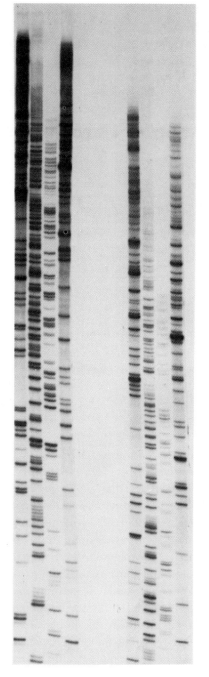

Fig. 4. Autoradiograph of a DNA sequencing gel. The origin is at the top and migration of the DNA chains, according to size, is downwards. The gel on the left has been run for 2 · 5 hr and that on the right for 5 hr with the same polymerisation mixtures.

Considerably longer sequences have been read off but these are usually less reliable.

One problem with the method is that it requires single-stranded DNA as template. This is the natural form of the DNA in the bacteriophages φX174 and G4, but most DNA is double-stranded and it is frequently difficult to separate the two strands. One way of overcoming this was devised by A.J.H. Smith (14). If the double-stranded linear DNA is treated with exonuclease III (a double-strand specific 3′ exonuclease) each chain is degraded from its 3′ end, as shown in Fig. 5, giving rise to a structure that is largely single-stranded and can be used as template for DNA polymerase with suitable small primers. This method is particularly suitable for use with fragments cloned in plasmid vectors and has been used extensively in the work on human mitochondrial DNA.

CLONING IN SINGLE-STRANDED BACTERIOPHAGE

Another method of preparing suitable template DNA that is being more widely used is to clone fragments in a single-stranded bacteriophage vector (15–17). This approach is summarised in Fig. 6. Various vectors have been described. We have used a derivative of bacteriophage M13 developed by Gronenborn & Messing (16) which contains an insert of the β-galactosidase gene with an *Eco*RI restriction enzyme site in it. The presence of β-galactosidase in a plaque can be readily detected by using a suitable colour-forming substrate (X-gal). The presence of an insert in the *Eco*RI site destroys the β-galactosidase gene, giving rise to a colourless plaque.

Besides being a simple and general method of preparing single-stranded DNA this approach has other advantages. One is that it is possible to use the same primer on all clones. Heidecker *et al.* (18) prepared a 96-nucleotide long restriction fragment derived from a position in the M13 vector flanking the *Eco*RI site (see Fig. 6). This can be used to prime into, and thus determine, a sequence of about 200 nucleotides in the inserted DNA. Smaller synthetic primers have now been prepared (19,20) which allow longer sequences to be determined. The approach that we have used is to prepare clones at random from restriction enzyme digests and determine the sequence with the flanking primer. Computer programmes (21) are then used to store, overlap and arrange the data.

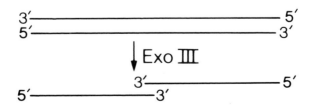

Fig. 5. Degradation of double-stranded DNA with exonuclease III.

Another important advantage of the cloning technique is that it is a very efficient and rapid method of fractionating fragments of DNA. In all sequencing techniques, both for proteins and nucleic acids, fractionation has been an important step and major progress has usually been dependent on the development of new fractionation methods. With the new rapid methods for DNA sequencing fractionation is still important and as the sequencing procedure itself is becoming more rapid more of the work has involved fractionation of the restriction enzyme fragments by electrophoresis on acrylamide. This becomes increasingly difficult as larger DNA molecules are studied and may involve several successive fractionations before pure fragments are obtained. In the

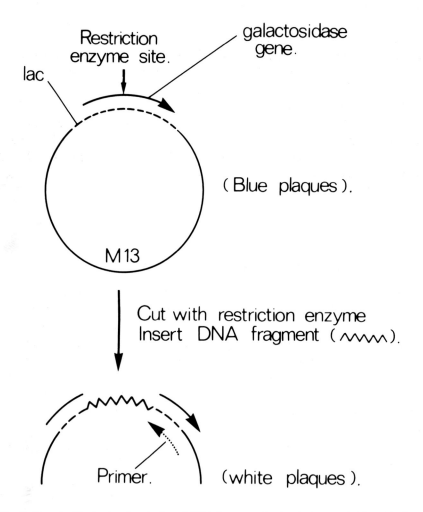

Fig. 6. Diagram illustrating the cloning of DNA fragments in the single-stranded bacteriophage vectorM13mp2 (16) and sequencing the insert with a flanking primer.

new method these fractionations are replaced by a cloning procedure. The mixture is spread on an agar plate and grown. Each clone represents the progeny of a single molecule and is therefore pure, irrespective of how complex the original mixture was. It is particulary suitable for studying large DNAs. In fact there is no theoretical limit to the size of DNA that could be sequenced by this method.

We have applied the method to fragments from mitochondrial DNA (22,23) and to bacteriophage λ DNA. Initially new data can be accumulated very quickly (under ideal conditions at about 500−1,000 nucleotides a day). However at later stages much of the data produced will be in regions that have already been sequenced, and progress then appears to be much slower. Nevertheless we find that most new clones studied give some useful data, either for correcting or confirming old sequences. Thus in the work with bacteriophage λ DNA we have about 90 % of the molecule identified in sequences and most of the new clones we study contribute some new information. In most studies on DNA one is concerned with identifying the reading frames for protein genes, and to do this the sequence must be correct. Errors can readily occur in such extensive sequences and confirmation is always necessary. We usually consider it necessary to determine the sequence of each region on both strands of the DNA.

Although in theory it would be possible to complete a sequence determination solely by the random approach, it is probably better to use a more specific method to determine the final remaining nucleotides in a sequencing study. Various methods are possible (22,24), but all are slow compared with the random cloning approach.

BACTERIOPHAGE φX174 DNA

The first DNA to be completely sequenced by the copying procedures was from bacteriophage φX174 (10,12)−a single-stranded circular DNA, 5,386 nucleotides long, which codes for ten genes. The most unexpected finding from this work was the presence of 'overlapping' genes. Previous genetic studies had suggested that genes were arranged in a linear order along the DNA chains, each gene being encoded by a unique region of the DNA. The sequencing studies indicated however that there were regions of the φX DNA that were coding for two genes. This is made possible by the nature of the genetic code. Since a sequence of three nucleotides (a codon) codes for one amino acid, each region of DNA can theoretically code for three different amino acid sequences, depending on where translation starts. This is illustrated in Fig. 7. The reading frame or phase in which translation takes place is defined by the position of the initiating ATG codon, following which nucleotides are simply read off three at a time. In φX there is an initiating ATG within the gene coding for the D protein, but in a different reading frame. Consequently this initiates an entirely

```
    Gly      His      Ala      Cys      Arg      Arg
  G G C . C A C . G C A . T G C . A G G . C G G
```

```
      Ala      Thr      His      Ala      Gly
  G . G C C . A C G . C A T . G C A . G G C . G G
```

```
        Pro      Arg      Met      Gln      Ala
  G G . C C A . C G C . A T G . C A G . G C G . G
```

Fig. 7. Diagram illustrating how one DNA sequence can code for three different amino acid sequences. The dots indicate the positions of triplet codons coding for the amino acids.

different sequence of amino acids, which is that of the E protein. Fig. 8 shows the genetic map of φX. The E gene is completely contained within the D gene and the B gene within the A.

Further studies (25) on the related bacteriophage, G4, revealed the presence of a previously unidentified gene, which was called K. This overlaps both the A and C genes, and there is a sequence of four nucleotides that codes for part of all three proteins, A, C and K, using all of its three possible reading frames.

It is uncertain whether overlapping genes are a general phenomenon or whether they are confined to viruses, whose survival may depend on their rate of replication and therefore on the size of the DNA: with the overlapping genes more genetic information can be concentrated in a given sized DNA.

Further details of the sequence of bacteriophage φX174 DNA have been published elsewhere (10,12).

MAMMALIAN MITOCHONDRIAL DNA

Mitochondria contain a small double-stranded DNA (mtDNA) which codes for two ribosomal RNAs (rRNAs), 22—23 transfer RNAs (tRNAs) and about 10—13 proteins which appear to be components of the inner mitochondrial membrane and are somewhat hydrophobic. Other proteins of the mitochondria are encoded by the nucleus of the cell and specifically transported to the mitochondria. Using the above methods we have determined the nucleotide sequence of human mtDNA (23) and almost the complete sequence of bovine mtDNA. The sequence revealed a number of unexpected features which indicated that the transcription and translation machinery of mitochondria is rather different from that of other biological systems.

The genetic code in mitochondria

Hitherto it has been believed that the genetic code was universal. No differences were found in the *E. coli*, yeast or mammalian systems that had been studied. Our initial sequence studies were on human mtDNA. No amino acid sequence of the proteins that were encoded by human mtDNA were known. However Steffans & Buse (26) had determined the sequence of subunit II of cytochrome oxidase (COII) from bovine mitochondria, and Barrell, Bankier & Drouin (27) found that a region of the human mtDNA that they were studying had a sequence that would code for a protein homologous to this amino acid sequence – indicating that it most probably was the gene for the human COII. Surprisingly the DNA sequence contained TGA triplets in the reading frame of the homologous protein. According to the normal genetic code TGA is a termination codon and if it occurs in the reading frame of a protein the polypetide chain is terminated at that position. It was noted that in the positions where TGA occured in the human mtDNA sequence, tryptophan was found in the bovine protein sequence. The only possible conclusion seemed to be that in mitochondria TGA was not a termination codon but was coding for tryptophan. It was similarly concluded that ATA, which normally codes for isoleucine, was coding for methionine. As these studies were based on a

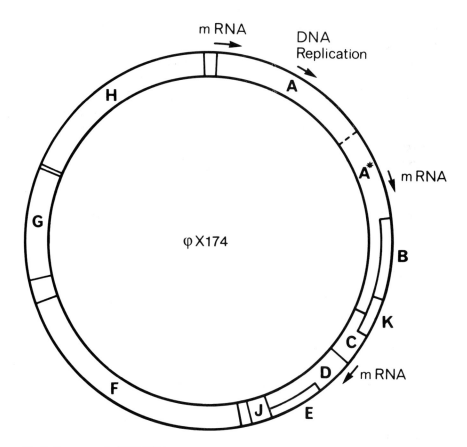

Fig. 8. Gene map of φX174 DNA.

comparison of a human DNA with a bovine protein, the possibility that the differences were due to some species variation, although unlikely, could not be completely excluded. For a conclusive determination of the mitochondrial code it was necessary to compare the DNA sequence of a gene with the amino acid sequence of the protein it was coding for. This was done by Young & Anderson (28) who isolated the bovine mtDNA, determined the sequence of its COII gene and confirmed the above differences.

Fig. 9 shows the human and bovine mitochondrial genetic code and the frequency of use of the different codons in human mitochondria. All codons are used with the exception of UUA and UAG, which are terminators, and AGA and AGG, which normally code for arginine. This suggests that AGA and AG are probably also termination codons in mitochondria. Further support for this is that no tRNA recognizing the codons has been found (see below) and that some of the unidentified reading frames found in the DNA sequence appear to end with these codons.

In parallel with our studies on mammalian mtDNA, Tzagoloff and his colleagues (29,30) were studying yeast mtDNA. They also found changes in the genetic code, but surprisingly they are not all the same as those found in mammalian mitochondria. These differences are summarised in Table 1.

SECOND LETTER

	U		C		A		G		
U	UUU UUC	Phe 77 140	UCU UCC	Ser 32 99	UAU UAC	Tyr 46 89	UGU UGC	Cys 5 17	U C
	UUA UUG	Leu 73 17	UCA UCG	83 7	UAA UAG	Ter – –	UGA UGG	Trp 93 10	A G
C	CUU CUC	Leu 65 167	CCU CCC	Pro 41 119	CAU CAC	His 18 79	CGU CGC	Arg 7 25	U C
	CUA CUG	276 45	CCA CCG	52 7	CAA CAG	Gln 81 9	CGA CGG	29 2	A G
A	AUU AUC	Ile 125 196	ACU ACC	Thr 51 155	AAU AAC	Asn 33 130	AGU AGC	Ser 14 39	U C
	AUA AUG	Met 166 40	ACA ACG	133 10	AAA AAG	Lys 85 10	AGA AGG	Ter – –	A G
G	GUU GUC	Val 30 49	GCU GCC	Ala 43 124	GAU GAC	Asp 15 55	GGU GGC	Gly 24 88	U C
	GUA GUG	71 18	GCA GCG	80 8	GAA GAG	Glu 64 24	GGA GGG	67 34	A G

FIRST LETTER (left margin) — THIRD LETTER (right margin)

Fig. 9. The human mitochondrial genetic code, showing the coding properties of the tRNAs (boxed codons) and the total number of codons used in the whole genome shown in Fig. 10. (One methionine tRNA has been detected, but as there is some uncertainty about the number present and their coding properties, these codons are not boxed.)

Table 1. Coding changes in mitochondria

| Codon | Amino acid coded | | |
	Normal	Mammalian mitochondria	Yeast mitochondria
UGA	Term	Trp	Trp
AUA	Ile	Met	Ile
CUN	Leu	Leu	Thr
AGA, AGG	Arg	Term?	Arg
CGN	Arg	Arg	Term?

Transfer RNAs

Transfer RNAs have a characteristic base-pairing structure which can be drawn in the form of the 'cloverleaf' model. By examining the DNA sequence for cloverleaf structures and using a computer programme (31), it was possible to identify genes coding for the mt-tRNAs.

Besides the cloverleaf structure, normal cytoplasmic tRNAs have a number of so-called 'invariable' features which are believed to be important to their biological function. Most of the mammalian mt-tRNAs are anomalous in that some of these invariable features are missing. The most bizarre is one of the serine tRNA in which a complete loop of the cloverleaf structure is missing (32,33), Nevertheless it functions as a tRNA.

In normal cytoplasmic systems at least 32 tRNAs are required to code for all the amino acids. This is related to the 'wobble' effect. Codon-anticodon relationships in the first and second positions of the codons are defined by the normal base-pairing rules, but in the third position G can pair with U. The result of this is that one tRNA can recognise two codons. There are many cases in the genetic code where all four codons starting with the same two nucleotides code for the same amino acid. These are known as 'family boxes'. The situation for the alanine family box is shown in Table 2, indicating that with the normal wobble system two tRNAs are required to code for the four alanine codons.

Table 2. Coding properties of alanine tRNAs

Codon	Anticodon (wobble)	Anticodon (mitochondria)
GCU		
	GGC	
GCC		
		UGC
CA	UGC	
GCG		

Note that the first position of the codon pairs with the third position of the anticodon and vice versa; e. g.

$^{5'}$ G C U $^{3'}$ (codon)

$_{3'}$ C G G $_{5'}$ (anticodon)

Only 22 tRNA genes could be found in mammalian mtDNA, and for all the family boxes there was only one, which had a T in the position corresponding to the third position of the codon (34). It seems very unlikely that none of the other predicated tRNAs would have been detected and the most feasible explanation is that in mitochondria one tRNA can recognise all four codons in a family box and that a U in the first position of the anticodon can pair with U, C, A or G in the third position of the codon. Clearly in boxes in which two of the codons code for one amino acid and two for a different one, there must be two different tRNAs and the wobble effect still applies. Such tRNAs are found, as expected, in the mitochondrial genes. The coding properties of the mt-tRNAs are shown in Fig. 9. Similar conclusions have been reached by Heckman *et al.* (35) and by Bonitz *et al.* (36), working respectively on neurospora and yeast mitochondria.

Distribution of protein genes

Mitochondrial DNA was known to code for three of the subunits of cytochrome oxidase, probably three subunits of the ATPase complex, cytochrome b, and a number of other unidentified proteins. In order to identify the protein-coding genes, the DNA was searched for reading frames; i.e. long stretches of DNA containing no termination codons in one of the phases and thus being capable of coding for long polypeptide chains. Such reading frames should start with an initiation codon, which in normal systems is nearly always ATG, and end with a termination codon. Fig. 10 summarises the distribution of the reading frames

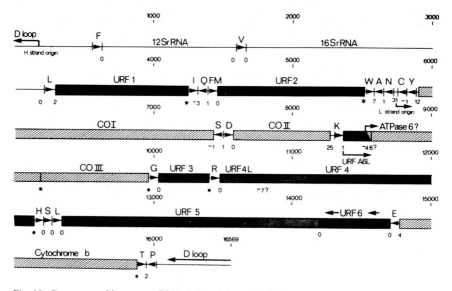

Fig. 10. Gene map of human mtDNA deduced from the DNA sequence. Boxed regions are the predicted reading frames for the proteins. URF = unidentified reading frame. tRNA genes are denoted by the one-letter amino acid code and are either L strand coded (▶) or H strand coded (◀). Numbers above the genes show the scale in nucleotides and below the predicted number between genes.

* Indicates that termination codons are created by polyadenylation of the mRNA.

on the DNA and these are believed to be the genes coding for the proteins. The gene for COII was identified from the amino acid sequence as described above, for subunit I of the cytochrome oxidase from amino acid sequence studies on the bovine protein by J. E. Walker (personal communication), and COIII, cytochrome b and, probably, ATPase 6 were identified by comparison with the DNA sequences of the corresponding genes in yeast mitochondria. Tzagoloff and his colleagues were able to identify these genes in yeast by genetic methods. It has not yet been possible to assign proteins to the other reading frames.

One unusual feature of the mtDNA is that it has a very compact structure. The reading frames coding for the proteins and the rRNA genes appear to be flanked by the tRNA genes with no, or very few, intervening nucleotides. This suggests a relatively simple model for transcription, in which a large RNA is copied from the DNA and the tRNAs are cut out by a processing enzyme, and this same processing leads to the production of the rRNAs and the messenger RNAs (mRNAs), most of which will be monocistronic. Strong support for this model comes from the work of Attardi (37,38) who has identified the RNA sequences at the 5′ and 3′ ends of the mRNAs, thus locating them on the DNA sequence. One consequence of this arrangement is that the initiation codon is at, or very near, the 5′ end of the mRNAs. This suggests that there must be a different mechanism of initiation from that found in other systems. In bacteria there is usually a ribosomal binding site before the initiating ATG codon, whereas in eucaryotic systems the 'cap' structure on the 5′ end of the mRNA appears to have a similar function and the first ATG following the cap acts as initiator. It seems that mitochondria may have a more simple, and perhaps more primitive, system with the translation machinery recognising simply the 5′ end of the mRNA. Another unique feature of mitochondria is that ATA and possibly ATT can act as initiator codons as well as ATG.

On the basis of the above model, some of the mRNAs will not contain termination codons at their 3′ ends after the tRNAs are cut out. However they have T or TA at the 3′ end. The mRNAs are generally found polyadenylated at their 3′ ends and this process will necessarily give rise to the codon TAA to terminate those protein reading frames.

The compact structure of the mammalian mitochondrial genome is in marked contrast to that of yeast, which is about five times as large and yet codes for only about the same number of proteins and RNAs. The genes are separated by long AT-rich stretches of DNA with no obvious biological function. There are also insertion sequences within some of the genes, whereas these appear to be absent from mammalian mtDNA.

REFERENCES

1. Holley, R. W., Les Prix Nobel, p. 183 (1968)
2. Sanger, F., Les Prix Nobel, p. 134 (1958)
3. Sanger, F., Brownlee, G. G. & Barrell, B. G., J. Mol. Biol. *13*, 373 (1965)
4. Robertson, H. D., Barrell, B. G., Weith, H. L. & Donelson, J. E., Nature, New Biol. *241*, 38 (1973)
5. Ziff, E. B., Sedat, J. W. & Galibert, F., Nature, New Biol. *241*, 34 (1973)
6. Billeter, M. A., Dahlberg, J. E., Goodman, H. M., Hindley, J. & Weissmann, C., Nature *224*, 1083 (1969)
7. Berg, P., Fancher, H. & Chamberlin, M., Symp. "Informational Macromolecules", pp 467–483, Academic Press: New York & London (1963)
8. Sanger, F., Donelson, J. E., Coulson, A. R., Kössel, H. & Fischer, D., Proc. Natl. Acad. Sci. USA *70*, 1209 (1973)
9. Sanger, F. & Coulson, A. R. J. Mol. Biol. *94*, 441 (1975)
10. Sanger, F., Air, G. M., Barrell, B. G., Brown, N. L., Coulson, A. R. Fiddes, J. C., Hutchison, C. A., Slocombe, P. M. & Smith, M., Nature *265*, 687 (1977)
11. Sanger, F., Nicklen, S. & Coulson, A. R., Proc. Natl. Acad. Sci. USA *74*, 5463 (1977)
12. Sanger, F., Coulson, A. R., Friedmann, T., Air, G. M., Barrell, B. G., Brown, N. L., Fiddes, J. C., Hutchison, C. A., Slocombe, P. M. & Smith, M., J. Mol. Biol. *125*, 225 (1978)
13. Godson, G. N., Barrell, B. G., Staden R. & Fiddes, J. C., Nature *276*, 236 (1978)
14. Smith, A. J. H., Nucl. Acids Res. *6*, 831 (1979)
15. Barnes, W. M., Gene *5*. 127 (1979)
16. Gronenborn, B. & Messing, J., Nature *272*, 375 (1978)
17. Herrmann, R., Neugebauer, K., Schaller, H. & Zentgraf, H., In "The Singlestranded DNA Phages" (Eds. Denhardt, D. T., Dressler, D. & Ray, D. S.) pp 473–476, Cold Spring Harbor Laboratory, New York (1978)
18. Heidecker, G., Messing, J. & Gronenborn, B., Gene *10*, 69 (1980)
19. Anderson, S., Gait, M. J., Mayol, L. & Young, I. G., Nucl. Acids. Res. *8*, 1731 (1980)
20. Gait, M. J., Unpublished results (1980)
21. Staden, R., Nucl. Acids Res. *8*, 3673 (1980)
22. Sanger, F., Coulson, A. R., Barrell, B. G., Smith, A. J. H. & Roe, B. A., J. Mol. Biol. *143*, 161 (1980)
23. Barrell, B. G., Anderson, S., Bankier, A. T., de Bruijn, M. H. L., Chen, E., Coulson, A. R., Drouin, J., Eperon, I. C., Nierlich, D. P., Roe, B. A., Sanger, F., Schreier, P. H., Smith, A. J. H., Staden, R. & Young, I. G., 31st Mosbach Colloq. "Biological Chemistry of organelle Formation", (Eds. Bücher, T., Sebald, W. & Weiss, H.) pp 11–25, Springer-Verlag, Berlin (1980)
24. Winter, G. & Fields, S., Nucl. Acids Res. *8*, 1965 (1980)
25. Shaw, D. C., Walker, J. E., Northrop, F. D., Barrell, B. G., Godson, G. N. & Fiddes, J. C., Nature *272*, 510 (1978)
26. Steffans, G. J. & Buse, G., Hoppe-Seyler's Z. Physiol. Chem. *360*, 613 (1979)
27. Barrell, B. G., Bankier, A. T. & Drouin, J., Nature *282*, 189 (1979)
28. Young, I. G. & Anderson, S., Gene, *12*, 257 (1980)
29. Macino, G., Coruzzi, G., Nobrega, F. G., Li, M. & Tzagoloff, A., Proc. Natl. Acad. Sci. USA *76*, 3784 (1979)
30. Macino, G. & Tzagoloff, A., Proc. Natl. Acad. Sci. USA *76*, 131 (1979)
31. Staden, R., Nucl. Acids Res. *8*, 817 (1980)
32. de Bruijn, M. H. L., Schreier, P. H., Eperon, I. C., Barrell, B. G., Chen, E. Y., Armstrong, P. W., Wong, J. F. H. & Roe, B. A., Nucl. Acids Res. *8*, 5213 (1980)
33. Arcari, P. & Brownlee, G. G., Nucl. Acids. Res. *8*, 5207 (1980)
34. Barrell, B. G., Anderson, S., Bankier, A. T., de Bruijn, M. H. L., Chen, E., Coulson, A. R., Drouin, J., Eperon, I. C., Nierlich, D. P., Roe, B. A., Sanger, F., Schreier, P. H., Smith, A. J. H., Staden, R. & Young, I. G., Proc. Natl. Acad. Sci. USA *77*, 3164 (1980)

35. Heckman, J. E., Sarnoff, J., Alzner-De Weerd, B., Yin, S. & RajBhandary, U. L., Proc. Natl. Acad. Sci. USA *77*, 3159 (1980)
36. Bonitz, S. G., Berlani, R., Coruzzi, G., Li, M., Macino, G., Nobrega, F. G., Nobrega, M. P., Thalenfeld, B. E. & Tzagoloff, A., Proc. Natl. Acad. Sci. USA, *77*, 3167 (1980)
37. Montoya, J., Ojala, D. & Attardi, G., Nature *290*, 465 (1981)
38. Ojala, D., Montoya, J. & Attardi, G., Nature *290*, 470 (1981)